Study Guide

for

Solomon, Berg, and Martin's

Biology

Eighth Edition

Ronald S. Daniel
The California State University

Sharon Callaway Daniel
Orange Coast College

Ronald L. Taylor
Author and Lecturer

THOMSON

BROOKS/COLE

Australia • Brazil • Canada • Mexico • Singapore • Spain • United Kingdom • United States

Printed in the United States of America
1 2 3 4 5 6 7 11 10 09 08 07

Printer: Thomson/West
Cover Image: Gail Shumway/Taxi/Getty Images

ISBN-13: 978-0-495-11415-4
ISBN-10: 0-495-11415-4

Thomson Higher Education
10 Davis Drive
Belmont, CA 94002-3098
USA

For more information about our products, contact us at:
Thomson Learning Academic Resource Center
1-800-423-0563

For permission to use material from this text or product, submit a request online at
http://www.thomsonrights.com.
Any additional questions about permissions can be submitted by email to **thomsonrights@thomson.com.**

PREFACE

□

This study guide provides a means to learn the content of biology, and to an extent, it also acquaints the beginning student with some of the experiential aspects of biology. It is designed for use with the eighth edition of the textbook *Biology* by Solomon, Berg, and Martin.

While it is of utmost importance that students learn the subject matter of biology, it is also important that they appreciate the ways that biologists think and perceive the world about them. Accordingly, we believe that students should be engaged in the processes of biology while learning its content. This study guide offers learning aids and study strategies that we have found effective in achieving these goals. The approach we have taken has also been influenced by reviewers of the previous editions of this book, and by feedback from our students about pedagogical methods they find valuable.

The variety of exercises we have included accommodates the different cognitive learning styles among students. Consequently, study guide chapters contain the following sections: *Chapter Overview, Reviewing Concepts, Building Words, Matching, Making Comparisons, Making Choices,* and *Visual Foundations.* Each section has a specific purpose, and the whole is designed to provide a well-rounded learning experience. *Answers* to questions are provided in the back of the book.

ACKNOWLEDGMENTS

We thank our students who helped us learn what does and does not work, especially our student editors at Orange Coast College who worked through the study guide and informed us of anything that was ambiguous or unclear. We also appreciate the contributions of the reviewers of our previous study guides, and we thank our colleagues and other instructors of biology whose creative ideas and dedication inspired us to prepare this study guide.

Student Editors
Leslie N. Bohner
Katherine M. Smith
Miranda M. Whitacre

Ronald S. Daniel, PhD
Professor Emeritus, Biological Sciences
California State Polytechnic University, Pomona

Sharon Callaway Daniel, MS
Professor and Chair of Biological Sciences
Orange Coast College, Costa Mesa, California

Ronald L. Taylor, PhD
Author and Lecturer

CONTENTS

□

TO THE STUDENT

□

We prepared this study guide to help you learn biology. We have included the ways of studying and learning that our most successful students find helpful. If you study the textbook, use this study guide regularly, and follow the guidelines provided here, you will learn more about biology and, in the process, you are likely to earn a better grade. It is our sincere desire to provide you with enjoyable experiences during your excursion into the world of life.

Study guide chapters contain the following sections: *Chapter Overview, Reviewing Concepts, Building Words, Matching, Making Comparisons, Making Choices,* and *Visual Foundations*. Although these sections provide a fairly comprehensive review of the material in each chapter, they may not cover all of the material that your instructor expects you to know. The ways that you can use these sections are described below.

CHAPTER OVERVIEW

Each chapter in this study guide begins with an overview of the chapter. It focuses on the major concepts of the chapter, and, when appropriate, relates them contextually to what you have learned in previous chapters. We believe you will find it beneficial to read this section carefully *before* you undertake any of the exercises in the chapter, and again *after* you have completed them. Review this material any time you find yourself losing sight of "the big picture."

REVIEWING CONCEPTS

This section outlines the textbook chapter. It contains important concepts and fill-in questions. Be sure that you understand them. Examine the textbook material to be sure that you have correctly filled in all of the blank spaces. Once completed, check your work in the *Answers* section at the back of the study guide.

BUILDING WORDS

The language of biology is fundamental to an understanding of the subject. Many terms used in biology have common word roots. If you know word roots, you can frequently figure out the meaning of a new term. This section provides you with a list of some of the prefixes and suffixes (word roots) that are used in the chapter.

Learn the meanings of these prefixes and suffixes, then use them to construct terms in the blank spaces next to the definitions provided. Once you have filled in the blanks for all of the definitions, check your work in the *Answers* section at the back of the study guide.

Learning the language of biology also means that you can communicate in its language, and effective communication in any language means that you can read, write, and speak the language. Use the glossary in your textbook to learn how to pronounce new terms. Say the words out loud. You will discover that pronouncing words aloud will not only teach you verbal communication, it will also help you remember them.

MATCHING

This section tests your knowledge of key terms used in the chapter. Check you answers in the *Answers* section.

MAKING COMPARISONS

This section uses tables to test your ability to recognize relationships between key elements in the chapter. Check your answers in the *Answers* section

MAKING CHOICES

The items in this section were selected for two purposes. First, and perhaps most importantly, they help you develop study skills. They provide a means to re-examine, integrate, and analyze selected textbook material. To answer many of them, you must understand how material presented in one part of a chapter relates to material presented in other parts of the chapter. This section provides a means to evaluate your mastery of the subject matter and to identify material that needs additional study.

Once you have studied the textbook material and your notes, close your textbook, put away your notes, and try answering the multiple choice items. Check your answers in the *Answers* section. Don't be discouraged if you miss some answers. Take each item that you missed one at a time, go back to the textbook and your notes, and determine what you missed and why. Do not merely copy down the correct answers — that defeats the purpose. If you dig the answers out of the textbook, you will find that your understanding of the material will improve. Some points that you thought you already knew will become clearer, and you will develop a better understanding of the relationships between different aspects of biology.

VISUAL FOUNDATIONS

Selected visual elements from the textbook are presented in this section. Coloring the parts of biological illustrations causes one to focus on the subject more carefully. Coloring also helps to clearly delineate the parts and reveals the relationship of parts to one another and to the whole.

Begin by coloring the open boxes (❏). This will provide you with a quick reference guide to the illustration when you study it later. Close your textbook while you are doing this section. After completing the *Visual Foundations* section, use your textbook to check your work. Make notes about anything you missed for future reference. Come back to these figures periodically for review. If an item to be colored is listed in the singular form, and there is more than one of them in the illustration, be sure to color all of them. For example, if you are asked to color the "mitochondrion," be sure to color all of the mitochondria shown in the illustration.

OTHER STUDY SUGGESTIONS

When possible, read material in the textbook *before* it is covered in class. Don't be concerned if you don't understand all of the material at first, and don't try to memorize every detail in this first reading. Reading before class will contribute greatly to your understanding of the material as it is presented in class, and it will help you take better notes. As soon as possible after the lecture, recopy all of your class notes. You should do this while the classroom presentation is still fresh in your mind. If you are prompt and disciplined about it, you will be able to recall additional details and include them in your notes. Use material in this study guide and your textbook to fill in the gaps.

What is the best way for you to study? That is something you have to figure out for yourself. Reading all of the chapters assigned by your instructor is basic, but it is not all there is to studying.

A technique that many successful students have found beneficial is to organize a study group of 4-6 students. Meet periodically, especially before examinations, and go over the material together. Bring along your textbooks, study guides, and expanded notes. Have each member of the group ask every other member several questions about the material you are reviewing. Discuss the answers as a group until the principles and details are clear to everyone. Use the combined notes of all members to fill in any gaps in your notes. Consider the reasons for correct and incorrect answers to questions and items in the *Reviewing Concepts* and *Making Choices* sections. This study technique can do wonders. There is probably no better way to find out how well you know the material than to verbalize it in front of a critical audience. Just ask any biology instructor!

If you follow the guidelines presented here and carefully complete the work provided in this study guide, you will make giant strides toward understanding the principles of biology. If you decide to specialize in some area of biology, you will find the general background you obtained in this course to be invaluable.

PART I

The Organization of Life

CHAPTER 1

A View of Life

Biology is the study of living things. The basic themes of biology include the evolution of life, the transmission of information, and the flow of energy through organisms. Evolution is responsible for the diversity of life on the planet. Organisms transmit information chemically, electrically, and behaviorally. They require a continuous input of energy to maintain the chemical transactions and cellular organization essential to their existence. Living things share a common set of characteristics that distinguishes them from nonliving things. They are organized at a variety of levels, each more complex than the previous, from the chemical level up through the ecosystem level. To make sense of the millions of organisms that have evolved on our planet, scientists have developed a binomial and hierarchical system of nomenclature. Organisms are studied by a system of observation, hypothesis, experiment, more observation, and revised hypothesis known as the scientific method. Increasingly, important ethical considerations arise from experiments conducted on living things.

REVIEWING CONCEPTS
Fill in the blanks.

INTRODUCTION
1. Knowledge of biological concepts is a vital tool for understanding the challenges that confront modern society. Some of the more important challenges are:
    (a)_____,
    (b)_____,
    (c)_____,
    (d)_____.

THREE BASIC THEMES
2. The three basic themes of biology are (a)_____,
    (b)_____, and (c)_____.

CHARACTERISTICS OF LIFE
3. The essential characteristics of life include:

    _____

    _____

    _____.

Organisms are composed of cells
4. The _____ asserts that all living things are composed of cells.

Organisms grow and develop
5. Living things grow by increasing _____ of cells.

6. _____ is a term that encompasses all the changes that occur during the life of an organism.

Organisms regulate their metabolic processes

7. Metabolism can be described as the sum of all the _____ that take place in the organism.

8. _____ is the tendency of organisms to maintain a constant internal environment.

Organisms respond to stimuli

9. Physical or chemical changes in an internal or external environment that evoke a response from all life forms are called _____.

Organisms reproduce

10. We know that organisms come from previously existing life as the result of scientists like Francesco Redi in the 17th century and _____ in the 19th century.

11. In general, living things reproduce in one of two ways: _____.

Populations evolve and become adapted to the environment

12. Inherited characteristics that enhance an organism's ability to survive in a particular environment are called _____.

LEVELS OF BIOLOGICAL ORGANIZATION

Organisms have several levels of organization

13. The basic structural and functional unit of life is the _____.

Several levels of ecological organization can be identified

14. All of the communities of organisms on Earth are collectively referred to as the _____.

15. All of the members of one species that live in the same geographic area make up a _____.

INFORMATION TRANSFER

DNA transmits information from one generation to the next

16. _____ are the units of hereditary material.

17. The two scientists credited with working out the structure of DNA are _____.

Information is transmitted by chemical and electrical signals

18. _____ are large molecules important in determining the structure and function of cells and tissues.

EVOLUTION: THE BASIC UNIFYING CONCEPT OF BIOLOGY

Biologists use a binomial system for naming organisms

19. The scientific name (binomial) of an organism consists of the _____ names for that organism.

20. The lowest category of classification is the _____.

21. The science of classifying and naming organisms is _____.

Taxonomic classification is hierarchical

22. The broadest of the taxonomic groups is the _____.

23. Families are grouped into _____.

The tree of life includes three domains and six kingdoms

24. The six kingdoms are: (a)_____ and _____, the two kingdoms of prokaryotes; (b)_____, unicellular organisms with distinct nuclei such as protozoa and algae; (c)_____, heterotrophic decomposers that include molds and yeasts; (d)_____, complex multicellular producers that can make food by photosynthesizing; and (e)_____, complex multicellular organisms that must obtain food from a "ready made" source.

25. A waxy covering that prevents water loss and stomata are present in members of the kingdom _____.

Species adapt in response to changes in their environment

26. Without the ability to adapt in different environments, many species would become _____.

Natural selection is an important mechanism by which evolution proceeds

27. Charles Darwin's famous book, "_____ _____," published in 1859, had and still has a profound influence on the biological sciences, as well as impacting on the thinking of society in general.

28. Darwin's theory of natural selection was based on the observations that individuals within species show (a)_____ significantly, and more individuals are produced than can possibly (b)_____, and individuals with the most advantageous characteristics are the most likely to (c)_____ and pass those characteristics on to their offspring.

Populations evolve as a result of selective pressures from changes in the environment

29. All the genes in a population make up its _____.

30. Natural selection favors organisms with traits that enable them to cope with _____.

THE ENERGY FOR LIFE

31. All the energy transformations and chemical processes that occur in organisms are referred to as _____.

32. _____ is the process by which molecular energy is released for the activities of life.

33. A self-sufficient ecosystem contains three major categories of organisms: _____. It must also have a continuous input of energy.

34. Organisms that can produce their own food from simple raw materials are called _____ or _____.

THE PROCESS OF SCIENCE

35. The _____ is a system of observation, hypothesis, experiment, more observation, and revised hypothesis.

Science requires systematic thought processes

36. With _____ reasoning, one begins with supplied information or premises and draws conclusions based on those premises.

37. With _____ reasoning, one draws a conclusion from specific observations.

Scientists make careful observations and ask critical questions

38. In 1928, bacteriologist _____ discovered the antibiotic, penicillin.

Chance often plays a role in scientific discovery

39. Significant discoveries are usually made by those who are in the habit of looking critically at nature and recognizing a _____.

A hypothesis is a testable statement

40. The characteristics of a good hypothesis include being
(a)_____, (b)_____, and
(c)_____.

Many predictions can be tested by experiment

41. An experimental group differs from a control group only with respect to the _____ being studied.

Researchers must avoid bias

42. In a _____, neither the patient nor the physician knows who is getting the experimental drug and who is getting the placebo.

Scientists interpret the results of experiments and make conclusions

43. Sampling errors can lead to inaccurate conclusions because

_____.

A theory is supported by tested hypotheses

44. A _____ is an integrated explanation of a number of hypotheses, each supported by consistent results from many observations or experiments.

Many hypotheses cannot be tested by direct experiment

45. Two examples of hypotheses that cannot be tested by direct experiment include

_____.

Paradigm shifts allow new discoveries

46. A _____ is a set of assumptions or concepts that constitutes a way of thinking about reality.

Systems biology integrates different levels of information

47. Biologists that integrate data from various levels of complexity with the goal of understanding the big picture are known as _____.

Science has ethical dimensions

48. Some of the more controversial areas of scientific inquiry with important ethical dimensions include _____

_____.

BUILDING WORDS

Use combinations of prefixes and suffixes to build words for the definitions that follow.

Prefixes	The Meaning		Suffixes	The Meaning
a-	without, not, lacking		-logy	"study of"
auto-	self, same		-stasis	equilibrium caused by opposing forces
bio-	life		-troph	nutrition, growth, "eat"
eco-	home		-zoa	animal
hetero-	different, other			
homeo-	similar, "constant"			
photo-	light			
proto-	first, earliest form of			

Prefix	Suffix	Definition
_____	_____	1. The study of life.
_____	_____	2. The study of groups of organisms interacting with one another and their nonliving environment.
_____	-sexual	3. Reproduction without sex.
_____	_____	4. Maintaining a constant internal environment.
_____	-system	5. A community along with its nonliving environment.
_____	-sphere	6. The planet earth including all living things.
_____	_____	7. An organism that produces its own food.
_____	_____	8. An organism that is dependent upon other organisms for food, energy, and oxygen.
_____	_____	9. Animal-like protists.
_____	-synthesis	10. The conversion of solar (light) energy into stored chemical energy by plants, blue-green algae, and certain bacteria.

MATCHING

Terms:

a.   Adaptation
b.   Cell
c.   Cilia
d.   Consumer
e.   Decomposer
f.   Domain
g.   Kingdom
h.   Order
i.   Phylum
j.   Population
k.   Species
l.   Stimulus
m.   Taxonomy
n.   Tissue

For each of these definitions, select the correct matching term from the list above.

_____ 1. A group of organisms of the same species that live in the same geographical area at the same time.

_____ 2. The broadest category of classification used.

_____ 3. The science of naming, describing, and classifying organisms.

_____ 4. Hairlike structures that project from the surface of some cells and are used by some organisms for locomotion.

_____ 5. A physical or chemical change in the internal or external environment of an organism potentially capable of provoking a response.

_____ 6. A group of closely associated, similar cells that work together to carry out specific functions.

_____ 7. An evolutionary modification that improves the chances of survival and reproductive success in a given environment.

_____ 8. A microorganism that breaks down organic material.

_____ 9. The basic structural and functional unit of life, which consists of living material bounded by a membrane.

_____10. A group of organisms with similar structural and functional characteristics that in nature breed only with each other and have a close common ancestry; a group of organisms with a common gene pool.

MAKING COMPARISONS

Fill in the blanks.

Level of Biological Organization	Examples of Components
Chemical level	Atoms, molecules
#1	Plasma membrane, organelles, nucleus
#2	Group of cells that have similar structure and function
#3	Functional structures made up of tissues
Organ system level	#4
Organism	#5

MAKING CHOICES

Place your answer(s) in the space provided. Some questions may have more than one correct answer.

_____ 1. In science, a tentative explanation for an observation or phenomenon is known as a/an
    a. supposition
    d. theory
    b. theory
    e. hypothesis
    c. belief

_____ 2. Organisms that can use sunlight energy to synthesize complex food molecules are known as
    a. heterotrophs.
    d. consumers.
    b. cyanobacteria.
    e. producers.
    c. autotrophs.

_____ 3. The kingdom(s) containing organisms that have a nucleus surrounded by a membrane ("true nucleus") and are unicellular is/are
    a. Bacteria.
    d. Plantae.
    b. Protista.
    e. Animalia.
    c. Fungi.

_____ 4. A snake is a member of the kingdom/phylum
    a. Prokaryotae/vertebrate.
    d. Animalia/chordata.
    b. Fungi/ascomycete.
    e. Plantae/angiosperm.
    c. Protista/Aves.

_____ 5. The kingdom(s) containing organisms that are both single-celled and have a chromosome that is not separated from the cytoplasm by a membrane is/are
    a. Fungi.
    d. Plantae.
    b. Animalia.
    e. Protista.
    c. Bacteria.

_____ 6. Maintenance of an internal balanced internal environment within an organism is known as
    a. homeostasis.
    d. adaptation.
    b. metabolism.
    e. cyclosis.
    c. response.

_____ 7. Prokaryotic organisms
    a. lack a nuclear membrane.
    d. are simple plants.
    b. include bacteria.
    e. comprise two kingdoms.
    c. contain membrane-bound organelles.

_____ 8. To comply with scientific method, a hypothesis must
    a. be true.
    d. withstand rigorous investigations.
    b. be testable.
    e. eventually reach the status of a principle.
    c. be popular and fundable.

_____ 9. The characteristic(s) common to all living things include(s)
    a. reproduction.
    d. growth.
    b. presence of a true nucleus.
    e. heterotrophic synthesis.
    c. adaptation.

The Federal Center for Disease Control tested 600 female prostitutes who regularly use condoms. They found that none of the women had the AIDS virus. Use this study and data to answer questions 10-12.

_____10. In terms of determining a relationship between use of condoms by women and the kind of sexual behavior that might lead to contraction of AIDS, this experiment

    a. could be called a theory.

    b. is well-designed.

    c. should be repeated with more subjects.

    d. is useless.

    e. should include male subjects.

_____11. The experiment

    a. proves that condoms are effective in preventing AIDS.

    b. indicates that AIDS is not transmitted through sexual activity.

    c. suggests that condoms may be effective in preventing AIDS.

    d. probably means that none of the partners of the female prostitutes had AIDS.

    e. shows that women are less prone to contract AIDS than men.

_____12. To determine the effectiveness of using condoms in preventing AIDS among women, this study was lacking in not having

    a. enough subjects.

    b. male subjects.

    c. controls.

    d. subjects who do not use condoms during sex.

    e. women who are not prostitutes.

_____13. The number of domains in the tree of life is

    a. one.

    b. two.

    c. three.

    d. four

    e. none.

_____14. One unifying principle in biology, known as the cell theory, states that the following consist of cells:

    a. animals, not plants

    b. eukaryotes, not prokaryotes

    c. prokaryotes, not eukaryotes

    d. plant and animal kingdoms, not other kingdoms

    e. all living things.

_____15. Which of the following is true concerning the levels of organization in multicellular organisms?

    a. tissues are comprised of cells

    b. organs are made up of tissues

    c. a group of tissues and organs make up an organ system

    d. molecules are comprised of atoms

    e. cells are made up of atoms, molecules, and organelles.

VISUAL FOUNDATIONS

Questions 1-3 refer to the diagram below. Some questions may have more than one correct answer.

_____ 1. The diagram represents:
   a. developmental theory.
   b. a controlled experiment.
   c. deductive reasoning.
   d. scientific method.
   e. all of the above.

_____ 2. Which of the following statements is/are supported by the diagram?
   a. A theory and a hypothesis are identical.
   b. Conclusions are based on results obtained from carefully designed experiments.
   c. Using results from a single sample can result in sampling error.
   d. A tested hypothesis may lead to new observations.
   e. Science is infallible.

_____ 3. Which of the following statements best describes the activity depicted by the box labeled #3 in the diagram?
   a. Make a prediction that can be tested.
   b. Generate a related hypothesis.
   c. Form generalized inductive conclusions.
   d. Develop a scientific principle.
   e. Test a large number of predicted cases.

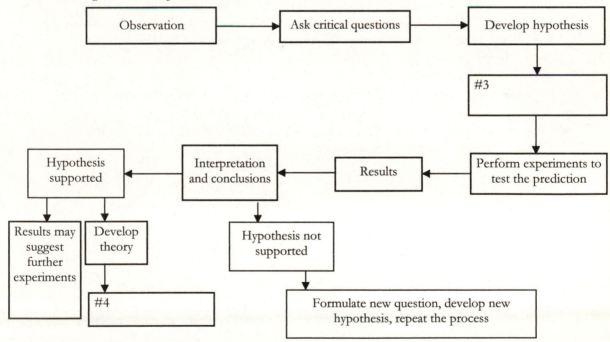

CHAPTER 2

Atoms and Molecules:
The Chemical Basis of Life

Living things, like everything else on our planet, are made of atoms and molecules. Therefore, to understand life is first to understand the interaction and activity of atoms and molecules. Living things share remarkably similar chemical compositions. About 96% of an organism's mass is made up of just four elements — oxygen, carbon, hydrogen, and nitrogen. The smallest portion of an element that retains its chemical properties is the atom. Atoms, in turn, consist of subatomic particles. Those most important to an understanding of life are the protons, neutrons, and electrons. The number and arrangement of electrons determine the chemical behavior of atoms. Molecules and compounds are formed by the joining of atoms. This chapter lays the foundation for understanding how the structure of atoms determines the way they form chemical bonds to produce compounds. The focus is on small, simple substances, known as inorganic compounds, that play critical roles in life's processes. One of these, water, is considered to have been essential to the origin of life, as well as to the continued survival and evolution of life on Earth.

REVIEWING CONCEPTS
Fill in the blanks.

INTRODUCTION

1. Among the biologically important groups of inorganic compounds are _____.

ELEMENTS AND ATOMS

2. An element is a substance that cannot be separated into simpler parts by _____.

3. Over 96% of an organism's weight is made up of the four elements _____

_____.

4. The three main parts of an atom are the (a)_____. The (b)_____ compose the atomic nucleus, and the (c)_____ occupy the mostly empty space surrounding the nucleus.

5. (a)_____ have one unit of positive charge, (b)_____ are neutral, and (c)_____ have a negative charge.

An atom is uniquely identified by its number of protons

6. The number of protons in an atom is called the _____.

7. The symbol $_8O_6$ indicates that oxygen has _____ protons.

Protons plus neutrons determine atomic mass

8. The total number of protons and neutrons in an atom is referred to as the

   _____ .

9. An atomic mass unit (amu) is also called a _____ .

Isotopes of an element differ in number of neutrons

10. The (a)_____ of isotopes vary, but
    the (b)_____ remains constant.

11. Unstable isotopes that emit radiation when they decay are termed

    _____ .

Electrons move in orbitals corresponding to energy levels

12. Electrons with the same energy comprise an _____ .

13. The electrons in the outermost energy level of an atom are called
    _____ electrons.

CHEMICAL REACTIONS

14. All elements in the same group on the periodic table have similar

    _____ .

Atoms form compounds and molecules

15. A chemical compound consists of _____
    _____ that are combined in a fixed ratio.

16. A _____ forms when two or more atoms are joined by chemical
    bonds.

Simplest, molecular, and structural chemical formulas give different information

17. A _____ formula simply uses chemical symbols to
    describe the chemical composition of a compound.

18. A _____ formula shows not only the types and numbers
    of atoms in a molecule but also their arrangement.

One mole of any substance contains the same number of units

19. The molecular mass of a compound is the sum of the _____
    _____ .

20. The amount of a compound in grams that is equal to its atomic mass is called one
    _____ .

Chemical equations describe chemical reactions

21. (a)_____ are substances that participate in the reaction,
    (b)_____ are substances formed by the reaction.

22. Arrows are used to indicate the direction of a reaction; reactants are placed to the
    (a)_____ of arrows, products are placed to the (b)_____ of arrows.

CHEMICAL BONDS

In covalent bonds electrons are shared

23. Covalently bonded atoms with similar electronegativities are called
    (a)_____ , and covalently bonded atoms with differing
    electronegativities are called (b)_____ .

Ionic bonds form between cations and anions

24. Ionic compounds are comprised of positively charged ions called (a)_____, and negatively charged ions called (b)_____.

25. The process known as _____ has occurred when cations and anions of an ionic compound are surrounded by the charged ends of water molecules.

Hydrogen bonds are weak attractions

26. Hydrogen bonds are individually (a)_____, but are collectively (b)_____ when present in large numbers.

Van der Waals interactions are weak forces

27. Van der Waals interactions are weak forces based on fluctuating _____.

REDOX REACTIONS

28. Reactions involving oxidation *and* reduction are called _____ reactions.

29. When iron rusts, it loses electrons and is said to be (a)_____, while at the same time oxygen gains electrons and is (b)_____.

WATER

30. In humans, water makes up about _____% of total body weight.

Hydrogen bonds form between water molecules

31. Cohesion of water molecules is due to _____ between the molecules.

32. The tendency of water to move through narrow tubes as the result of adhesion and cohesion is known as _____.

Water molecules interact with hydrophilic substances by hydrogen bonding

33. Substances that interact readily with water are said to be (a)_____, whereas those that don't are called (b)_____.

Water helps maintain a stable temperature

34. Water has a high _____, which helps marine organisms maintain a relatively constant internal temperature.

35. A _____ is a unit of heat energy.

ACIDS, BASES, AND SALTS

36. A base is a proton acceptor that generally dissociates in solution to yield (a)_____. An acid is a proton donor that dissociates in solution to yield (b)_____.

pH is a convenient measure of acidity

37. The pH scale covers the range (a)_____. Neutrality is at pH (b)_____.

38. A solution with a pH above 7 is a(an) (a)_____. A solution with a pH below 7 is a(an) (b)_____.

Buffers minimize pH change

39. A _____ includes a weak acid and a weak base.

An acid and a base react to form a salt

40. Dissociated ions called _____ can conduct an electric current when a salt, an acid, or a base is dissolved in water.

BUILDING WORDS

Use combinations of prefixes and suffixes to build words for the definitions that follow.

Prefixes	The Meaning	Suffixes	The Meaning
equi-	equal	-hedron	face
hydr(o)-	water	-philic	loving, friendly, lover
neutro-	neutral	-phobic	fearing
non-	not		
tetra-	four		

Prefix	Suffix	Definition
_____	-n	1. An uncharged subatomic particle.
_____	-librium	2. The condition of a chemical reaction when the forward and reverse reaction rates are equal.
_____	_____	3. The shape of a molecule when it appears as a three-dimensional pyramid.
_____	-ation	4. The process whereby ionic compounds combine chemically with water and dissolve.
_____	_____	5. Not attracted to water; insoluble in water.
_____	-electrolyte	6. A substance that does not form ions when dissolved in water, and therefore does not conduct an electric current.
_____	_____	7. Having a strong affinity for water; soluble in water.

MATCHING

Terms:

a. Anion
b. Atom
c. Atomic mass
d. Calorie
e. Cation

f. Covalent bond
g. Electrolyte
h. Electron
i. Hydrogen bond
j. Ion

k. Ionic bond
l. Oxidation
m. Proton
n. Reduction
o. Valence electrons

For each of these definitions, select the correct matching term from the list above.

____ 1. The loss of electrons, or the loss of hydrogen atoms from a compound.

____ 2. The total number of protons and neutrons in the atomic nucleus.

____ 3. Substances dissolved in water that can conduct electrical current.

____ 4. The smallest subdivision of an element that retains its chemical properties.

____ 5. Chemical bond that involves one or more shared pairs of electrons.

____ 6. The name of the outermost electrons.

____ 7. An ion with a negative electric charge.

____ 8. A particle bearing an electric charge, either positive or negative.

____ 9. An ion with a positive electric charge.

____10. The amount of heat required to raise 1 gram of a substance 1 degree Celsius.

MAKING COMPARISONS

Fill in the blanks.

Name	Formula
Sodium ion	Na^+
#1	CO_2
Calcium	#2
#3	Fe_2O_3
Methane	#4
#5	Cl^-
#6	K^+
Hydrogen ion	#7
#8	Mg
Ammonium ion	#9
#10	CO_2
Water	#11
#12	$C_6H_{12}O_6$
Bicarbonate ion	#13
#14	H_2CO_3
Sodium chloride	#15
#16	OH^-
Sodium hydroxide	#17

MAKING CHOICES

Place your answer(s) in the space provided. Some questions may have more than one correct answer.

_____ 1. Substances that cannot be broken down into simpler substances by ordinary chemical means are
   a. compounds.
   b. covalently bonded atoms.
   c. mixtures.
   d. molecules.
   e. elements.

_____ 2. Cohesion of water molecules results from
   a. attraction between water and other substances.
   b. hydrogen bonding between water molecules.
   c. arrested adhesion.
   d. covalent bonding.
   e. capillary action.

_____ 3. The symbol $_6C$ stands for
   a. calcium with six electrons.
   b. carbon with six protons.
   c. carbon with three protons and three neutrons.
   d. calcium with three protons and three electrons.
   e. an isotope of carbon.

_____ 4. Cations are
   a. negative ions.
   b. positive ions.
   c. isotopes.
   d. incomplete products.
   e. negative molecules.

Use the following chemical equation to answer questions 5-10.

$$2 H_2 + O_2 \rightarrow 2 H_2O \text{ (water)}$$

_____ 5. The product has a molecular mass of
   a. one.
   b. twelve.
   c. larger than either reactant alone.
   d. sixteen.
   e. eighteen.

_____ 6. The reactant(s) in the formula is/are
   a. hydrogen.
   b. oxygen.
   c. H_2O.
   d. water.
   e. the arrow.

_____ 7. How many oxygen atoms are involved in this reaction?
   a. one
   b. two
   c. three
   d. four
   e. none

_____ 8. The product(s) is/are
   a. hydrogen.
   b. oxygen.
   c. H_2O.
   d. water.
   e. the arrow.

_____ 9. The lightest reactant has a mass number of
   a. one.
   b. two.
   c. larger than the product.
   d. eight.
   e. ten.

_____10. The direction in which this reaction tends to occur is
a. right to left.
b. from products to reactants.
c. from hydrogen and oxygen to water.
d. left to right.
e. from reactants to products.

_____11. Oxidation has occurred when
a. an atom gains an electron.
b. a molecule loses an electron.
c. hydrogen bonds break.
d. an atom loses an electron.
e. an ion gains an electron.

_____12. When two atoms combine by sharing one pair of their electrons, the bond is a
a. divalent, single.
b. covalent, single.
c. divalent, double.
d. covalent, double.
e. covalent, polar.

_____13. A form of an element that is progressing toward a more stable state by emitting radiation is called a(n)
a. Dalton atom.
b. radioisotope.
c. enerton.
d. quantum unitron.
e. radionuclide.

_____14. The outer electron energy level of an atom
a. may contain two or more orbitals.
b. determines the element.
c. contains the same number of electrons as protons.
d. often contains at least one neutron.
e. determines chemical properties.

_____15. If the number of neutrons in an atom is changed, it will also change
a. atomic mass.
b. atomic number.
c. the element.
d. the number of electrons.
e. the elements chemical properties.

_____16. Dissociation of which of the following into ions may result in the formation of electrolytes?
a. salts
b. acids
c. hydrogen
d. bases
e. alcohols

_____17. A substance containing equal amounts of hydrogen (H^+) and hydroxyl (OH^-) ions would be
a. acidic, pH below 7.
b. alkaline, pH below 7.
c. acidic, pH above 7.
d. alkaline, pH above 7.
e. neutral, pH 7.

_____18. H−O−H is/are
a. a structural formula.
b. water.
c. a chemical formula.
d. a molecule.
e. a structure with the molecular mass 18.

_____19. The atomic nucleus contains
a. protons.
b. neutrons.
c. electrons.
d. positrons.
e. isotopes.

VISUAL FOUNDATIONS

Color the parts of the illustration below as indicated.

RED ☐ electron

YELLOW ☐ neutron

BLUE ☐ proton

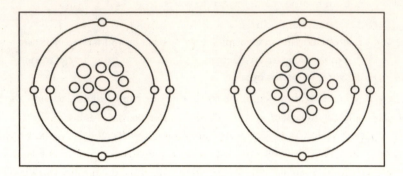

Questions 1-2 refer to the illustration above. Two atoms are represented in this illustration. Some questions may have more than one correct answer.

1. The atom on the left is called (a)_____. The atom on the right is called (b)_____.

____ 2. These atoms:

   a. are ions.

   b. have the same atomic mass.

   c. have the same mass.

   d. are from different elements.

   e. have the same atomic number.

CHAPTER 3

❑

The Chemistry of Life: Organic Compounds

In the previous chapter, you acquired an understanding as to how the structure of atoms determines the way they form chemical bonds to produce compounds. The focus was on the smaller, simpler substances known as inorganic compounds. The focus in this chapter is on the larger and more complex organic compounds. Most of the chemical compounds in organisms are organic compounds. Carbon is the central component of organic compounds, probably because it forms bonds with a greater number of different elements than any other type of atom. The major groups of organic compounds are the carbohydrates, lipids, proteins, and nucleic acids. Carbohydrates serve as energy sources for cells, and as structural components of fungi, plants, and some animals. Lipids function as biological fuels, as structural components of cell membranes, and as hormones. Proteins serve as structural components of cells and tissues and as the facilitators of thousands of different chemical reactions. Nucleic acids transmit hereditary information and determine what proteins a cell manufactures. They are made up of subunits called nucleotides. Some single and double nucleotides are central to cell function.

REVIEWING CONCEPTS
Fill in the blanks.

INTRODUCTION

1. _____ are the four major "classes" of macromolecules in living systems.

2. Macromolecules are built from smaller molecules. Proteins, for example, are built from _____.

CARBON ATOMS AND MOLECULES

3. Carbon has _____ (#?) valence electrons.

4. Carbon atoms can form _____(#?) bonds with a large variety of other atoms.

Isomers have the same molecular formula, but different structures

5. _____ isomers are compounds that differ in the covalent arrangements of their atoms.

6. _____ isomers are compounds that are different in the spatial arrangement of atoms or groups of atoms.

7. _____ isomers are molecules that are mirror images of one another.

Functional Groups change the properties of organic molecules

8. Each class of organic compound is characterized by specific
_____ groups with distinctive properties.

9. R — OH is the symbol for a _____ group.

10. R — CHO is the symbol for a _____ group.

11. R — COOH is the symbol for a _____ group.

Many biological molecules are polymers

12. Polymers are macromolecules produced by linking _____ together.

13. The process by which polymers are degraded into their subunit components is
_____.

CARBOHYDRATES

14. Carbohydrates contain the elements _____
_____ in a ratio of about 1:2:1.

Monosaccharides are simple sugars

15. The simple sugar _____ is an important and nearly ubiquitous fuel molecule in living cells.

Disaccharides consist of two monosaccharide units

16. Disaccharides form when two monosaccharides are bonded by a
_____ linkage.

Polysaccharides can store energy or provide structure

17. Long chains of _____ link together to form a polysaccharide.

18. _____ is the main storage carbohydrate of plants.

19. Starch occurs in two forms. The simpler form, _____, is unbranched.

20. Plant cells store starch in organelles called _____.

21. _____ is the main storage carbohydrate of animals.

22. _____ is the most abundant carbohydrate on earth.

Some modified and complex carbohydrates have special roles

23. _____ is a complex carbohydrate that forms the exoskeleton of arthropods.

24. A carbohydrate combined with a protein forms (a)_____;
carbohydrates combined with lipids are called (b)_____.

LIPIDS

25. Lipids are composed of the elements _____.

Triacylglycerol is formed from glycerol and three fatty acids

26. A triacylglycerol consists of one (a)_____ molecule combined with one to three (b)_____ molecules.

Saturated and unsaturated fatty acids differ in physical properties

27. Saturated fatty acids are solid due to _____.

Phospholipids are components of cell membranes

28 A phospholipid molecule assumes a distinctive configuration in water because of its _____ property, which means that one end of the molecule is hydrophilic and the other end is hydrophobic.

Carotenoids and many other pigments are derived from isoprene units

29. Carotenoids consist of five-carbon hydrocarbon monomers known as _____.

Steroids contain four rings of carbon atoms

30. A steroid consists of _____ atoms arranged in four attached rings, three of which contain _____ carbon atoms and the fourth contains _____.

Some chemical mediators are lipids

31. Some chemical mediators are produced by the modification of _____ that have been removed from phospholipids.

PROTEINS

Amino acids are the subunits of proteins

32. All amino acids contain (a)_____ groups but vary in the (b)_____ group.

33. _____ bonds join amino acids.

34. The bonds that join the subunits of proteins are called _____ bonds.

35. Proteins are large, complex molecules made of subunits called _____.

Proteins have four levels of organization

36. The levels of organization distinguishable in protein molecules are _____.

37. The amino acid sequence is the primary structure of a _____.

38. Secondary structure results from _____ bonding involving the backbone.

39. Tertiary structure depends on interactions among _____.

40. Quaternary structure results from interactions among _____.

The amino acid sequence of a protein determines its conformation

41. The conformation of a protein determines its _____.

NUCLEIC ACIDS

42. Nucleic acids consist of _____subunits.

43. Each nucleotide subunit in a nucleic acid is composed of a (a)_____ _____ nitrogenous base, a five-carbon (b)_____, and an inorganic (c)_____ group.

Some nucleotides are important in energy transfers and other cell functions

44. Energy for life functions is supplied mainly by the nucleotide _____.

IDENTIFYING BIOLOGICAL MOLECULES

BUILDING WORDS

Use combinations of prefixes and suffixes to build words for the definitions that follow.

Prefixes	**The Meaning**		**Suffixes**	**The Meaning**
amphi-	two, both, both sides		-lysis	breaking down, decomposition
amylo-	starch		-mer	part
di-	two, twice, double		-plast	formed, molded, "body"
hex-	six		-saccharide	sugar
hydr(o)-	water			
iso-	equal, "same"			
macro-	large, long, great, excessive			
mono-	alone, single, one			
poly-	much, many			
tri-	three			

Prefix	**Suffix**	**Definition**
_____	_____	1. A compound that has the same molecular formula as another compound but a different structure and different properties.
_____	-molecule	2. A large molecule consisting of thousands of atoms.
_____	_____	3. A single organic compound that links with similar compounds in the formation of a polymer.
_____	_____	4. The degradation of a compound by the addition of water.
_____	_____	5. A simple sugar.
_____	-ose	6. A sugar that consists of six carbons.
_____	_____	7. Two monosaccharides covalently bonded to one another.
_____	_____	8. A macromolecule consisting of repeating units of simple sugars.
_____	_____	9. A starch-forming granule.
_____	-acylglycerol	10. A compound formed of one fatty acid and one glycerol molecule.
_____	- acylglycerol	11. A compound formed of two fatty acids and one glycerol molecule.
_____	- acylglycerol	12. A compound formed of three fatty acids and one glycerol molecule.
_____	-pathic	13. Having one hydrophilic end and one hydrophobic end.
_____	-peptide	14. A compound consisting of two amino acids.
_____	-peptide	15. A compound consisting of a long chain of amino acids.

MATCHING

Terms:

a.	Amino acid	f.	Glucose	k.	Phospholipid
b.	Amino group	g.	Hydrocarbon	l.	Protein
c.	Carotenoid	h.	Lipid	m.	Starch
d.	Condensation	i.	Nucleotide	n.	Steroid
e.	Glycogen	j.	Phosphate group	o.	Tertiary structure

For each of these definitions, select the correct matching term from the list above.

_____ 1. A molecule composed of one or more phosphate groups, a 5-carbon sugar, and a nitrogenous base.

_____ 2. A group of yellow to orange pigments in plants.

_____ 3. A large complex organic compound composed of chemically linked amino acid subunits.

_____ 4. Complex lipid molecules containing carbon atoms arranged in four interlocking rings.

_____ 5. An organic compound containing an amino group and a carboxyl group.

_____ 6. The three-dimensional structure of a protein molecule.

_____ 7. An organic compound that is made up of only carbon and hydrogen atoms.

_____ 8. Fatlike substance in which there are two fatty acids and a phosphorus-containing group attached to glycerol.

_____ 9. Polysaccharide used by plants to store energy.

_____10. The principle carbohydrate stored in animal cells.

MAKING COMPARISONS

Fill in the blanks.

Functional Group	Abbreviated Formula	Class of Compounds Characterized by Group	Description
Carboxyl	R — COOH	Carboxylic acids (organic acids)	Weakly acidic; can release a H^+ ion
#1	R — PO_4H_2	#2	Weakly acidic; one or two H^+ ions can be released
Methyl	#3	Component of many organic compounds	#4
Amino	#5	#6	Weakly basic; can accept an H^+ ion
Sulfhydryl	R — SH	#7	Helps stabilize internal structure of proteins
Hydroxyl	#8	#9	#10

MAKING CHOICES

Place your answer(s) in the space provided. Some questions may have more than one answer.

_____ 1. The secondary structure of a protein results from
    a. ionic bonds.
    b. hydrogen bonds.
    c. covalent bonds.
    d. polymerization.
    e. peptide bonds.

_____ 2. Hydrolysis results in
    a. smaller molecules.
    b. larger molecules.
    c. reduced ambient water.
    d. increase in polymers.
    e. increase in monomers.

_____ 3. An isomer can be described as one of a group of molecules with the same molecular formulas but different
    a. elements.
    b. compounds.
    c. structures.
    d. physical properties.
    e. chemical properties.

_____ 4. The complex carbohydrate storage molecule in plants is
    a. glycogen.
    b. starch.
    c. usually in an extracellular position.
    d. composed of simple glucose subunits.
    e. more soluble than the animal storage molecule.

_____ 5. Nucleotides
    a. are polypeptide polymers.
    b. are polymers of simple carbohydrates.
    c. contain a base, phosphate, and sugar.
    d. comprise complex proteins.
    e. contain genetic information.

_____ 6. Proteins contain
    a. glycerol.
    b. polypeptides.
    c. peptide bonds.
    d. amino acids.
    e. nucleotides.

_____ 7. The main structural components of cells and tissues are
    a. carbon-based molecules.
    b. carbohydrates, lipids, and DNA.
    c. basically strings of covalently bonded carbons.
    d. organic compounds.
    e. DNA, RNA, and ATP.

_____ 8. Compared to unsaturated fatty acids, saturated fatty acids
    a. contain more hydrogens.
    b. contain more double bonds.
    c. are more like oils.
    d. are more prone to be solids.
    e. contain more carbohydrates.

_____ 9. Glycogen is
    a. a molecule in animals.
    b. a molecule in plants.
    c. stored mainly in the liver and muscle of vertebrates.
    d. composed of simple carbohydrate subunits.
    e. more soluble than starch.

_____10. A monomer is to a polymer as a
    a. chain is to a link.
    b. link is to a chain.
    c. monosaccharide is to a polysaccharide.
    d. necklace is to a bead.
    e. bead is to a necklace.

_____11. $C_6H_{12}O_6$ is the formula for
    a. glucose.
    b. a type of monosaccharide.
    c. cellulose.
    d. fructose.
    e. a type of disaccharide.

_____12. Proteins
    a. are a major source of energy.
    b. contain carbon, hydrogen, and oxygen.
    c. can combine with carbohydrates.
    d. are found in both plants and animals.
    e. are polypeptide polymers.

_____13. Phospholipids

    a. are found in cell membranes.   d. are polar.

    b. are in the amphipathic lipid group.   e. are polypeptide polymers.

    c. contain glycerol, fatty acids, and phosphate groups.

_____14. Polymers are

    a. made from monomers.   d. macromolecules.

    b. formed by dehydration synthesis.   e. smaller than monomers.

    c. found in monomers.

_____15. –COOH

    a. is a carboxyl group.   d. coverts to $-COO^+$.after accepting a proton.

    b. is an amino group.   e. is a polymer.

    c. becomes ionized $-COO^-$.after donating a proton

VISUAL FOUNDATIONS

Color the parts of the illustration below as indicated.

    RED   ❏   saturated fatty acid

    GREEN   ❏   glycerol

    YELLOW   ❏   phosphate group

    BLUE   ❏   choline

    ORANGE   ❏   unsaturated fatty acid

Questions 1-4 refer to the illustration above. Some questions may have more than one correct answer.

_____ 1. The illustration represents

    a. a polypeptide.   d. lecithin.

    b. a phospholipid.   e. starch.

    c. glycogen.

_____ 2. The hydrophobic end of this molecule is composed of

    a. amino acids.   d. phosphate groups.

    b. fatty acids.   e. steroids.

    c. glucose.

_____ 3. This kind of molecule contributes to the formation of

    a. neutral fat.   d. cellulose.

    b. cell membranes.   e. complex proteins.

    c. RNA.

_____ 4. This molecule is amphipathic, meaning that

    a. it contains nonpolar groups.   d. it is found in all life forms.

    b. it is very large.   e. it can act as an acid.

    c. it contains hydrophobic and hydrophilic regions.

Organization of the Cell

Chapters 2 and 3 introduced you to the inorganic and organic materials that are critical to an understanding of the cell, the basic unit of life. In this chapter and those that follow, you will see how cells utilize these chemical materials. Because all cells come from preexisting cells, they have similar needs and therefore share many fundamental features. Most cells are microscopically small because of limitations on the movement of materials across the plasma membrane. Cells are studied by a combination of methods, including microscopy and cell fractionation. Unlike prokaryotic cells, eukaryotic cells are complex, containing a variety of membranous organelles. Although each organelle has its own particular structure and function, all of the organelles of a cell work together in an integrated fashion. The cytoskeleton provides mechanical support to the cell and functions in cell movement, the transport of materials within the cell, and cell division. Most cells are surrounded by an extracellular matrix that is secreted by the cells themselves.

REVIEWING CONCEPTS

Fill in the blanks.

INTRODUCTION

1. _____ are the building blocks of complex multicellular organisms.

THE CELL THEORY

2. The two components of the cell theory are _____
_____
_____.

CELL ORGANIZATION AND SIZE

The organization of all cells is basically similar

3. All cells are enclosed by a _____.

4. The internal structures of cells that carry on life's activities are called
_____.

Cell size is limited

5. In general, the smaller the cell, the larger the _____.

6. A micrometer is 1/1,000,000 of a meter and therefore _____ of a millimeter.

Cell size and shape are related to function

7. The _____ of cells are related to the functions they perform.

METHODS FOR STUDYING CELLS

Light microscopes are used to study stained or living cells

8. (a)_____ is the ratio of microscopic size to actual size of an object, and (b)_____ is the capacity to distinguish fine detail in an image.

Electron microscopes provide a high-resolution image that can be greatly magnified

9. Powerful electron microscopes are used to study the _____ of cells.

Biologists us biochemical techniques to study cell components

10. Cell components are separated by a method known as

_____.

PROKARYOTIC AND EUKARYOTIC CELLS

11. _____ cells are relatively lacking in complexity, and their genetic material is not enclosed by membranes.

12. _____ cells are relatively complex and possess both membrane-bound organelles and a "true" nucleus.

13. In eukaryotic cells, DNA is contained in the _____.

CELL MEMBRANES

14. Membranes divide the eukaryotic cell into _____.

15. The internal membrane system in the eukaryotic cell is known as the

_____.

THE CELL NUCLEUS

16. The _____ consists of two concentric membranes that separate the nuclear contents from the surrounding cytoplasm.

17. DNA is associated with proteins, forming a complex known as

_____.

ORGANELLES IN THE CYTOPLASM

Ribosomes manufacture proteins

18. Ribosomes are tiny particles found free in the cytoplasm or attached to certain membranes; they consist of (a)_____ and are synthesized by the (b)_____.

The endoplasmic reticulum is a network of internal membranes

19. The _____ is a complex of intracytoplasmic membranes that compartmentalize the cytoplasm.

20. Rough RER is studded with _____ that are involved in protein synthesis.

21. Some proteins constructed on RER are transported by _____ for secretion to the outside or insertion in other membranes.

The Golgi complex processes, sorts, and modifies proteins

22. In some cells the Golgi consists of flattened membranous sacs called

_____.

Lysosomes are compartments for digestion

23. Lysosomes are small sacs containing _____ that can break down (lyse) complex molecules, foreign substances, and "dead" organelles.

24. Primary lysosomes are formed by budding from the _____.

Vacuoles are large, fluid-filled sacs with a variety of functions

25. The membrane of a vacuole is called the _____.

Peroxisomes metabolize small organic compounds

26. Peroxisomes get their name from the fact that they produce _____ during oxidation reactions.

Mitochondria and chloroplasts are energy-converting organelles

27. The chemical reactions that convert food energy to ATP (cellular respiration) take place in organelles called _____.

Mitochondria make ATP through cellular respiration.

28. Mitochondria negatively affect health and aging through the release of _____.

Chloroplasts convert light energy to chemical energy through photosynthesis

29. _____ are organelles that contain green pigments that trap light energy for photosynthesis.

30. _____ membranes contain chlorophyll that traps sunlight energy and converts it to chemical energy in ATP.

31. The matrix within chloroplasts, where carbohydrates are synthesized, is called the (a)_____, while the ATP that supplies energy to drive the process is synthesized on membranes called (b)_____.

THE CYTOSKELETON

32. The cytoskeleton is a network of protein filaments that are responsible for the _____ of cells.

33. _____ and _____ are both made up of globular protein subunits that can rapidly assemble and disassemble.

Microtubules are hollow cylinders

34. _____ may play a role in some types of microtubule formation.

35. Microtubules consist of the two proteins _____ and _____.

Microfilaments consist of intertwined strings of actin

36. In muscle cells actin is associated with the protein _____ to form fibers associated with muscle contraction.

Intermediate filaments help stabilize cell shape

37. Keratin and neurofilaments are examples of tough, flexible fibers called _____ that stabilize cell shape.

CELL COVERINGS

38. In most eukaryotic cells polysaccharide side chains of proteins and lipids form a
_____, or cell coat, that is part of the plasma membrane.

39. Plant cell walls are composed primarily of _____ and smaller
quantities of other polysaccharides.

BUILDING WORDS

Use combinations of prefixes and suffixes to build words for the definitions that follow.

Prefixes	The Meaning		Suffixes	The Meaning
chloro-	green		-karyo(te)	nucleus
chromo-	color		-plast(id)	formed, molded, "body"
cyto-	cell		-some	body
eu-	good, well, "true"			
leuko-	white (without color)			
lyso-	loosening, decomposition			
micro-	small			
myo-	muscle			
pro-	"before"			

Prefix	Suffix	Definition
_____	-phyll	1. A green pigment that traps light for photosynthesis.
_____	_____	2. Organelles containing pigments that give fruits and flowers their characteristic colors.
_____	-plasm	3. Cell contents exclusive of the nucleus.
_____	-skeleton	4. A complex network of protein filaments within the cell.
glyoxy-	_____	5. A microbody containing enzymes used to convert stored fats in plant seeds to sugars.
_____	_____	6. An organelle that is not pigmented and is found primarily in roots and tubers, where it is used to store starch.
_____	_____	7. An organelle containing digestive enzymes.
_____	-filaments	8. Small, solid filaments, 7 nm in diameter, that make up part of the cytoskeleton of eukaryotic cells.
_____	-tubules	9. Small, hollow filaments, 25 nm in diameter, that make up part of the cytoskeleton of eukaryotic cells.
_____	-villi	10. Small, finger-like projections from cell surfaces that increase surface area.
_____	-sin	11. A muscle protein that, together with actin, is responsible for muscle contraction.
peroxi-	_____	12. An organelle containing enzymes that split hydrogen peroxide, rendering it harmless.
_____	_____	13. Precursor organelles.
_____	-karyotes	14. Organisms that evolved before organisms with nuclei.

_____   -sol   15. Cell contents exclusive of the nucleus and organelles; the fluid component of the cytoplasm.

_____   _____   16. An organism with a distinct nucleus surrounded by nuclear membranes.

MATCHING

Terms:

a. Actin
b. Centriole
c. Chloroplast
d. Endoplasmic reticulum
e. Granum

f. Mitochondrion
g. Nuclear envelope
h. Nucleolus
i. Plasma membrane
j. Ribosome

k. Secretory vesicle
l. Stroma
m. Vacuole
n. Vesicle

For each of these definitions, select the correct matching term from the list above.

_____ 1. One of a pair of small, cylindrical organelles lying at right angles to each other near the nucleus.

_____ 2. Site of ribosome synthesis.

_____ 3. An intracellular organelle that is the site of aerobic respiration.

_____ 4. A fluid-filled, membrane-bounded sac found within the cytoplasm; may function in storage, digestion, or water elimination.

_____ 5. Any small sac, especially a small spherical membrane-bounded compartment, within the cytoplasm.

_____ 6. A chlorophyll-bearing intracellular organelle of some plant cells.

_____ 7. A stack of thylakoids within a chloroplast.

_____ 8. The fluid region of the chloroplast.

_____ 9. An interconnected network of intracellular membranes.

_____10. An organelle that is part of the protein synthesis machinery.

MAKING COMPARISONS

Fill in the blanks.

Cell Structure	Location of Structure	Function of Structure	Kind of Cell Containing This Structure
Nuclear area containing a single strand of circular DNA	Cytoplasm	Inheritance, control center	Prokaryotic
#1	Cytoplasm and RER	Synthesis of polypeptides	#12
Golgi complex	#3	#4	Most eukaryotic
Mitochondria	#5	#6	#7
Complex chromosomes	#8	#9	#10
Centrioles	#11	#12	#13

Cell Structure	Location of Structure	Function of Structure	Kind of Cell Containing This Structure
Cell wall	#14	Protection, support	#15
#16	#17	Encloses cellular contents, regulates passage of materials into and out of cell	#18
Chloroplasts	#19	#20	Eukaryotic (primarily plants)

MAKING CHOICES

Place your answer(s) in the space provided. Some questions may have more than one correct answer.

_____ 1. A nanometer (nm) is

    a. one billionth of a meter.   d. one millionth of a micrometer.

    b. one millionth of a centimeter.   e. one thousandth of a micrometer.

    c. one millionth of a millimeter.

_____ 2. Which of the following is in the nucleolus, but not normally found in the rest of chromatin?

    a. DNA   d. RNA

    b. protein   e. ribosomes

    c. chromosomes

_____ 3. Cell membrane functions include

    a. isolation of chemical reactions.   d. maintaining cell shape.

    b. selective permeability.   e. concentration of reactants.

    c. moderating between a cell's internal and external environments.

_____ 4. The high resolution attained by the electron microscope is attributed to its use of

    a. fixed specimens.   d. grid patterns.

    b. electromagnetic lenses.   e. electrons instead of light.

    c. short wavelengths.

_____ 5. Cells are small at least in part because as size increases the surface-to-volume ratio

    a. doubles.   d. decreases.

    b. decreases to half.   e. reduces efficiency of cell activities.

    c. increases.

_____ 6. Inner membranous folds in mitochondrion

    a. are called cristae.   d. contain enzymes.

    b. increase membrane surface area.   e. contain structural proteins.

    c. extend into the intermembrane space.

_____ 7. The Golgi complex functions to
    a. modify proteins.
    b. process proteins.
    c. packages glycoproteins.
    d. sort molecules.
    e. break down large carbohydrates.

_____ 8. The membranes that comprise the endomembrane system include
    a. Golgi complex.
    b. lysosomes.
    c. endoplasmic reticulum.
    d. transport vesicles.
    e. plasma membrane.

_____ 9. Lysosomes
    a. contain digestive enzymes.
    b. contain nucleic acids.
    c. possess a membrane.
    d. break down complex molecules.
    e. break down organelles.

_____10. Which of the following cells contain plastids?
    a. animal
    b. plant
    c. some eukaryotic
    d. some prokaryotic
    e. algae

_____11. The "cytoskeleton" of eukaryotic cells
    a. changes constantly.
    b. includes microfilaments.
    c. includes some DNA.
    d. functions in cell movements.
    e. includes protein.

_____12. The cell theory states that
    a. new cells come from preexisting cells.
    b. all cells are descended from ancient cells.
    c. cells divide.
    d. living things are composed of cells.
    e. cells contain genetic material.

_____13. Chloroplasts and mitochondria both
    a. are found in plant cells.
    b. have two membranes.
    c. contain DNA.
    d. are found in animal cells.
    e. contain a matrix.

_____14. Ribosomal RNA is synthesized in the
    a. cytoplasm.
    b. nuclear matrix.
    c. cristae.
    d. mitochondria.
    e. nucleolus.

_____15. Actin, myosin, and tubulin are
    a. proteins.
    b. in chromatin.
    c. primarily in neurons.
    d. components of filaments.
    e. components of the plasma membrane.

VISUAL FOUNDATIONS

Color the parts of the illustration below as indicated.

RED ☐ plasma membrane
GREEN ☐ nuclear area
YELLOW ☐ cell wall
PINK ☐ capsule
BLUE ☐ flagellum
ORANGE ☐ pili

Questions 1 and 2 pertain to the illustration above.

1. Is this a prokaryotic cell or a eukaryotic cell? _____

2. Which of the colored parts is/are unique to this kind of cell? _____

Color the parts of the illustration below as indicated.

RED ❑ plasma membrane

ORANGE ❑ nucleus

YELLOW ❑ cell wall

BLUE ❑ prominent vacuole

GREEN ❑ chloroplast

BROWN ❑ mitochondrion

TAN ❑ endomembrane system (Note: The plasma membrane and the outer portion of the nuclear envelop were assigned other colors. Use tan to identify the remaining structures in this system.)

Questions 3-5 pertain to the illustration below.

3. What kind of cell is illustrated by this generalized diagram? _____

4. Which of the colored parts is/are considered components of the cytoplasm of this cell? _____

_____

5. Which of the colored parts is/are characteristic of this kind of cell and not generally associated with other types of cells? _____

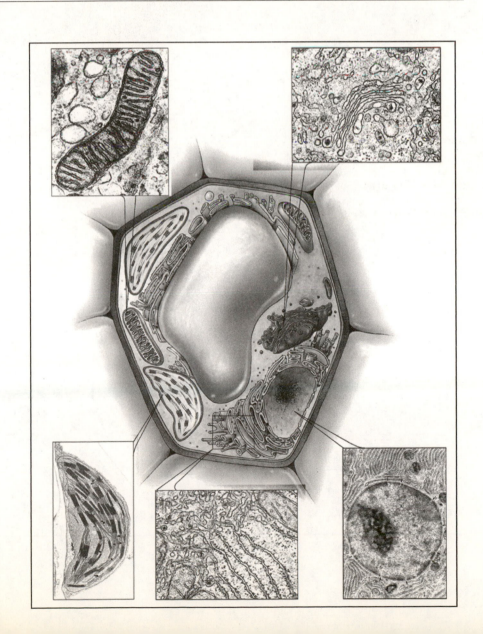

Color the parts of the illustration below as indicated.

RED ☐ plasma membrane

ORANGE ☐ nucleus

YELLOW ☐ centriole

BROWN ☐ mitochondrion

BROWN ☐ endomembrane system (Note: The plasma membrane and the outer portion of the nuclear envelope have already been assigned colors. Use tan to identify the remaining structures in this system.)

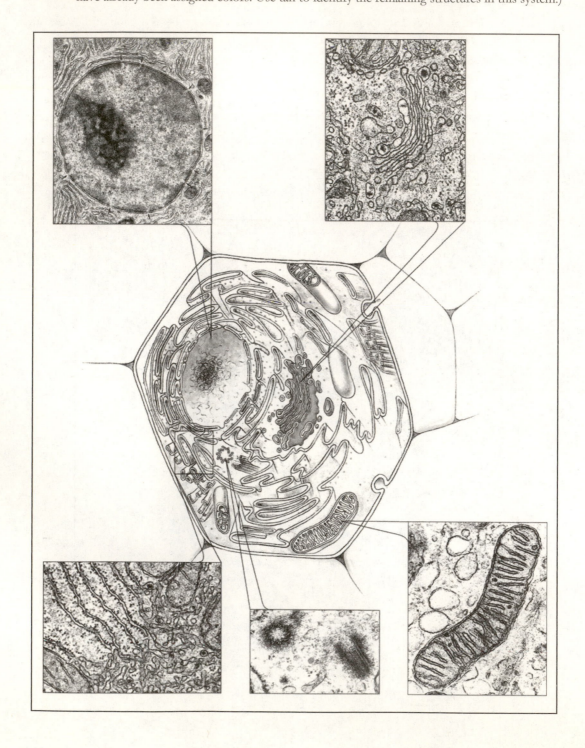

Questions 6-8 pertain to the illustration above.

6. What kind of cell is illustrated by this generalized diagram? _____

7. Which of the colored parts is/are considered components of the cytoplasm of this cell? _____

_____

8. Which of the colored parts is/are characteristic of this kind of cell and not generally associated with other types of

cells? _____

CHAPTER 5

Biological Membranes

Biological membranes serve to enclose cells, separating the interior world of the cell from the exterior. In eukaryotes, they also enclose a variety of internal structures and form extensive internal membrane systems. Additionally, membranes function to provide work surfaces for many chemical reactions, to regulate movement of materials in and out of the cell, to transmit signals and information between the environment and the interior of the cell, and to act as an essential part of energy transfer and storage systems. Membranes are phospholipid bilayers with both internal and peripheral proteins associated with them. Membranes have the ability to seal themselves, round up and form closed vesicles, and fuse with other membranes. Molecules pass through membranes in a variety of ways. In multicellular organisms, cell membranes have specialized structures associated with them that allow neighboring cells to form strong connections or to establish rapid communication.

REVIEWING CONCEPTS
Fill in the blanks.

INTRODUCTION

1. All cells are physically separated from the external environment by a

   _____ .

2. Membrane proteins known as _____ are responsible
   for calcium-dependent adhesion between cells in multicellular sheets.

THE STRUCTURE OF BIOLOGICAL MEMBRANES
Phospholipids form bilayers in water

3. The physical properties of biological membranes are due primarily to their

   _____ .

4. The headgroups of phospholipid molecules are said to be (a)_____
   because they readily associate with water. On the other hand, the
   (b)_____ ends turn away from water and associate with each other.

5. The cylindrical shape and strongly _____ character of
   phospholipid molecules are features responsible for the formation of the bilayer.

Current data support a fluid mosaic model of membrane structure

6. The theory of membrane structure known as the (a)_____
   model holds that membranes consist of a dynamic, fluid (b)_____
   and embedded or otherwise associated globular (c)_____ .

Biological membranes are two-dimensional fluids

7. The crystal-like properties of many phospholipid bilayers is due to the orderly
   arrangement of (a)_____ on the outside and
   (b)_____ on the inside.

8. The two-dimensional fluid property of the lipid bilayer is due to the constant motion of _____.

Biological membranes fuse and form closed vesicles

9. Products are secreted from cells by the fusion of vesicles with the

_____.

Membrane proteins include integral and peripheral proteins

10. _____ have hydrophobic regions that interact with the fatty acid tails of the membrane phospholipids.

11. _____ are not embedded in the lipid bilayer.

Proteins are oriented asymmetrically across the bilayer

12. Membrane proteins that will become part of the inner surface of the plasma membrane are manufactured by (a)_____ and moved to the membrane through the (b)_____.

Membrane proteins function in transport, in information transfer, and as enzymes

13. A variety of membrane functions is made possible by the diversity of _____ molecules in membranes.

PASSAGE OF MATERIALS THROUGH CELL MEMBRANES

Biological membranes present a barrier to polar molecules

14. Small _____ molecules easily permeate plasma membranes.

Transfer proteins transfer molecules across membranes

15. _____ are channel proteins that facilitate the transport of water through the plasma membrane in response to osmotic gradients.

PASSIVE TRANSPORT

Diffusion occurs down a concentration gradient

16. Diffusion rate is affected by temperature and the _____ _____ of the moving particles.

Osmosis is diffusion of water across a selectively permeable membrane

17. The movement of water through a selectively permeable membrane from a region of higher concentration to a region of lower concentration is called

_____.

18. The _____ of a solution is determined by the amount of dissolved substances in the solution.

Facilitated diffusion occurs down a concentration gradient

19. In facilitated diffusion, the net movement is always from a region of (a)_____ solute concentration to a region of (b)_____ solute concentration.

ACTIVE TRANSPORT

Active transport systems "pump" substances against their concentration gradients

20. Because of the sodium-potassium pump, there are _____ (fewer or more) potassium ions inside the cell relative to the sodium ions outside

21. The distribution of sodium and potassium across cell membranes causes the inside of the cell to be _____ charged relative to the outside.

Carrier proteins can transport one or two solutes

22. (a)_____ transport one type of substance in one direction and (b)_____ move two types of substances in one direction.

Cotransport systems indirectly provide energy for active transport

23. The process whereby the movement of one solute down its concentration gradient provides the energy to transport another solute up its concentration gradient is known as _____.

EXOCYTOSIS AND ENDOCYTOSIS

In exocytosis, vesicles export large molecules

24. Exocytosis is the exportation of waste materials or specific products by the fusion of a _____ with the plasma membrane of a cell.

In endocytosis, the cell imports materials

25. In (a)_____ a cell ingests large, solid particles. In (b)_____ a cell takes in dissolved materials.

CELL JUNCTIONS

26. Three types of junctions, the _____ _____, are specialized structures associated with the plasma membrane of animal cells.

Anchoring junctions connect cells of an epithelial sheet

27. The two common types of anchoring junctions are (a)_____ that connect with microfilaments of the cytoskeleton, and (b)_____ that are anchored to intermediate filaments in the cell.

Tight junctions seal off intercellular spaces between some animal cells

28. The _____, composed of tight junctions, prevents many substances in the blood from passing into the brain.

Gap junctions allow the transfer of small molecules and ions

29. Gap junctions consist of a cluster of (a)_____ that form (b)_____ that connect the cytoplasm of adjacent cells.

Plasmodesmata allow certain molecules and ions to move between plant cells

30. Plasmodesmata in plant cells are functionally equivalent to _____ in animal cells.

31. Most plasmodesmata contain a cylinder-shaped structure called the _____.

BUILDING WORDS

Use combinations of prefixes and suffixes to build words for the definitions that follow.

Prefixes	The Meaning		Suffixes	The Meaning
desm(o)-	bond		-cyto(sis)	cell
endo-	within		-desm(a)	bond

exo-	outside, outer, external	-some	body
hyper-	over		
hypo-	under		
iso-	equal, "same"		
phago-	eat, devour		
pino-	drink		

Prefix	**Suffix**	**Definition**
_____	_____	1. A process whereby materials are taken into the cell.
_____	_____	2. The process whereby waste or secretion products are ejected from a cell by fusion of a vesicle with the plasma membrane.
_____	-tonic (-osmotic)	3. Having an osmotic pressure or solute concentration that is greater than a standard solution.
_____	-tonic (-osmotic)	4. Having an osmotic pressure or solute concentration that is less than a standard solution.
_____	-tonic (-osmotic)	5. Having an osmotic pressure or solute concentration that is the same as a standard solution.
_____	_____	6. The ingestion of large solid particles, such as bacteria and food, by a cell.
_____	_____	7. A type of endocytosis whereby fluid is engulfed by vesicles originating at the cell surface.
_____	_____	8. A button-like plaque (body) present on two opposing cell surfaces, that holds (bonds) the cells together by means of protein filaments that span the intercellular space.
plasmo-	_____	9. A cytoplasmic channel connecting (bonding) adjacent plant cells and allowing for the movement of small molecules and ions between cells.

MATCHING

Terms:

a. Active transport
b. Concentration gradient
c. Cotransport
d. Dialysis
e. Diffusion
f. Facilitated diffusion
g. Fluid-mosaic model
h. Gap junction
i. Osmosis
j. Signal transduction
k. Selectively permeable membrane
l. Tight junction
m. Turgor pressure

For each of these definitions, select the correct matching term from the list above.

_____ 1. The modern picture of membranes in which protein molecules float in a phospholipid bilayer.

_____ 2. The diffusion of water across a selectively permeable membrane.

_____ 3. The transport of ions or molecules across a membrane by a specific carrier protein.

_____ 4. Regions in a system of differing concentration, such as exist in a cell and its environment, that cause molecules to move from areas of higher concentration to lower concentration.

_____ 5. Energy-requiring transport of a molecule across a membrane from a region of low concentration to a region of high concentration.

_____ 6. A membrane that allows some substances to cross it more easily than others.

_____ 7. The random movement of molecules from a region of higher concentration to one of lower concentration of that substance.

_____ 8. The internal pressure in a plant cell caused by the diffusion of water into the cell.

_____ 9. A specialized structure between some animal cells, producing a tight seal that prevents materials from passing through the spaces between the cells.

_____10. Structure consisting of specialized regions of the plasma membranes of two adjacent cells containing numerous pores that allow passage of certain small molecules and ions between them.

MAKING COMPARISONS

Fill in the blanks.

Transport Mechanism	Description of the Transport Mechanism
Diffusion	Net movement of a substance from an area of high concentration to an area of low concentration
#1	Cell ingests large particles such as bacteria and other food; literally means "cell eating"
#2	The passive transport of solutes down a concentration gradient aided by a specific protein in a membrane
#3	The active transport of substances into the cell by the formation of invaginated regions or "folds" of the plasma membrane that pinch off and become cytoplasmic vesicles
#4	Transfer of solutes by proteins located within the membrane
#5	The active transport of solutes against a concentration gradient by a specific carrier protein in a membrane
Exocytosis	#6
#7	Diffusion of water across a selectively permeable membrane; a passive transport mechanism
#8	Transport of substances across a membrane requiring the expenditure of energy by the cell
Pinocytosis	#9
#10	Points of attachment between cells that hold cells together

MAKING CHOICES

Place your answer(s) in the space provided. Some questions may have more than one correct answer.

_____ 1. Facilitated diffusion

a. moves molecules against a gradient.

b. does not occur in prokaryotes.

c. requires use of ATP.

d. involves protein channels.

e. moves substances both into and out of cells.

_____ 2. Active transport

a. can move molecules against a gradient.

b. does not occur in prokaryotes.

c. requires use of ATP.

d. occurs in animal cells, not in plant cells.

e. moves substances both into and out of cells.

_____ 3. A molecule is called amphipathic when it

a. prevents free passage of substances.

b. is in a membrane.

c. has hydrophobic and hydrophilic regions.

d. is a lipid.

e. is embedded in a bilipid layer.

_____ 4. If membranes were not fluid and dynamic, which of the following might still occur normally?

a. active transport

b. facilitated diffusion

c. simple diffusion

d. endo- and exocytosis

e. osmosis

_____ 5. Membrane fusion enables

a. diversity of proteins.

b. exocytosis.

c. endocytosis.

d. fusion of vesicles and plasma membrane.

e. formation of Golgi bodies.

_____ 6. Cell walls

a. are large plasma membranes.

b. occur in plant cells.

c. are more rigid than membranes.

d. are fluid.

e. contain desmosomes.

Red blood cells are placed in three beakers containing the following solutions: beaker A, distilled water; beaker B, isotonic solution; beaker C, 13% salt solution. Use this information to answer questions 7-9.

_____ 7. The cells in beaker C

a. shriveled.

b. swelled.

c. plasmolyzed.

d. were unaffected.

e. probably eventually burst.

_____ 8. The cells in beaker A

a. shriveled.

b. swelled.

c. plasmolyzed.

d. were unaffected.

e. probably eventually burst.

_____ 9. The cells in beaker B

a. shriveled.

b. swelled.

c. plasmolyzed.

d. were unaffected.

e. probably eventually burst.

_____ 10. Plasma membranes of eukaryotic cells have a large amount of

a. cholesterols.

b. phospholipid.

c. rigidity.

d. fluidity.

e. protein.

_____11. Diffusion rate depends on
    a. the flow of water.   d. the plasma membrane.
    b. concentration gradient.   e. kinetic energy.
    c. energy from the cell.

_____12. Endocytosis may include
    a. secretion vacuoles.   d. a combination of inbound particles with proteins.
    b. pinocytosis.   e. receptor-mediation.
    c. phagocytosis.

_____13. Plasmodesmata
    a. are channels in cytoplasm.   d. are the same as several plasmodesma.
    b. connect plant cells.   e. connect ER of adjacent cells.
    c. are plant cell structures equivalent to animal cell desmosomes.

_____14. The principal cell adhesion molecules in vertebrates are known as
    a. cadherins.   d. plasmodesma.
    b. integral proteins.   e. β-pleated molecules.
    c. glycoproteins.

_____15. Sugars may be added to proteins
    a. and porins.   d. in the lumen of ER.
    b. to form gated channels.   e. in the extracellular space.
    c. resulting in a glycoprotein.

VISUAL FOUNDATIONS

Color the parts of the illustration below as indicated. Label the interior and exterior of the cell, the carbohydrate chains, and the lipid bilayer.

RED ☐ alpha helix
GREEN ☐ hydrophilic region of transmembrane proteins
YELLOW ☐ hydrophobic region of transmembrane proteins
BLUE ☐ glycolipid
ORANGE ☐ glycoprotein
BROWN ☐ cholesterol
TAN ☐ peripheral protein

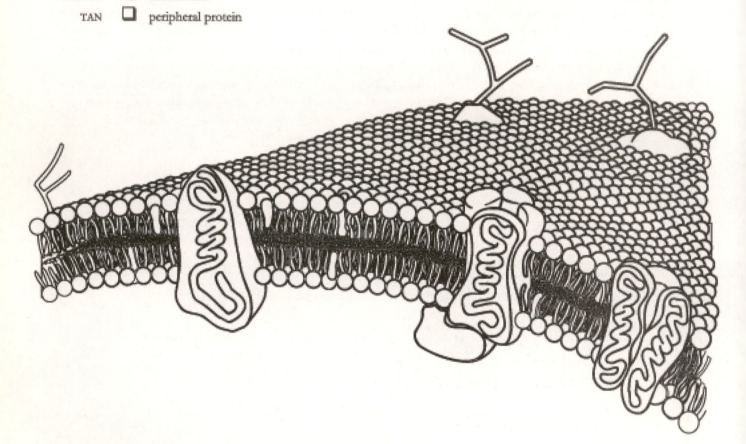

1. What is the name of the model for membrane structure illustrated above?_____

2. Which of the colored parts is most important in forming junctions between adjacent cells? _____

CHAPTER 6

Cell Communication

No study of the organization of life, the subject matter of Part I of the textbook, would be complete without an examination of the way cells communicate with one another, i.e., how cells send and receive signals, how information crosses the plasma membrane and is transmitted through the cell, and finally, how cells respond to the signals they receive. To maintain homeostasis, cells must continuously receive and respond to information from the outside environment, some of which is sent by other cells. In multicellular organisms, cells must also communicate with one another, sometimes directly and sometimes indirectly. Organisms of different species, and even different kingdoms and domains, communicate with one another. Over billions of years, elaborate systems of cell communication have evolved. Most organisms communicate with other members of their species by secreting chemical signals. Faulty signals can cause or contribute to a variety of diseases, such as diabetes and cancer.

REVIEWING CONCEPTS
Fill in the blanks.

INTRODUCTION

1. To maintain homeostasis, cells must continuously receive and respond to information from_____.

2. Prokaryotes, protists, fungi, plants, and animals all communicate with other members of their species by _____.

3. A community of microorganisms attached to a solid surface is known as a _____.

CELL SIGNALING: AN OVERVIEW

4. The term _____ refers to the mechanisms by which cells communicate with one another.

5. The cells that can respond to a signaling molecule are called _____.

6. _____ is the process by which a cell converts an extracellular signal into an intracellular signal that results in a response.

SENDING SIGNALS

7. _____ is a process in which a signaling molecule diffuses through the interstitial fluid and acts on nearby cells.

8. _____ are paracrine regulators that modify cAMP levels and interact with other signaling molecules to regulate metabolic activities.

9. Most neurons signal one another by releasing chemical compounds called _____.

RECEPTION

10. A _____ is a large protein or glycoprotein that binds with a signaling molecule.

11. A signaling molecule that binds to a specific receptor is called a _____.

12. Pigments in plants, some algae, and some animals that absorb blue light and play a role in biological rhythms are called _____.

13. Receptor down-regulation occurs in response to (a)_____ hormone concentrations and involves (b)_____ the number of receptors.

14. Receptor up-regulation occurs in response to (a)_____ hormone concentrations and involves (b)_____ the number of receptors.

Three types of receptors occur on the cell surface

15. The three main types of receptors on the cell surface are (a)_____, (b)_____, and (c)_____.

16. _____ are transmembrane proteins with a binding site for a signaling molecule outside the cell and an enzyme component inside the cell.

Some receptors are intracellular

17. Intracellular proteins that regulate the expression of specific genes are called _____.

SIGNAL TRANSDUCTION

18. Many regulatory molecules transmit information to the cell's interior without physically crossing the _____.

Ion channel-linked receptors open or close channels

19. Gamma-aminobutyric acid is a neurotransmitter that binds to _____ in neurons, thereby inhibiting neural signaling.

G-protein-linked receptors initiate signal transduction

20. G proteins are a group of regulatory proteins important in many _____ pathways.

Second messengers are intracellular signaling agents

21. A chain of molecules in the cell that relays a signal is called a _____.

22. In many signaling cascades in prokaryotic and animal cells, the second messenger is _____.

23. Adenylyl cyclase is an enzyme on the _____ side of the plasma membrane.

24. When adenylyl cyclase is activated, it catalyzes the formation of (a)_____, which in turn activates (b)_____, leading to the phosphorylation of (c)_____, resulting in some response in the cell.

25. Ion pumps in the plasma membrane normally maintain a _____ calcium ion concentration in the cytosol compared to its concentration in the extracellular fluid.

Enzyme-linked receptors function directly

26. Most enzyme-linked receptors are (a)_____ , enzymes that (b)_____ the amino acid tyrosine in proteins.

Many activated intracellular receptors are transcription factors

27. Some hydrophobic signaling molecules diffuse across the membranes of target cells and bind with _____ in the cytosol or in the nucleus.

28. Nitric oxide is a gaseous signaling molecule that passes into target cells binding directly with the enzyme (a)_____ , thereby catalyzing the production of (b)_____ , and resulting in the relaxation of the smooth muscle in (c)_____ .

Scaffolding proteins increase efficiency

29. Scaffolding proteins organize groups of (a)_____ into (b)_____ , ensuring that signals are relayed accurately, rapidly, and more efficiently.

Signals can be transmitted in more than one direction

30. Transmembrane proteins that connect the cell to the extracellular matrix are called _____ .

RESPONSES TO SIGNALS

31. In both plants and animals, _____ regulate development by causing changes in the expression of specific genes.

32. When a plant is exposed to drought conditions, gene activation and protein synthesis lead to increased concentration of _____ in its leaves.

The response to a signal is amplified

33. The process of magnifying the strength of a signaling molecule is called _____ .

Signals must be terminated

34. Signal termination returns the receptor and each of the components of the _____ pathway to their _____ states.

EVOLUTION OF CELL COMMUNICATION

35. Evidence suggests that cell communication first evolved in _____ and continued to change over time as new types of organisms evolved.

BUILDING WORDS

Use combinations of prefixes and suffixes to build words for the definitions that follow.

Prefixes	The Meaning		Suffixes	The Meaning
di-	two, twice, double		-chrome	color
intra-	within		-crine	to secrete
neuro-	nerve			
para-	beside, near			
phyto-	plant			
tri-	three			

Prefix	Suffix	Definition
_____	_____	1. Pertains to cell secretions (signaling molecules) that diffuse through the interstitial fluid and act on nearby cells.
_____	-transmitter	2. Substance used by neurons to transmit impulses across a synapse.
_____	_____	3. A blue-green, proteinaceous pigment involved in photoperiodism and a number of other light-initiated physiological responses of plants.
_____	-cellular	4. Pertains to the inside of a cell.
_____	-phosphate	5. A chemical compound with three phosphate groups.
_____	-phosphate	6. A chemical compound with two phosphate groups.
_____	-acylglycerol	7. A compound formed of two fatty acids and one glycerol molecule.

MATCHING

Terms:

a.	Adenylyl cyclase	g.	Intracellular receptor	m.	Signal termination
b.	Calmodulin	h.	Ligand	n.	Signal transduction
c.	Enzyme-linked receptor	i.	Neurotransmitter	o.	Transcription factor
d.	G protein-linked receptor	j.	Nitric oxide	p.	Tyrosine kinase
e.	Guanosine diphosphate	k.	Ras protein		
f.	Hormone	l.	Signal amplification		

For each of these definitions, select the correct matching term from the list above.

____ 1. A signaling molecule that binds to a specific receptor.

____ 2. The process in which a receptor converts an extracellular signal into an intracellular signal that causes some change in the cell.

____ 3. A gaseous signaling molecule that passes into target cells.

____ 4. A signaling molecule released by neurons to signal one another.

____ 5. A chemical messenger secreted by endocrine glands.

____ 6. A transmembrane protein composed of seven alpha helices connected by loops that extend into the cytosol or outside the cell.

____ 7. A transmembrane protein with a binding site for a signaling molecule outside the cell and a binding site for an enzyme inside the cell.

____ 8. A receptor located in the cytosol or in the nucleus.

____ 9. A molecule similar to ADP but containing the base guanine instead of adenine.

_____ 10. An enzyme-linked receptor that phosphorylates specific tyrosines in certain signaling proteins inside the cell.

_____ 11. One of a group of small G proteins that, when activated, trigger a cascade of reactions.

_____ 12. A receptor located in the nucleus that activates or represses specific genes.

_____ 13. An enzyme on the cytoplasmic side of the plasma membrane that catalyzes the formation of cyclic AMP from ATP.

_____ 14. A protein that when combined with calcium ions affects the activity of protein kinases and protein phosphatases.

_____ 15. The process of enhancing signal strength as a signal is relayed through a signal transduction pathway.

_____ 16. The process of inactivating the receptor and each component of the signal transduction pathway once they have done their jobs.

MAKING COMPARISONS

Refer to this illustration of signal transduction to fill in blanks in the table below.

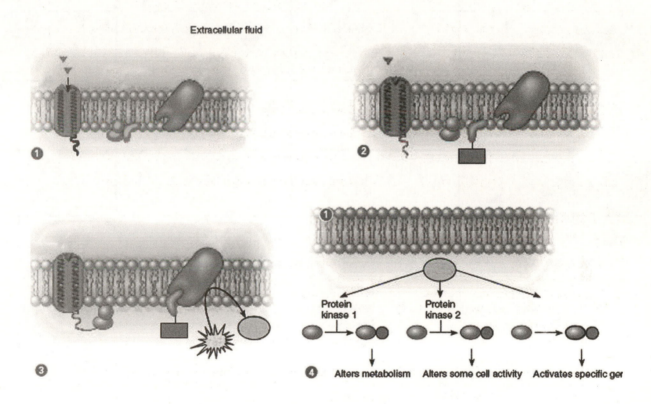

Identify the structures in the table and fill in the blanks.

Structure	Name of Structure	Structure	Name of Structure
entire structure	receptor molecule	at the arrow	#1

Structure	Name of Structure	Structure	Name of Structure
at the arrow	#2	at the arrow	#3
at the arrow	#4	entire structure	#5
entire structure	#6	entire structure	#7
at the arrow	#8	at the arrow	#9
at the arrow	#10		

MAKING CHOICES

Place your answer(s) in the space provided. Some questions may have more than one correct answer.

_____ 1. A community of bacteria attached to a solid surface is a

a. clone.
b. quorum.
c. biofilm.
d. communication population.
e. colony.

_____ 2. Cells in multicellular organisms that respond to a chemical signal from other cells in the same organism are called

a. transmitter cells.
b. signaling cells.
c. responders.
d. secretory cells.
e.. target cells

_____ 3. The process in cells of converting an extracellular signal into an intracellular signal to which a response occurs is known as

a. optimization.
b. transduction.
c. internal regulation.
d. chemical transposition.
e. chemical signal reception.

_____ 4. The process whereby a signaling molecule diffuses through interstitial fluid and affects nearby cells is

    a. paracrine regulation.
    d. interstitial modification.

    b. cellular reception.
    e. endocrine facilitation.

    c. interstitial reception.

_____ 5. The signaling chemicals transferred across the gap between neurons are called

    a. neuroreceptors.
    d. neuronal enhancers.

    b. synaptic transmitters.
    e. neuron communicators.

    c. neurotransmitters.

_____ 6. Endocrine glands

    a. have no ducts.
    d. secrete chemical messengers into interstitial fluid.

    b. secrete hormones.
    e. secrete hormones into capillaries.

    c. communicate directly with neurons.

_____ 7. Proteins and glycoproteins that bind with ligands are called

    a. enzymes.
    d. bonding elements.

    b. messengers.
    e. receptors.

    c. signaling molecules.

_____ 8. Receptor down-regulation may involve

    a. destruction of receptors by lysosomes.
    d. increased intake of insulin by cells.

    b. decreased numbers of receptors.
    e. degradation of receptors.

    c. regulation of blood glucose levels.

_____ 9. Receptor up-regulation occurs in response to

    a. low hormone concentrations.
    d. increase in protein membrane complexes.

    b. high hormone concentrations.
    e. integration.

    c. receptor down-regulation.

_____ 10. Receptors on the cell surface that convert chemical signals into electrical signals are called

    a. G protein-linked receptors.
    d. protein complex receptors.

    b. ion channel-linked receptors.
    e. ligand-gated channels.

    c. enzyme-linked receptors.

_____ 11. Proteins that regulate the expression of genes are

    a. cell surface proteins.
    d. intracellular receptors.

    b. cell membrane glycoproteins.
    e. intranuclear integrators.

    c. transcription factors.

_____ 12. When gamma-aminobutyric acid (GABA) binds to a ligand-gated chloride ion channel in a neuron

    a. the channel opens.
    d. neural signaling is inhibited.

    b. adjacent channels close.
    e. the neuron's electrical potential changes.

    c. chloride ions flow into the neuron.

_____ 13. Second messengers

    a. are ions or small molecules.
    d. are enzymes.

    b. may amplify a neuronal signal.
    e. are hormones.

    c. are found mainly in interstitial fluids.

_____ 14. Calcium ion messengers are involved in

    a. microtubule disassembly.
    d. immune system cell activation.

    b. muscle contraction.
    e. initiation of development.

    c. blood clotting.

_____ 15. Calmodulin

    a. combines to only one specific enzyme.   d. is found in eukaryotic cells.

    b. produces the binding ezyme calmodulinase.   e. is a Ca^{2+} binding protein.

    c. changes the shape of the 4- Ca^{2+} binding complex

_____ 16. Ras proteins

    a. are activated when bound to GTP.   d. trigger a series of reactions called the Ras pathway.

    b. are large proteins.   e. phosphorylate glucosamine.

    c. are a group of G proteins.

_____ 17. Integrins

    a. are transmembrane proteins.   d. are enzymes that catalyze signaling reactions.

    b. are intracellular signaling transmitters.   e. may be involved in "inside-out" signaling.

    c. interface between the cell and the extracellular matrix.

_____ 18. The components of signal transduction pathways have been identified in

    a. protists.   d. plants.

    b. fungi.   e. yeasts.

    c. animals.

VISUAL FOUNDATIONS

Color the parts of this illustration of the mechanism of intracellular reception as indicated.

RED   ❑   extracellular signaling molecules

GREEN   ❑   signaling molecules moving through the cytosol

YELLOW   ❑   intranuclear receptor

BLUE   ❑   transcription factor (activated receptor)

ORANGE   ❑   DNA

BROWN   ❑   m-RNA

PINK   ❑   ribosomes

TAN   ❑   protein that alters cell activity

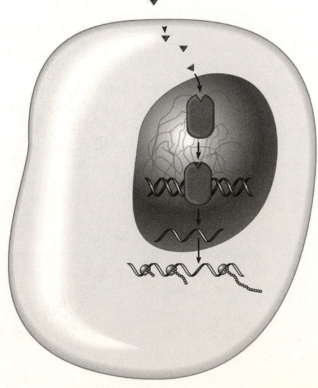

Color the parts of this illustration of types of cell signaling as indicated.

RED	❏	generalized signaling cells
GREEN	❏	generalized target cells
YELLOW	❏	receptors
BLUE	❏	signaling molecules
ORANGE	❏	signaling neuron
BROWN	❏	receptor neuron

PART II

□

Energy Transfer Through Living Systems

CHAPTER 7

Energy and Metabolism

Having concluded Part I of the textbook, you have an understanding of the chemical materials that function in life's processes and of the cell and its membranes. You now have the foundation to examine one of the basic themes of biology – the flow of energy through organisms. Life depends on a continuous input of energy. The myriad chemical reactions of cells that enable them to function involve energy transformations. These transformations are governed by the laws of thermodynamics that explain why organisms cannot produce energy but must continuously capture it from somewhere else, and why in every energy transaction, some energy is dissipated as heat. Some chemical reactions occur spontaneously, releasing free energy that is then available to perform work. Other reactions require an input of free energy before they can occur. Chemical reactions in organisms are regulated by enzymes, substances that lower the amount of energy needed to activate reactions. The effectiveness of some drugs is due to their ability to inhibit enzymes that are critical to the normal functioning of certain pathogenic organisms.

REVIEWING CONCEPTS
Fill in the blanks.

INTRODUCTION

1. Life depends on a continuous supply of outside energy. Fortunately, _____ capture outside sources of energy and incorporate it in chemical bonds (food).

2. Plants convert radiant energy to _____ energy.

BIOLOGICAL WORK

3. Energy is the capacity to do _____.

Organisms carry out conversions between potential energy and kinetic energy

4. Energy is in one of two forms: (a)_____ is "stored energy" and (b)_____ is "energy of motion."

THE LAWS OF THERMODYNAMICS

5. The study of energy and its transformations is called _____.

The total energy in the universe does not change

6. The first law of thermodynamics states that energy can be neither (a)_____ nor _____, however, it can be (b)_____ and changed in form.

The entropy of the universe is increasing

7. In every energy conversion or transfer some energy is dissipated as _____.

8. The term entropy refers to the _____ in the universe.

ENERGY AND METABOLISM

Enthalpy is the total potential energy of a system

9. The energy required to break a chemical bond is referred to as _____.

10. The total potential energy of a chemical reactions, or enthalpy, equals the _____ of the reactants and products.

Free energy is available to do cell work

11. An increase in _____ leads to a decrease in the amount of free energy.

Chemical reactions involve changes in free energy

12. The change in free energy during a reaction is equal to the change in (a)_____ minus the product of the absolute temperature multiplied by the change in (b)_____.

Free energy decreases during an exergonic reaction

13. _____ are spontaneous and they release energy that can perform work.

14. The mathematical/chemical symbol for change in free energy is _____.

Free energy increases during an endergonic reaction

15. In an endergonic reaction, free energy has a _____ value.

Diffusion is an exergonic process

16. In a concentration gradient, energy moves from a region of (a)_____ concentration to a region of (b)_____ concentration.

Free-energy changes depend on the concentrations of reactants and products

17. In a state of _____ , the rate of the reverse reaction equals the rate of the forward reaction.

Cells drive endergonic reactions by coupling them to exergonic reactions

18. Exergonic reactions (a)_____[release or require input of?] free energy; endergonic reactions (b)_____[release or require input of?] free energy.

19. In the reaction C → D, where the value of $\triangle G$ is negative, the reactant has _____(more or less?) free energy than the free energy of the product.

ATP, THE ENERGY CURRENCY OF THE CELL

20. The three main parts of the ATP molecule are _____, _____, and _____.

ATP donates energy through the transfer of a phosphate group

21. Phosphate bonds in ATP are transferred by the process known as _____.

ATP links exergonic and endergonic reactions

22. Exergonic reactions are generally part of (a)_____ pathways and endergonic reactions are generally part of (b)_____ pathways.

The cell maintains a very high ratio of ATP to ADP

23. The cell cannot store _____ quantities of ATP.

ENERGY TRANSFER IN REDOX REACTIONS

Most electron carriers transfer hydrogen atoms

24. Stripped electrons and the energy they possess are transferred to an _____.

ENZYMES

All reactions have a required energy of activation

25. The activation energy of a reaction begins the reaction by using energy to _____.

An enzyme lowers a reaction's activation energy

26. Enzymes lower the activation energy to _____ the rate of the reaction.

An enzyme works by forming an enzyme-substrate complex

27. Enzymes lower activation energy by forming an unstable intermediate called the _____.

28. When the substrate binds to the enzyme molecule, it causes a change, known as _____.

29. Most enzyme names end in _____.

Enzymes are specific

30. Most enzymes are specific because the shape of the _____ is closely related to the shape of the substrate.

Many enzymes require cofactors

31. An organic cofactor is called a _____.

32. Some enzymes have two components, the cofactor and a protein referred to as the _____.

Enzymes are most effective at optimal conditions

33. Factors that affect enzyme activity include (list three) _____ _____.

Enzymes are organized into teams in metabolic pathways

34. When enzymes work in teams, the (a)_____ from one enzyme-substrate reaction becomes the (b)_____ for the next enzyme-substrate reaction.

35. A series of reactions can be illustrated as A $\rightarrow$ B $\rightarrow$ C ... etc. where each reaction ($\rightarrow$) is carried out by a specific enzyme. Such a series is referred to as a _____.

The cell regulates enzymatic activity

36. Feedback inhibition is a type of enzyme regulation in which the formation of a product _____ an earlier reaction in a sequence.

Enzymes can be inhibited by certain chemical agents

37. Inhibition is _____ when the inhibitor-enzyme bond is weak.

38. _____ inhibition occurs when the inhibitor competes with the normal substrate for binding to the active site of the enzyme.

39. _____ inhibition occurs when the inhibitor binds to the enzyme at a site other than the active site.

Some drugs are enzyme inhibitors

40. Because sulfa drugs have a chemical structure similar to _____, they are able to selectively affect bacteria.

BUILDING WORDS

Use combinations of prefixes and suffixes to build words for the definitions that follow.

Prefixes	The Meaning		Suffixes	The Meaning
allo-	other, "another"		-calor(ie)	heat
ana-	up		-ergonic	work, "energy"
cata-	down		-steric	"space"
end(o)-	within			
ex(o)-	outside, outer, external			
kilo-	thousand			

Prefix	Suffix	Definition
_____	_____	1. Heat energy; the amount of heat required to raise the temperature of 1000 grams (1 kg) of water from 14.5° C to 15.5° C.
_____	-bolism	2. In living organisms, the "building up" (synthesis) of more complex substances from simpler ones.
_____	-bolism	3. In living organisms, the "breaking down" of more complex substances into simpler ones.
_____	_____	4. A spontaneous reaction that releases free energy and can therefore perform work.
_____	_____	5. A reaction that requires an input of free energy from the surroundings.
_____	_____	6. Refers to a receptor site on some region of an enzyme molecule other than the active site.

MATCHING

Terms:

a. Catalyst
b. Coenzyme
c. Energy
d. Enthalpy
e. Entropy
f. Enzyme
g. Free energy
h. Heat energy
i. Kinetic energy
j. Potential energy
k. Substrate
l. Thermodynamics

For each of these definitions, select the correct matching term from the list above.

_____ 1. A quantitative measure of the amount of randomness or disorder of a system.

_____ 2. An organic substance that is required for a particular enzymatic reaction to occur.

_____ 3. A substance on which an enzyme acts.

_____ 4. Energy in motion.

_____ 5. The study of energy and its transformations.

_____ 6. A substance that increases the speed at which a chemical reaction occurs without being used up during the reaction.

_____ 7. An organic catalyst that greatly increases the rate of a chemical reaction without being consumed by that reaction.

_____ 8. The total potential energy of a system.

_____ 9. Stored energy.

_____10. The capacity or ability to do work.

MAKING COMPARISONS

Fill in the blanks.

Chemical Reaction	Type of Metabolic Pathway	Energy Required (Endergonic) or Released (Exergonic)
Synthesis of ATP	Anabolic	Endergonic
Hydrolysis	#1	#2
Phosphorylation	#3	#4
ATP → ADP	#5	#6
Oxidation	#7	#8
Reduction	#9	#10
FAD → FADH$_2$	#11	#12

MAKING CHOICES

Place your answer(s) in the space provided. Some questions may have more than one correct answer.

_____ 1. Enzymes

a. are lipoproteins.

b. lower required activation energy.

c. speed up biological chemical reactions.

d. become products after complexing with substrates.

e. are regulated by genes.

_____ 2. Enzyme activity may be affected by

a. cofactors.

b. temperature.

c. pH.

d. substrate concentration.

e. genes.

_____ 3. A kilocalorie (kcal) is
    a. a measure of heat energy.   d. the temperature of water.
    b. an energetic electron.   e. an essential nutrient.
    c. a way to measure energy generally.

Use the following formula to answer question 4:

$$ATP + H_2O \rightarrow ADP + P\Delta G = -7.3 \text{ kcal/mole}$$

_____ 4. This reaction
    a. hydrolyzes ATP.   d. is endergonic.
    b. loses free energy.   e. is exergonic.
    c. produces adenosine triphosphate.

_____ 5. Compliance with the second law of thermodynamics presumes that
    a. disorder is increasing.   d. all energy will eventually be useless to life.
    b. maintaining order in a system requires input of energy.   e. heat dissipates in all systems.
    c. entropy will decrease as order in organisms increases.

_____ 6. Kinetic energy is
    a. doing work.   d. energy of motion.
    b. stored energy.   e. energy of position or state.
    c. in chemical bonds.

_____ 7. In the formula $\Delta G = \Delta H - T\Delta S$
    a. free energy decreases as entropy decreases.   d. temperature increase decreases free energy.
    b. entropy and free energy are inversely related.   e. increasing enthalpy increases free energy.
    c. change in enthalpy is greater than change in free energy.

_____ 8. Chemical energy in molecules is
    a. kinetic energy.   d. stored in chemical bonds.
    b. potential energy.   e. energy in motion.
    c. released in an endergonic reaction.

_____ 9. The sum of all chemical activities in an organism is known as
    a. anabolism.   d. entropy.
    b. catabolism.   e. enthalpy.
    c. metabolism.

_____ 10. An exergonic reaction
    a. releases energy.   d. reduces total free energy.
    b. can be spontaneous.   e. produces a negative value for ΔG
    c. is a "downhill" reaction.

_____ 11. In the reaction $A \rightleftharpoons B$
    a. "**A**"is the product molecule.   d. free energy is reduced.
    b. "**B**"is the reactant molecule.   e. a tendency to accumulate "**B**" is indicated.
    c. double arrows indicate a reversible reaction.

_____ 12. According to the first law of thermodynamics
    a. energy conversions are never 100% efficient.   d. energy cannot be created or destroyed.
    b. during chemical reactions some energy is lost as heat   e. total energy in the universe is increasing.
    c. disorganization in the universe is decreasing

VISUAL FOUNDATIONS

Color the parts of the illustration below as indicated.

RED ☐ active sites

GREEN ☐ substrates

YELLOW ☐ enzyme

BLUE ☐ allosteric site

ORANGE ☐ regulator

BROWN ☐ cyclic AMP

TAN ☐ enzyme-substrate complex

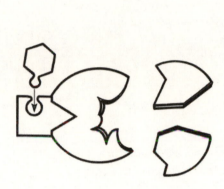
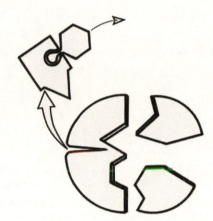
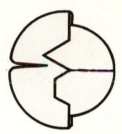

CHAPTER 8

How Cells Make ATP:
Energy-Releasing Pathways

This chapter continues your study of the flow of energy through living systems. Cells break down nutrients one step at a time, releasing energy from chemical bonds and transferring it to ATP where it is available for cellular work. There are three major pathways by which cells extract energy from nutrients: aerobic respiration, anaerobic respiration, and fermentation. Aerobic respiration involves a series of reactions in which hydrogen is transferred from a nutrient to oxygen, resulting in the formation of water. In anaerobic respiration, fuel molecules are broken down in the absence of oxygen, and an inorganic compound serves as the final hydrogen (electron) acceptor. Fermentation is a type of anaerobic respiration in which the final electron acceptor is an organic compound derived from the initial nutrient.

REVIEWING CONCEPTS
Fill in the blanks.

INTRODUCTION

1. _____ is the exergonic aspect of metabolism involving breakdown of complex molecules.

2. _____ is the endergonic aspect of metabolism involving the synthesis of complex molecules.

3. Most anabolic reactions are endergonic and require _____ or some other energy source to drive them.

REDOX REACTIONS

4. Most cells that live in environments where oxygen is plentiful use the catabolic process called _____ to extract free energy from nutrients.

5. In aerobic respiration, a fuel molecule is oxidized, yielding the by-products (a)_____ with the release of the essential (b)_____ that is required for life's activities.

THE FOUR STAGES OF AEROBIC RESPIRATION

6. The four stages of aerobic respiration of glucose are _____, _____, _____, and _____.

In glycolysis, glucose yields two pyruvates

7. Glycolysis reactions take place in the _____ of the cell.

8. In the first phase of glycolysis, two ATP molecules are consumed and glucose is split into two _____ molecules.

9. In the second phase of glycolysis, each of the molecules resulting from splitting of glucose is oxidized and transformed into a _____ molecule.

10. Glycolysis *nets* _____ (#?) ATPs.

Pyruvate is converted to acetyl CoA

11. (a)_____(#?) NADH molecules form during glycolysis and another (b)_____(#?) form during the formation of acetyl CoA from pyruvate.

The citric acid cycle oxidizes acetyl CoA

12. The eight step citric acid cycle completes the oxidation of glucose. For each acetyl group that enters the citric acid cycle, (a)_____ (#?) NAD^+ are reduced to NADH, (b)_____ (#?) molecules of CO_2 are produced, and (c)_____ (#?) hydrogen atoms are removed.

13. _____(#?) acetyl CoAs are completely degraded with two turns of the citric acid cycle.

The electron transport chain is coupled to ATP synthesis

14. The hydrogens removed during glycolysis, acetyl CoA formation, and the citric acid cycle are first transferred to the primary hydrogen acceptors (a)_____, then they are processed through (b)_____.

15. The (a)_____ consists of a series of electron acceptors embedded in the inner membrane of mitochondria. (b)_____ is the final acceptor in the chain.

Aerobic respiration of one glucose yields a maximum of 36 to 38 ATPs

16. Mitochondrial shuttle systems harvest the electrons of _____ produced in the cytosol.

Cells regulate aerobic respiration

17. Aerobic respiration requires a steady input of fuel molecules and _____.

ENERGY YIELD OF NUTRIENTS OTHER THAN GLUCOSE

18. Human beings and many other animals usually obtain most of their energy by oxidizing _____. Amino acids are also used.

19. Amino groups are metabolized by a process called _____.

20. The _____ components of a triacylglycerol are used as fuel.

21. Fatty acids are oxidized and split into acetyl groups by the process of _____. The acetyl molecules enter the citric acid cycle.

ANAEROBIC RESPIRATION AND FERMENTATION

22. _____ is a means of extracting energy that produces inorganic end products and does not require oxygen.

23. Fermentation is an anaerobic process in which the final acceptor of electrons from NADH is an _____.

Alcohol fermentation and lactate fermentation are inefficient

24. When hydrogens from NAD are transferred to acetaldehyde, _____ is formed.

25. When hydrogens from NAD are transferred to pyruvate, _____ is formed.

26. Fermentation yields a net gain of only (a)_____(#?) ATPs per glucose molecule, compared with about (b)_____(#?) ATPs per glucose molecule in aerobic respiration.

BUILDING WORDS

Use combinations of prefixes and suffixes to build words for the definitions that follow.

Prefixes	The Meaning	Suffixes	The Meaning
aero-	air	-be (bios)	life
an-	without, not, lacking	-lysis	breaking down, decomposition
de-	indicates removal, separation		
glyco-	sweet, "sugar"		

Prefix	Suffix	Definition
_____	_____	1. An organism that requires air or free oxygen to live.
_____	-aerobe	2. An organism that does not require air or free oxygen to live.
_____	-hydrogenation	3. A reaction in which hydrogens are removed from the substrate.
_____	-carboxylation	4. A reaction in which a carboxyl group is removed from a substrate.
	-amination	5. A reaction in which an amino group is removed from a substrate.
_____	_____	6. A sequence of reactions that breaks down a molecule of glucose (a sugar) to two molecules of pyruvate.

MATCHING

Terms:

a. Aerobic respiration
b. Anaerobic respiration
c. Chemiosmosis
d. Citric acid cycle
e. Cytochromes
f. Electron transport chain
g. Ethyl alcohol
h. Facultative anaerobe
i. Fermentation
j. Lactic acid
k. Oxidation
l. Phosphorylation
m. Pyruvic acid
n. Reduction

For each of these definitions, select the correct matching term from the list above.

_____ 1. The process by which a proton gradient drives the formation of ATP.

_____ 2. Aerobic series of chemical reactions in which acetyl Co-A is completely degraded to carbon dioxide and water with the release of ATP.

_____ 3. An organism that can live in either the presence or absence of oxygen.

_____ 4. Anaerobic respiration that utilizes organic compounds both as electron donors and acceptors.

_____ 5. Oxygen-requiring pathway by which organic molecules are broken down and energy is released that can be used for biological work.

_____ 6. Iron-containing proteins of the electron transport system that are alternately oxidized and reduced.

_____ 7. The loss of electrons or hydrogen atoms from a substance.

_____ 8. The introduction of a phosphate group into an organic molecule.

_____ 9. A series of chemical reactions during which hydrogens or their electrons are passed along from one receptor molecule to another, with the release of energy.

_____10. The gain of electrons or hydrogen atoms by a substance.

MAKING COMPARISONS

Fill in the blanks.

Reactions in Glycolysis	Catalyzing Enzyme	ATPs Used (−) or Produced (+)
glucose $\longrightarrow$ glucose-6-phosphate	hexokinase	− 1 ATP
glucose-6-phosphate $\longrightarrow$ #1_____	Phospho-glucoisomerase	zero
#1_____ $\longrightarrow$ fructose-1,6-biphosphate	#2	#3
fructose-1,6-biphosphate $\longrightarrow$ dihydroxyacetone phosphate and #4_____	Aldolase	#5
#4_____ $\longrightarrow$	Glyceraldehydes-3-phosphate dehydrogenase	#6
$\longrightarrow$ #7_____	Phosphoglycero-kinase	#8
#7_____ $\longrightarrow$ #9_____	Phosphoglycero-mutase	zero
#9_____ $\longrightarrow$ phosphoenolpyruvate	Enolase	#10
Phosphoenolpyruvate $\longrightarrow$ #11_____	Pyruvate kinase	#12

MAKING CHOICES

Place your answer(s) in the space provided. Some questions may have more than one correct answer.

_____ 1. Complete aerobic metabolism yields

    a. 36 to 38 ATPs.   d. 4 ATPs from the citric acid cycle.

    b. 2 ATPs from glycolysis.   e. 2 pyruvates.

    c. 32 to 34 ATPs from electron transport and chemiosmosis.

_____ 2. Anaerobic respiration

    a. does not involve an electron transport chain.   d. involves chemiosmosis.

    b. is performed by certain prokaryotes.   e. may use nitrate as the final hydrogen acceptor.

    c. uses an inorganic substance as the final hydrogen acceptor.

_____ 3. A facultative anaerobe

    a. is capable of carrying out aerobic respiration.   d. is capable of producing CO_2.

    b. is capable of carrying out alcohol fermentation.   e. requires oxygen.

    c. is capable of producing ethanol.

_____ 4. During oxidative phosphorylation

    a. ATP converts to ADP.   d. the electron transport chain is directly involved.

    b. ATP forms ADP.   e. chemiosmosis is directly involved.

    c. most of the ATP from aerobic metabolism is produced .

_____ 5. Production of acetyl CoA from pyruvate

    a. is anabolic.   d. takes place in endoplasmic reticulum.

    b. takes place in mitochondria.   e. yields CO_2 and NADH

    c. takes place in cytoplasm.

_____ 6. The citric acid cycle yields

    a. 2 H_2O and 4 CO_2.   d. one $FADH_2$.

    b. 6 NADH.   e. 2 acetyl CoA.

    c. 4 ATP.

_____ 7. Glycolysis

    a. generates a net profit of two ATPs.   d. produces two pyruvates.

    b. is more efficient than aerobic respiration.   e. reduces glucose to H_2O and CO_2.

    c. takes place in the cristae of mitochondria.

_____ 8. ATP synthase

    a. is a cytochrome.   d. is a transmembrane protein.

    b. converts ATP to ADP.   e. couples protons to electrons to form water.

    c. forms channels across the inner mitochondrial membrane.

_____ 9. Chemiosmosis involves

    a. flow of protons down an electrical gradient.   d. proton channels composed of electrons.

    b. flow of protons down a concentration gradient.   e. the production of ATP.

    c. pumping of protons into the mitochondrial matrix.

_____10. In the electron transport chain

    a. water is the final electron acceptor.   d. glucose is a common carrier molecule.

    b. cytochromes carry electrons.   e. electrons gain energy with each transfer.

    c. the final electron acceptor has a negative redox potential.

_____11. Cells may obtain energy from large molecules by means of

    a. catabolism.

    b. aerobic respiration.

    c. anaerobic respiration.

    d. fermentation.

    e. the Kreb's cycle.

_____12. Complete aerobic metabolism ultimately yields

    a. 36 to 38 ATPs.

    b. 2 ATPs from glycolysis.

    c. 32 to 34 ATPs from electron transport and chemiosmosis.

    d. 4 ATPs from the citric acid cycle.

    e. 2 pyruvates.

_____13. Anaerobic respiration

    a. does not involve an electron transport chain.

    b. is performed by certain prokaryotes.

    c. uses an inorganic substance as the final hydrogen acceptor.

    d. involves chemiosmosis.

    e. may use nitrate as the final hydrogen acceptor.

_____14. A facultative anaerobe

    a. is capable of carrying out aerobic respiration.

    b. is capable of carrying out alcohol fermentation.

    c. is capable of producing ethanol.

    d. is capable of producing CO_2.

    e. requires oxygen.

VISUAL FOUNDATIONS

Enter the appropriate enzyme in the spaces provided.

1. _____   5. _____

2. _____   6. _____

3. _____   7. _____

4. _____   8. _____

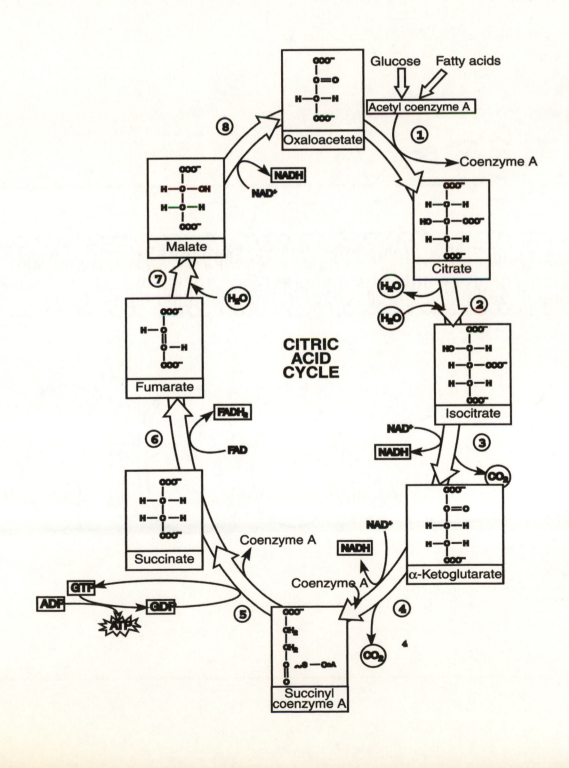

Color the parts of the illustration below as indicated.

RED ☐ electron path
GREEN ☐ complex I
YELLOW ☐ complex II
BLUE ☐ complex III
ORANGE ☐ complex IV
BROWN ☐ inner mitochondrial membrane
TAN ☐ matrix
PINK ☐ intermembrane space
VIOLET ☐ outer mitochondrial membrane

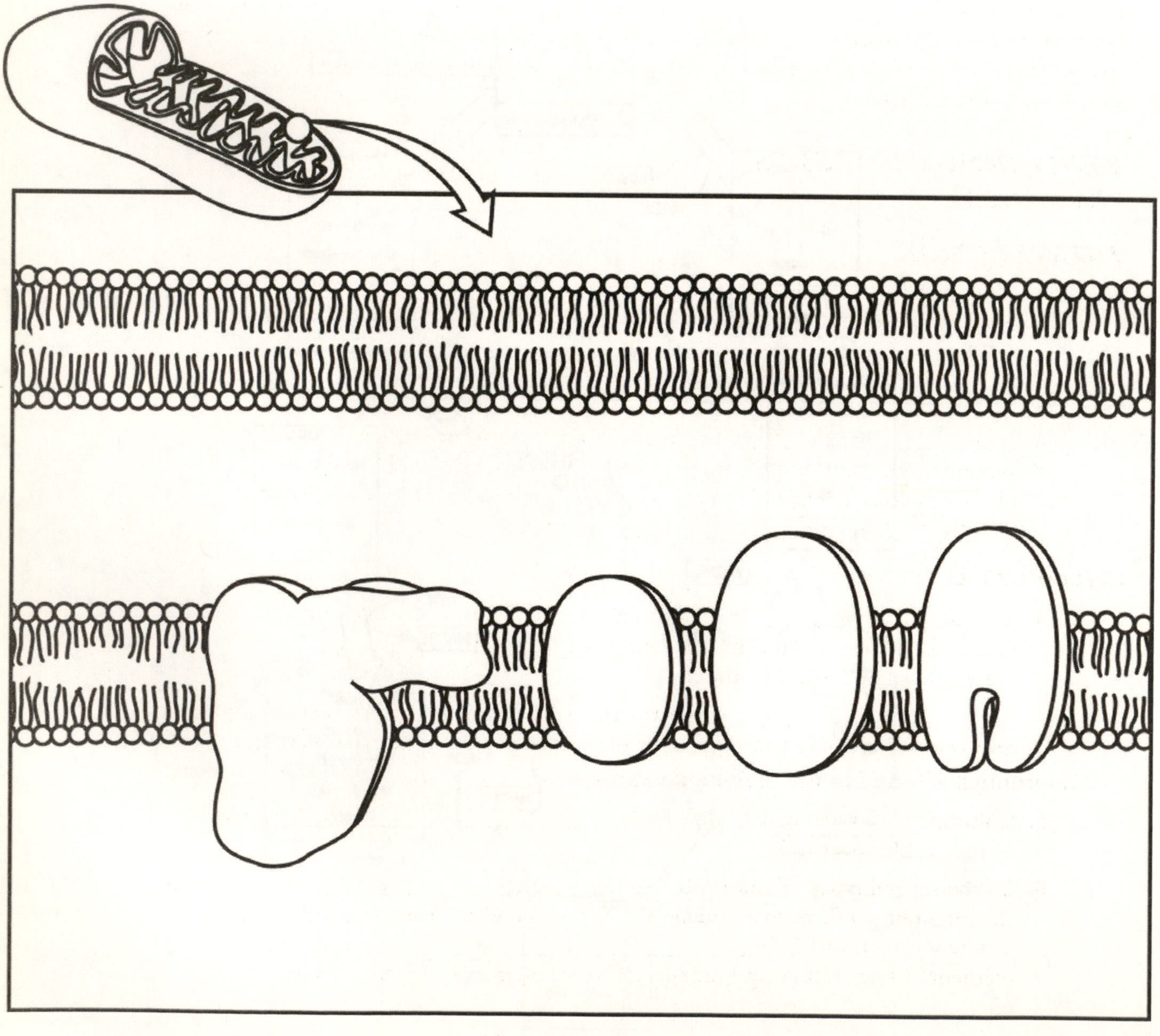

CHAPTER 9

❑

Photosynthesis: Capturing Energy

This chapter concludes the overview of the flow of energy through living systems by examining the process by which energy is captured from the environment and converted into the chemical bond energy of carbohydrates. This energy is what fuels the metabolic reactions that you studied in the previous chapter, i.e., the reactions that sustain all life. Photosynthesis, by far the most prevalent and important of the energy-capturing processes, captures solar energy and uses it to manufacture organic compounds from carbon dioxide and water. In photosynthetic prokaryotes, the process occurs in thylakoids, and in eukaryotes, the process occurs in organized structures within chloroplasts. Photosynthesis consists of both light-dependent and light-independent reactions. There are three different pathways by which carbon dioxide is assimilated into plants.

REVIEWING CONCEPTS

Fill in the blanks.

INTRODUCTION

1. Using the basic raw materials (a) _____,
   photosynthetic autotrophs use (b) _____ energy to make
   ATP and other molecules that hold (c)_____ energy.

LIGHT

2. (a)_____ occurs when energized electrons return to the
   ground state, emitting their excess energy in the form of visible light. In
   photosynthesis, energized electrons leave atoms and pass to an
   (b)_____ molecule.

CHLOROPLASTS

3. The (a)_____ is the fluid-filled region within the inner membrane of
   chloroplasts that contains most of the enzymes for photosynthesis. Most
   chloroplasts are located in the (b)_____ cells of leaves.

4. _____ are flat, disk-shaped membranes in chloroplasts that
   are arranged in stacks called grana. Chlorophyll is located in these membranes.

Chlorophyll is found in the thylakoid membrane

5. Chlorophyll absorbs light mainly in the _____ portions
   of the visible spectrum.

6. Of the several types of chlorophyll in plants, (a)_____ is
   the bright green form that initiates the light-dependent reactions, and the
   yellowish-green form, (b)_____, is an accessory
   pigment. Other yellow and orange accessory pigments in plant cells are
   (c)_____.

Chlorophyll is the main photosynthetic pigment

7. An (a)_____ is a graph that illustrates the relative absorption of different wavelengths of light by a given pigment. It is obtained with an instrument called a (b)_____.

8. The (a)_____ of photosynthesis is a measurement of the relative effectiveness of different wavelengths of light in affecting photosynthesis. It may be greater than can be accounted for by the absorption of chlorophyll alone, the difference accounted for by (b)_____ that transfer energy absorbed from the green wavelengths to chlorophyll.

OVERVIEW OF PHOTOSYNTHESIS

ATP and NADPH are the products of the light-dependent reactions: an overview

9. Light-dependent reactions provide useful chemical energy for _____ of photosynthetic products.

Carbohydrates are produced during the carbon fixation reactions: an overview

10. Light-independent reactions transfer the energy from (a)_____ to the bonds in (b)_____ molecules.

THE LIGHT-DEPENDENT REACTIONS

Photosystems I and II each consist of a reaction center and multiple antenna complexes

11. The light-dependent reactions of photosynthesis begin when _____ and/or accessory pigments absorb light.

12. Chlorophylls *a* and *b* and accessory pigment molecules are organized with pigment-binding proteins in the thylakoids membrane into units called _____.

13. Each antenna complex absorbs light energy and transfers it to the _____, which consists of chlorophyll molecules and proteins.

14. Light energy is converted to chemical energy in the reaction centers by a series of _____.

15. The reaction center of photosystem II is made up of a chlorophyll *a* molecule with an absorption peak of about 680 nm and is referred to as _____.

16. When a pigment molecule absorbs light energy, that energy is passed from one pigment molecule to another until it reaches the _____.

Noncyclic electron transport produces ATP and NADPH

17. In noncyclic electron transport, the energized electron is passed along an electron transport chain from one electron acceptor to another, until it is passed to _____, and iron-containing protein.

18. Like photosystem I, photosystem II is activated when a pigment molecule in an _____ absorbs a photon of light energy.

19. _____ is a process that not only yields electrons, but is also the source of almost all the oxygen in the Earth's atmosphere.

Cyclic electron transport produces ATP but no NADPH

20. As electrons pass from one acceptor to another in cyclic electron transport, they lose energy, some of which is used to pump protons across the

_____.

21. It is generally believed that ancient bacteria used _____ to produce ATP from light energy.

ATP synthesis occurs by chemiosmosis

22. The chemiosmotic model explains the coupling of ATP synthesis and

_____.

THE CARBON FIXATION REACTIONS

23. ATP and NADPH generated in the light-dependent reactions are used to reduce carbon dioxide to a carbohydrate, a process known as _____.

Most plants use the Calvin cycle to fix carbon

24. The Calvin cycle begins with the combination of one CO_2 molecule and one five-carbon sugar, (a)_____, to form a six-carbon molecule. This molecule instantly splits into two three-carbon molecules called (b)_____.

25. To produce one six-carbon carbohydrate, the light-independent reactions utilize six molecules of (a)_____, hydrogen obtained from (b)_____, and energy from (c)_____.

Photorespiration reduces photosynthetic efficiency

26. Photorespiration occurs mainly during bright, hot days when plant stomata are closed. It reduces photosynthetic efficiency because CO_2 cannot enter the system, causing the enzyme (a)_____ to bind RuBP to (b)_____ instead of CO_2.

27. Photorespiration is negligible in (a)_____(C_3 or C_4?) plants and relatively common among (b)_____ (C_3 or C_4?) plants.

The initial carbon fixation step differs in C_4 plants and in CAM plants

28. The C_4 pathway efficiently fixes _____ at low concentrations.

29. CAM plants fix _____ at night.

30. C_4 plants initially fix CO_2 into the four-carbon molecule _____.

31. The pores in leaves through which gases pass are called _____.

METABOLIC DIVERSITY

32. (a)_____ are organisms that cannot make their own food, unlike (b)_____ that are self nourishing.

PHOTOSYNTHESIS IN PLANTS AND IN THE ENVIRONMENT

33. Molecular oxygen is constantly replenished by the _____ that releases the oxygen that all aerobic organisms require for respiration.

BUILDING WORDS

Use combinations of prefixes and suffixes to build words for the definitions that follow.

Prefixes	The Meaning		Suffixes	The Meaning
auto-	self, same		-lysis	breaking down, decomposition
hetero-	different, other		-plast	formed, molded, "body"
meso-	middle		-troph	nutrition, growth, "eat"
photo-	light			
chloro-	green			

Prefix	Suffix	Definition
_____	-phyll	1. Tissue in the middle of a leaf specialized for photosynthesis.
_____	-synthesis	2. The conversion of solar (light) energy into stored chemical energy by plants, blue-green algae, and certain bacteria.
_____	_____	3. The breakdown (splitting) of water under the influence of light energy trapped by chlorophyll.
_____	-phosphorylation	4. Phosphorylation that uses light as a source of energy.
_____	_____	5. A membranous organelle containing green photosynthetic pigments.
_____	-phyll	6. A green photosynthetic pigment.
_____	_____	7. An organism that fixes carbon, producing the organic compounds it needs.
_____	_____	8. An organism that is dependent upon the organic molecules produced by other organisms as the building blocks from which it synthesizes the carbon compounds it needs.

MATCHING

Terms:

a. C_3 pathway
b. C_4 pathway
c. Calvin cycle
d. CAM
e. Carotenoid
f. Chemoautotroph
g. Chemoheterotroph
h. Granum
i. Photoheterotroph
j. Photon
k. Photosystem
l. Stoma
m. Stroma
n. Thylakoid

For each of these definitions, select the correct matching term from the list above.

_____ 1. The fluid region of the chloroplast surrounding the thylakoids.

_____ 2. Interconnected system of flattened sac-like membranous structures inside the chloroplast where light energy is converted into chemical energy.

_____ 3. A stack of thylakoids within a chloroplast.

_____ 4. A yellow to orange plant pigment.

_____ 5. The usual pathway for fixing carbon dioxide in the synthesis reactions of photosynthesis.

_____ 6. A cyclic series of reactions occurring during the light-independent phase of photosynthesis.

_____ 7. A highly organized cluster of photosynthetic pigments and electron/hydrogen carriers embedded in the thylakoid membranes of chloroplasts.

_____ 8. An autotrophic organism that obtains energy from inorganic compounds.

_____ 9. A particle or packet of electromagnetic radiation.

_____10. A metabolic pathway that fixes carbon in desert plants.

MAKING COMPARISONS

Fill in the blanks.

Category	Photosynthesis	Respiration
Eukaryotic cellular site	Chloroplasts	Cytosol and mitochondria
Type of metabolic pathway (anabolic or catabolic)	#1	#2
Site of electron transport chain	#3	#4
Source of hydrogen (electrons)	#5	#6
Terminal hydrogen acceptor	#7	#8

MAKING CHOICES

Place your answer(s) in the space provided. Some questions may have more than one correct answer.

Use the following formula to answer questions 1-3:

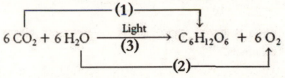

$$6 CO_2 + 6 H_2O \xrightarrow[\text{(3)}]{\text{Light}} C_6H_{12}O_6 + 6 O_2$$

_____ 1. What kind of reaction is represented by number one (1)?

    a. reduction   d. substitution

    b. oxidation   e. replacement

    c. neutral

_____ 2. What kind of reaction is represented by number two (2)?

    a. reduction   d. substitution

    b. oxidation   e. replacement

    c. neutral

_____ 3. What missing component is represented by number three (3)?

    a. carotene   d. palisade layer of a leaf

    b. sunlight   e. chlorophyll

    c. wavelengths in the range of 422-492 nm or 647-760 nm

_____ 4. Some of the requirements for the light-dependent reactions of photosynthesis include

    a. water.   d. NADP.

    b. photons of light.   e. oxygen.

    c. carbohydrates.

_____ 5. Chloroplasts in prokaryotic cells

    a. possess a CF_0–CF_1 complex.   d. contain chlorophyll.

    b. have double membranes.   e. carry out photosynthesis II.

    c. do not exist.

_____ 6. The light-independent reactions of photosynthesis

    a. generate PGA.   d. require ATP.

    b. generate oxygen.   e. require NADPH.

    c. take place in the stroma.

_____ 7. Thylakoids
    a. comprise grana.
    b. are continuous with the plasma membrane.
    c. possess the basic fluid mosaic membrane structure.
    d. are located in the stroma.
    e. carry out the light-independent reactions.

_____ 8. Chlorophyll
    a. contains magnesium in a porphyrin ring.
    b. dissolves in water.
    c. is green because it absorbs green portions of the light spectrum.
    d. is the only pigment found in most plants.
    e. is found mostly in the stroma.

_____ 9. The leaf tissue in which most of photosynthesis occurs is the
    a. periderm.
    b. epidermis.
    c. mesophyll.
    d. stomata layer.
    e. grana.

_____10. C_4 fixation
    a. replaces C_3 fixation.
    b. produces PGA.
    c. produces oxaloacetate.
    d. takes place in bundle sheath cells.
    e. supplements C_3 fixation.

_____11. The reactions of photosystem II
    a. include splitting water with light photons.
    b. produce H_2O.
    c. assist in producing an electrochemical gradient across the thylakoid membrane.
    d. use $NADP^+$ as a final electron acceptor.
    e. produce O_2.

_____12. Most producers are
    a. chemosynthetic heterotrophs.
    b. chemosynthetic autotrophs.
    c. photoautotrophs.
    d. photosynthetic heterotrophs.
    e. photosynthetic consumers.

_____13. In plants that have chloroplasts, chlorophyll is found mainly in the
    a. thylakoid lumen.
    b. thylakoid membranes.
    c. stomata.
    d. stroma.
    e. intermembrane spaces.

_____14. In photosynthesis, electrons in atoms
    a. are excited.
    b. produce fluorescence.
    c. are accepted by a reducing agent when they escape.
    d. are pushed to higher energy levels by photons of light.
    e. release absorbed energy as another wavelength of light.

_____15. Reactions occurring during electron flow in respiration and photosynthesis are
    a. both endergonic.
    b. both exergonic.
    c. exergonic and endergonic respectively.
    d. endergonic and exergonic respectively.
    e. neither endergonic nor exergonic.

VISUAL FOUNDATIONS

Color the parts of the illustration below as indicated. Also label the entry site of CO_2 and the structure that transports water.

RED ☐ stoma   GREEN ☐ chloroplasts

ORANGE ☐ palisade mesophyll   TAN ☐ upper epidermis

YELLOW ☐ spongy mesophyll

BLUE ☐ vein

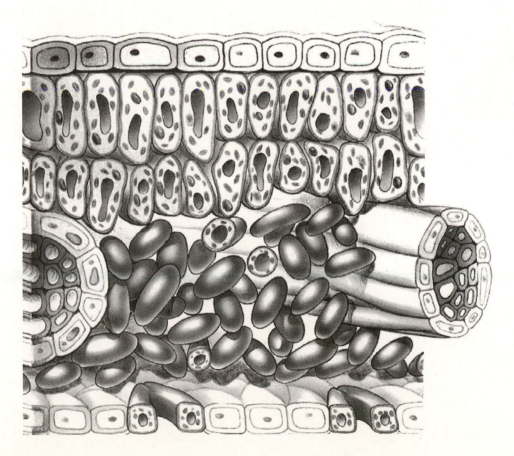

PART III

The Continuity of Life: Genetics

❑

CHAPTER 10

Chromosomes, Mitosis, and Meiosis

This chapter begins an examination of one of the major themes of biology – the transmission of information, specifically, the transmission of information from one generation of cell or organism to the next. In prokaryotes, the information is contained in a single circle of DNA. In eukaryotes, it is carried in the chromosomes contained within the cell nucleus. Chromosomes are made up of DNA and protein. The DNA is organized into informational units, or genes, that determine the characteristics of the organism. Each species is unique due to the information specified by its genes. Genes are passed from parent cell to daughter cell by a process that ensures that each new nucleus receives the same number and types of chromosomes as were present in the original nucleus. There are two basic types of reproduction — asexual and sexual. In asexual reproduction, a single parent cell usually splits, buds, or fragments into two or more individuals. In sexual reproduction, sex cells, or gametes, are produced by a process that halves the number of chromosomes in the resulting cells. When two gametes fuse, the resulting cell contains the same number of chromosomes as the parent cells.

REVIEWING CONCEPTS

Fill in the blanks.

INTRODUCTION

1. Cells contain a massive amount of precisely coded genetic information in the form of DNA, collectively called the organism's _____.

EUKARYOTIC CHROMOSOMES

2. DNA and associated proteins form a complex, the _____, that make up chromosomes.

DNA is organized into informational units called genes

3. According to the Human Genome Project, humans have about _____ genes that code for proteins.

DNA is packaged in a highly organized way in chromosomes

4. Positively charged histones associate with DNA forming structures called _____.

Chromosome number and informational content differ among species

5. Most human body cells have exactly _____ (#?) chromosomes.

THE CELL CYCLE AND MITOSIS

6. The period from the beginning of one cell division to the beginning of the next cell division is the _____.

Chromosomes duplicate during interphase

7. Interphase is divided into the G_1 phase, which stands for (a)_____ phase, the S phase, or (b)_____ phase, and the G_2 phase, or (c)_____ phase.

During prophase, duplicated chromosomes become visible with the microscope

8. Each chromatid includes a constricted region called the _____.

9. Attached to each centromere is a _____, a structure formed from proteins to which microtubules can bind.

10. Some of the microtubules radiating from each pole elongate toward the chromosome, forming the _____, which separates the chromosomes during anaphase.

Prometaphase begins when the nuclear envelope breaks down

11. In prometaphase, sister chromatids of each duplicated chromosome become attached to _____ at opposite poles of the cell.

Duplicated chromosomes line up on the midplane during metaphase

12. During metaphase, all the cell's chromosomes align at the cell's midplane, or _____.

During anaphase, chromosomes move toward the poles

13. Kinetochore microtubules shorten, or _____, during anaphase.

During telophase, two separate nuclei form

14. Telophase is characterized by the _____ _____ and a return to interphase-like condition.

Cytokinesis forms two separate daughter cells

15. In plant cells, cytokinesis occurs by forming a _____, a partition constructed in the equatorial region of the spindle and growing laterally toward the cell wall.

Mitosis produces two cells genetically identical to the parent cell

16. The regularity of the process of cell division ensures that each daughter nucleus receives exactly the same number and kinds of _____ the parent cell had.

Lacking nuclei, prokaryotes divide by binary fission

17. The reproduction process in which one cell divides into two offspring cells is named _____.

REGULATION OF THE CELL CYCLE

18. _____ are a group of plant hormones that promote mitosis both in normal growth and in wound healing.

SEXUAL REPRODUCTION AND MEIOSIS

19. Offspring inherit traits that are virtually identical to those of their single parent when reproduction is _____.

20. In (a)_____ reproduction, offspring receive genetic information from two parents. Haploid (n) gametes from the parents fuse to form a single (b)_____ (2n) cell called the (c)_____.

21. Homologous chromosomes are members of a pair of chromosomes that are similar in _____.

Meiosis produces haploid cells with unique gene combinations

22. Meiosis typically consists of two nuclear and cytoplasmic divisions referred to as _____.

Prophase I includes synapsis and crossing-over

23. During prophase I, homologous chromosomes come to lie lengthwise side by side in a process known as _____.

24. Homologous chromosomes exchange genetic material (crossing over) during the first meiotic (a)_____, providing more (b)_____ among gametes and offspring.

During meiosis I, homologous chromosomes separate

25. The haploid condition is established as the members of each pair of homologous chromosomes separate during the first meiotic _____.

Chromotids separate in meiosis II

26. In metaphase I, chromatids are arranged in bundles of (a)_____, and in metaphase II, chromatids are in groups of (b)_____.

Mitosis and meiosis lead to contrasting outcomes

27. (a)_____ results in two daughter cells identical to the original cell. (b)_____ results in four genetically different, haploid daughter cells.

The timing of meiosis in the life cycle varies among species

28. Sex cells (sperm and eggs or ova) are known as (a)_____, therefore, the formation of sex cells is referred to as (b)_____.

29. The formation of sperm is called _____.

30. The formation of eggs or ova is called _____.

BUILDING WORDS

Use combinations of prefixes and suffixes to build words for the definitions that follow.

Prefixes	The Meaning	Suffixes	The Meaning
centro-	center	-gen(esis)	production of
chromo-	color	-mere	part
dipl-	double, in pairs	-phyte	plant
gameto-	sex cells, eggs and sperm	-some	body
hapl-	single		
inter-	between		
oo-	egg		
spermato-	seed, "sperm"		
sporo-	spore		

Prefix	Suffix	Definition
_____	_____	1. A dark staining body within the cell nucleus containing genetic information.
_____	-phase	2. The stage in the life cycle of a cell that occurs between successive cell divisions.
_____	-oid	3. An adjective pertaining to a single set of chromosomes.
_____	-oid	4. An adjective pertaining to a double set of chromosomes.
_____	_____	5. The constricted part or region of a chromosome (often near the center) to which a spindle fiber is attached.
_____	_____	6. The process by which gametes (sex cells) are produced.
_____	_____	7. The process whereby sperm are produced.
_____	_____	8. The process whereby eggs are produced.
_____	_____	9. The stage in the life cycle of a plant that produces gametes by mitosis.
_____	_____	10. The stage in the life cycle of a plant that produces spores by meiosis.

MATCHING

Terms:

a.	Anaphase	f.	Kinetochore	k.	Prophase
b.	Chromatin	g.	Meiosis	l.	S phase
c.	Crossing over	h.	Metaphase	m.	Synapsis
d.	Cytokinesis	i.	Mitosis	n.	Telophase
e.	Interkinesis	j.	Polyploid		

For each of these definitions, select the correct matching term from the list above.

_____ 1. The last stage of mitosis and meiosis.

_____ 2. The phase in interphase during which DNA and other chromosomal components are synthesized.

_____ 3. Portion of the chromosome centromere to which the mitotic spindle fibers attach.

_____ 4. Process whereby genetic material is exchanged between homologous chromatids during meiosis.

_____ 5. The phase in mitosis during which the chromosomes line up along the equatorial plate.

_____ 6. DNA-protein fibers that condense to form chromosomes during prophase.

_____ 7. Process during which a diploid cell undergoes two successive nuclear divisions resulting in four haploid cells.

_____ 8. Having more than two sets of chromosomes per nucleus.

_____ 9. Division of the cell nucleus resulting in two daughter cells with the identical number of chromosomes as the parental cell.

_____10. Stage of cell division in which the cytoplasm divides into two daughter cells.

MAKING COMPARISONS

Fill in the blanks.

Event	Mitosis	Meiosis
Chromosome compaction (condensation)	Prophase	Prophase I, prophase II
Cytokinesis	#1	#2
Homologous chromosomes move to opposite poles	#3	#4
Tetrads form	#5	#6
Cytokinesis occurs	#7	#8
Chromatids separate	#9	#10
Chromosomes line up at the midplane of the cell	#11	#12
Duplication of DNA	#13	#14

MAKING CHOICES

Place your answer(s) in the space provided. Some questions may have more than one correct answer.

_____ 1. Typical gametes include

    a. somatic cells.
    b. 2n cells.
    c. sperm.
    d. ova.
    e. cells resulting from oogenesis.

_____ 2. In mitosis, cells with 16 chromosomes produce daughter cells with

    a. 32 chromosomes.
    b. 8 chromosomes.
    c. 16 chromosomes.
    d. 8 pairs of chromosomes.
    e. 4 pairs of chromosomes.

_____ 3. In meiosis, cells with 16 chromosomes produce daughter cells with

    a. 32 chromosomes.
    b. 8 chromosomes.
    c. 16 chromosomes.
    d. 8 pairs of chromosomes.
    e. 4 pairs of chromosomes.

_____ 4. Gametogenesis typically involves (n=single chromosomes [haploid]; 2n=paired chromosomes [diploid])

    a. 2n to 2n.
    b. 2n to n.
    c. reduction division.
    d. mitosis.
    e. meiosis.

_____ 5. Eukaryotic chromosomes

    a. contain DNA and protein.
    b. possess asters.
    c. are distinctly visible in a light microscope in interphase.
    d. are uncoiled in interphase.
    e. align in the equator during prophase.

Use this list to answer questions 6-15 about meiosis:

a. interphase	d. anaphase I	g. metaphase II
b. prophase I	e. telophase I	h. anaphase II
c. metaphase I	f. prophase II	i. telophase II

_____ 6. Nuclear membranes break down.

_____ 7. Homologous chromosomes synapse.

_____ 8. Chromatids separate.

_____ 9. Members of tetrad separate.

_____10. Crossing over takes place.

_____11. Pairs of chromosomes align at equatorial plane.

_____12. Reduction of chromosome number from 2n to n.

_____13. Tetrads form.

_____14. Diploid to haploid.

_____15. Single chromosomes align at equatorial plane.

Use this list to answer questions 16-25 about mitosis:

a. interphase	e. telophase	h. G_2 phase
b. prophase	f. T phase	i. S phase
c. metaphase	g. G_1 phase	j. M phase
d. anaphase		

_____16. The time between the synthesis phase and prophase.

_____17. Chromosomes begin to condense by coiling.

_____18. Condensed chromosomes uncoil.

_____19. DNA replicates.

_____20. Active synthesis and growth.

_____21. Chromosomes are lined up in a central plane.

_____22. All stages of mitosis collectively and cytokinesis.

_____23. Nuclear envelope breaks down.

_____24. The time between mitosis and start of the synthesis phase.

_____25. Kinetochores divide.

VISUAL FOUNDATIONS

Color the parts of the illustration below as indicated. Also label the phase in which DNA replicates and the phase in which cells that are not dividing are arrested.

RED   ❏   M phase

YELLOW   ❏   G1

BLUE   ❏   S

ORANGE   ❏   G2

VIOLET   ❏   Interphase

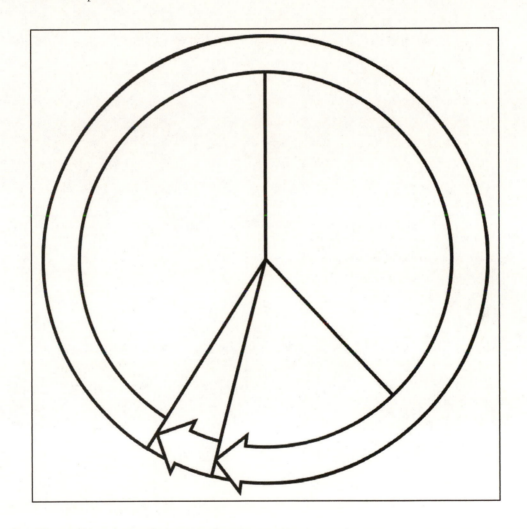

CHAPTER 11

$\Box$

The Basic Principles of Heredity

The previous chapter provided you with a basic understanding of chromosomal behavior during sexual reproduction, thereby preparing you for an examination of the process whereby genetic information is transferred from parent to offspring. Genes that occupy corresponding loci on homologous chromosomes govern variations of the same characteristic. Such genes, called alleles, exist in pairs in diploid organisms. A monohybrid cross is a cross between two individuals that carry different alleles for a single gene locus. Similarly, a dihybrid cross is a cross between two individuals that carry different alleles at each of two gene loci. The results of monohybrid and dihybrid crosses illustrate the two basic principles of genetics, namely, segregation and independent assortment. The laws of probability are used to predict the results of a cross between two individuals. Special chromosomes called sex chromosomes determine the sex of most animal species. The relationship between a single pair of alleles at a gene locus and the characteristic it controls may be simple, or it may be complex. Genetically identical individuals may develop differently in different environments.

REVIEWING CONCEPTS

Fill in the blanks.

INTRODUCTION

1. In the mid-19th century, the study of inheritance as a modern branch of science began with the work of _____.

MENDEL'S PRINCIPLES OF INHERITANCE

2. Offspring from the P generation cross are heterozygous; they are called the F_1 or
   (a)_____ generation. Offspring from the F_1 cross are called
   the (b)_____ generation.

3. A (a)_____ gene may mask the expression of a (b)_____ gene.

Alleles separate before gametes are formed: the principle of segregation

4. During meiosis, members of paired genes at each allele _____ so
   that each gamete contains only one allele of each pair.

Alleles occupy corresponding loci on homologous chromosomes

5. The site of a gene on a chromosome is called its _____.

6. The two chromosomes that make up a pair are called _____
   chromosomes.

A monohybrid cross involves individuals with different alleles of a given locus

7. An individual is said to be _____ for a particular
   feature when the two alleles it carries for that feature are different.

8. A _____ predicts the ratios of genotypes and phenotypes of the offspring of a cross.

9. The phenotype of an individual does not always reveal its _____.

10. A test cross is a cross between an individual of unknown genetic composition and a _____ individual.

A dihybrid cross involves individuals that have different alleles at two loci

11. A dihybrid cross is a cross between individuals that differ with respect to their alleles at _____.

Alleles on nonhomologous chromosomes are randomly distributed into gametes: the principle of independent assortment

12. The mechanics of _____ are the basis for independent assortment.

Recognition of Mendel's work came during the early 20th century

13. The chromosome theory of inheritance can be explained by assuming that genes are _____ in specific locations along the chromosomes.

USING PROBABILITY TO PREDICT MENDELIAN INHERITANCE

14. The _____ predicts the combined probabilities of independent events.

15. Probability can range from (a)_____(impossible) to (b)_____(certain).

16. The probability of two independent events occurring together is the _____ of the probabilities of each occurring separately.

17. The _____ predicts the combined probabilities of mutually exclusive events.

The rules of probability can be applied to a variety of calculations

18. If events are truly independent, (a)_____ have no influence on the probability of the occurrence of (b)_____.

MENDELIAN INHERITANCE AND CHROMOSOMES

Linked genes do not assort independently

19. Genes in the same chromosome are said to be _____ and do not assort independently.

20. Linked genes are recombined when chromatids exchange genetic material, a process known as _____ that occurs during meiotic prophase.

Calculating the frequency of crossing-over reveals the linear order of linked genes on a chromosome

21. A chromosome can be genetically mapped by determining the frequency of _____ among genes.

Sex is generally determined by sex chromosomes

22. The sex or gender of many animals is determined by the X and Y sex chromosomes. The other chromosomes in a given organism's genome are called _____.

23. When a Y-bearing sperm fertilizes an ovum, the result is a(n) (a)_____, and fertilization by an X-bearing sperm produces a(n) (b)_____.

24. The effect of X-linked genes is made equivalent in males and females by dose compensation, which is accomplished by a (a)_____ X-chromosome in the male or (b)_____ of one X-chromosome in the female.

25. A dense, metabolically inactive X chromosome at the edge of the nucleus in female mammalian cells is known as the _____.

EXTENSIONS OF MENDELIAN GENETICS

Dominance is not always complete

26. In genetic crosses involving _____, the genotypic and phenotypic ratios are identical.

Multiple alleles for a locus may exist in a population

27. Multiple alleles are (a)_____ (#?) different alleles that can occupy the same (b)_____.

A single gene may affect multiple aspects of the phenotype

28. _____ refers to the many different effects that can often result from a given gene.

Alleles of different loci may interact to produce a phenotype

29. _____ is when one allele of a gene pair determines whether alleles of other gene pairs are expressed.

Polygenes act additively to produce a phenotype

30. It is called _____ when two or more independent pairs of genes have similar and additive effects on a phenotype.

Genes interact with the environment to shape phenotype

31. The range of phenotypic possibilities that can develop from a single genotype under different environmental conditions is known as the _____.

BUILDING WORDS

Use combinations of prefixes and suffixes to build words for the definitions that follow.

Prefixes	The Meaning
di-	two, twice, double
hemi-	half
hetero-	different, other
homo-	same
mono-	alone, single, one
poly-	much, many

Prefix	Suffix	Definition
_____	-gene	1. Two or more pair of genes that affect the same trait in an additive fashion.
_____	-zygous	2. Having the same (identical) members of a gene pair.

_____	-zygous	3. Having dissimilar (different) members of a gene pair.
_____	-hybrid	4. Pertaining to the mating of individuals differing in two specific pairs of genes.
_____	-hybrid	5. Pertaining to the mating of individuals differing in one pair of genes.
_____	-zygous	6. Having only half (one) of a given pair of genes.

MATCHING

Terms:

a. Allele
b. Barr body
c. Dominant allele
d. Epistasis
e. Genotype

f. Inbreeding
g. Incomplete dominance
h. Linkage
i. Locus
j. Overdominance

k. Phenotype
l. Pleiotropy
m. Recessive allele

For each of these definitions, select the correct matching term from the list above.

_____ 1. Condition in which certain alleles at one locus can alter the expression of alleles at a different locus.

_____ 2. An alternative form of a gene.

_____ 3. The physical or chemical expression of an organism's genes.

_____ 4. Condition in which a single gene produces two or more phenotypic effects.

_____ 5. A condensed and inactivated X-chromosome appearing as a distinctive dense spot in the nucleus of certain cells of female mammals.

_____ 6. The allele that is not expressed in the heterozygous state.

_____ 7. The place on a chromosome at which the gene for a given trait occurs.

_____ 8. Condition in which both alleles of a locus are expressed in a heterozygote.

_____ 9. The allele that is always expressed when it is present.

MAKING COMPARISONS

Fill in the blanks.   T = tall; t = short (complete dominance); Y = yellow; y = green (complete dominance)

	T Y	**T y**	**t Y**	**t y**
T Y	tall plant with yellow seeds (TT YY)	tall plant with yellow seeds (TT Yy)	tall plant with yellow seeds (Tt YY)	tall plant with yellow seeds (Tt Yy)
T y	#1	#2	#3	#4
t Y	#5	#6	#7	#8
t y	#9	#10	#11	#12

MAKING CHOICES

Place your answer(s) in the space provided. Some questions may have more than one correct answer.

R and **r** are genes for flower color. Homozygous dominant and heterozygous genotypes both have red flowers; the homozygous recessive genotype has white flowers. **T** and **t** are genes that control plant height. Homozygous dominant plants are tall, heterozygous plants are medium height, and homozygous recessive plants are short. Genes for flower color and height are on different chromosomes. Use these data and the following list to answer questions 1-9. Construct Punnett Squares as needed.

a. RRTT	f. rrtt	k. 1:2:1:2:4:2:1:2:1	p. pink, medium
b. RrTt	g. 1:1	l. red, tall	q. pink, short
c. Rrtt	h. 1:2:1	m. red, medium	r. white, tall
d. rrTT	i. 9:3:3:1	n. red, short	s. white, medium
e. rrTt	j. 1:2:1:1:2:1	o. pink, tall	t. white, short

Plants with the genotypes Rrtt and rrTT are mated. Their offspring (F₁) are then mated to produce another generation (F₂). Questions 1-5 pertain to the F₂ generation.

_____ 1. Which of the genotypes listed above are found among offspring?

_____ 2. What would the phenotypic ratio among offspring be if both flower color and height genes had exhibited incomplete dominance?

_____ 3. What are the genotypic and phenotypic ratios among offspring?

_____ 4. What are the genotypes found among the offspring that are not in the list?

_____ 5. What are all of the phenotypes among offspring?

Questions 6-9 pertain to the following cross: Rrtt x rrTT.

_____ 6. What is the phenotypic ratio among offspring?

_____ 7. What are the genotypes and phenotypes of the parents?

_____ 8. What is the genotypic ratio among offspring?

_____ 9. What are the genotypes and phenotypes of the offspring?

_____10. Which of the following is/are consistent with Mendel's principle of dominance?

    a. All F₁ offspring express the dominant trait.   d. Both P generation parents are true breeding.

    b. All F₂ offspring express the dominant trait.   e. Only one P generation parent is true breeding.

    c. Both organisms in the P generation are homozygous.

_____11. Phenotypic expression

    a. always involves only one pair of genes.   d. refers specifically to the composition of genes.

    b. may involve many pairs of alleles.   e. refers to the appearance of a trait.

    c. is partly a function of environmental influences.

_____12. Pleiotropy means that a gene

    a. exhibits incomplete dominance.   d. has multiple effects.

    b. masks expression of other genes.   e. is sex-influenced.

    c. has the same effect as another pair of genes.

_____13. Which of the following would be true if hairy toes happened to be a recessive X-linked trait?

    a. All men would have hairy toes.   d. More women than men would have hairy toes.

    b. No women would have hairy toes.   e. More men than women would have hairy toes.

    c. Parents with hairy toes could have a child without hairy toes.

_____14. Linked genes are

    a. inseparable.

    b. generally on the same chromosome.

    c. in separate homologous chromosomes.

    d. in different chromatids.

    e. always separated during crossing over.

_____15. Given the following information about crossing over, determine the relative positions of three loci (X, Y, Z) on one chromosome: X and Y = 8%, Y and Z = 5%, X and Z = 3 map units.

    a. X, Y, Z

    b. X, Z, Y

    c. Z, Y, X

    d. Y, Z, X

    e. Z, Y, X

_____16. Mendel's "factors"

    a. are haploid in gametes.

    b. interact in gametes.

    c. segregate into separate gametes.

    d. are genes.

    e. affect flower color, but not seed color.

_____17. Epistasis means that alleles

    a. are sex-linked.

    b. have multiple effects.

    c. have the same effect as another pair of genes.

    d. mask the expression of another pair of genes.

    e. exhibit incomplete dominance.

_____18. Which of the following is/are true?

    a. XX is usually female.

    b. XY is usually male.

    c. birds and butterflies do not have sex chromosomes.

    d. hermaphrodites are XX or XY.

    e. XXY is usually male.

_____19. Which of the following applies if two pure-breeding P generation plants are used to ultimately produce an F_2 generation with flower colors in a ratio of 1 red: 2 pink: 1 white? (R is a gene for red; r = white)

    a. All F_1 plants were Rr.

    b. F_1 plants were red and white.

    c. At least one P generation parent was pink.

    d. F_2 genotypic ratio is 3:1.

    e. F_2 genotypic ratio is 1:2:1.

_____20. Polygenic inheritance means that a pair of genes

    a. exhibits incomplete dominance.

    b. affects expression of other genes.

    c. has a similar and additive effect as another independent pair of genes.

    d. has multiple effects.

    e. is sex-influenced.

VISUAL FOUNDATIONS

Use the cells in Row 1 as a reference. In Rows 2 and 3, draw in color the following structures: centrioles (yellow), spindle (black), properly assorted chromosomes in their appropriate colors (red and blue), and centrosomes (green).

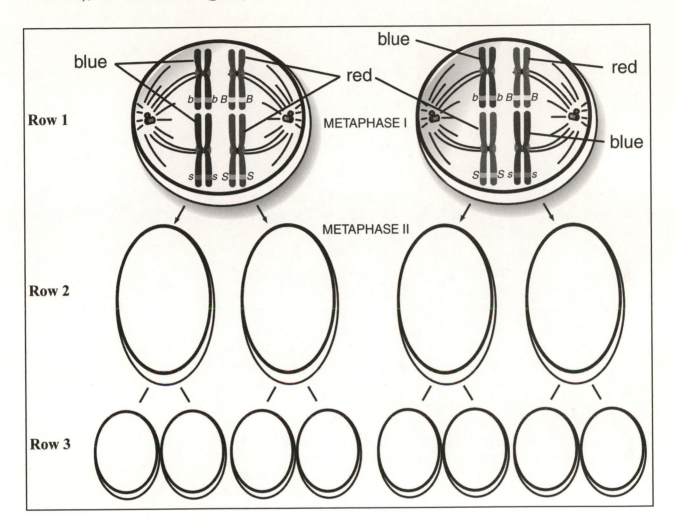

DNA: The Carrier of Genetic Information

In the previous chapter, you learned how genes are arranged in chromosomes and how they are transmitted from generation to generation. This chapter addresses the chemical nature of genes and how they function. Genes are made of deoxyribonucleic acid (DNA). Each DNA molecule consists of two polynucleotide chains arranged in a coiled double helix. The sequence of bases in DNA provides for the storage of genetic information. When DNA replicates, the hydrogen bonds between the two polynucleotide chains break, and the chains unwind and separate. Each half-helix then pairs with complementary nucleotides, replacing its missing partner. The result is two DNA double helices, each identical to the original and consisting of one original strand from the parent molecule and one newly synthesized complementary strand.

REVIEWING CONCEPTS
Fill in the blanks.

INTRODUCTION

1. DNA is a nucleic acid which carries the molecular basis of
_____.

EVIDENCE OF DNA AS THE HEREDITARY MATERIAL
DNA is the transforming principle in bacteria

2. A type of permanent genetic change in which the properties of one genetic type are conferred on another genetic type is known as _____.

DNA is the genetic material in certain viruses

3. _____ can reproduce by injecting only their DNA into cells, indicating that DNA is the genetic material.

THE STRUCTURE OF DNA
Nucleotides can be covalently linked in any order to form long polymers

4. Nucleotides consist of _____
_____.

5. The bases in nucleotides include the two purines, (a)_____
_____, and the two pyrimidines (b)_____
_____.

DNA is made of two polynucleotide chains intertwined to form a double helix

6. _____ studies by Franklin and Wilkins showed that DNA has a helical structure with nucleotide bases stacked like rungs of a ladder.

7. Watson and Crick devised a DNA model based for the most part on existing data. Their model suggested that DNA was formed from two

_____

arranged in a coiled double helix.

In double-stranded DNA, hydrogen bonds form between A and T and between G and C

8. (a)_____ hydrogen bonds form between adenine and thymine, and (b)_____ hydrogen bonds form between guanine and cytosine.

DNA REPLICATION

9. Replication of DNA is considered semiconservative because each "old" strand serves as a _____ for the formation of a new strand.

Meselson and Stahl verified the mechanism of semiconservative replication

10. Using density gradient centrifugation, scientists can separate large molecules like DNA on the basis of differences in their _____.

Semiconservative replication explains the perpetuation of mutations

11. A change in the sequence of bases in a strand of DNA that results in a genetic change is known as a _____.

DNA replication requires protein "machinery"

12. _____ catalyze the linking together of nucleotide subunits.

13. DNA synthesis proceeds in a (a)____' → (b)____' direction. One strand is copied continuously, the other adds (c)_____ fragments discontinuously.

Enzymes proofread and repair errors in DNA

14. The three enzymes involved in nucleotide excision repair are

_____.

Telomeres cap eukaryotic chromosome ends

15. Programmed cell death is known as _____.

16. The shortening of telomeres may contribute to _____.

BUILDING WORDS

Use combinations of prefixes and suffixes to build words for the definitions that follow.

Prefixes	The Meaning
a-	without, not, lacking
anti-	against, opposite of
semi-	half
tel(o)-	end

Prefix	Suffix	Definition
_____	-virulent	1. Not lethal; lacking in virulence.
_____	-parallel	2. The arrangement of the two polynucleotide chains in a DNA molecule, viz., "running" in opposite directions to each other.
_____	-conservative	3. The manner in which DNA replicates; half of the original DNA strand is conserved in each new double helix.
_____	-mere	4. The end of a eukaryotic chromosome.

MATCHING

Terms:

a. Chromatin
b. Deoxyribose
c. DNA ligase
d. DNA polymerase
e. Double helix

f. Histone
g. Mutation
h. Nucleotide
i. Purine
j. Pyrimidine

k. Replication fork
l. Transformation

For each of these definitions, select the correct matching term from the list above.

_____ 1. The incorporation of genetic material by a cell that causes it to change its phenotype.

_____ 2. The enzyme in DNA replication responsible for base pairing.

_____ 3. The region of the DNA where the molecule "unzips" for replication to begin.

_____ 4. The bases adenine and guanine.

_____ 5. Random heritable changes in DNA.

_____ 6. A molecule composed of one or more phosphate groups, a 5-carbon sugar, and a nitrogenous base.

_____ 7. A pentose sugar lacking a hydroxyl group on carbon-2.

_____ 8. The shape of the DNA molecule.

MAKING COMPARISONS

Fill in the blanks.

Researcher(s)	Contribution to the Body of Knowledge About DNA
Watson and Crick	Developed a model that demonstrated how DNA can both carry information and serve as its own template for duplication
#1	Found substance in heat-killed bacteria that "transformed" living bacteria
#2	Only the DNA of a bacteriophage is necessary for the reproduction of new viruses
Chargaff	#3
#4	Inferred from x-ray crystallographic films of DNA patterns that nucleotide bases are stacked like rungs in a ladder
#5	Chemically identified the transforming principle described by Griffith as DNA

MAKING CHOICES

Place your answer(s) in the space provided. Some questions may have more than one correct answer.

_____ 1. Okazaki fragments are
    a. joined by DNA ligase.
    b. comprised of at least 3000 nucleotides.
    c. linked by 3' hydroxyl and 5' phosphate.
    d. assembled on ribosomes.
    e. joined to a growing DNA strand by a phosphodiester linkage.

_____ 2. The two new strands in replicating DNA "grow" as follows:
    a. Leading continuously, lagging as fragments.
    b. Leading and lagging continuously. .
    c. Leading toward replication fork, lagging away from fork.
    d. Both strands "grow" toward replication fork.
    e. Both strands "grow" away from replication fork.

_____ 3. The molecules that "proofread" base pairing in replicated DNA, and correct any errors found are
    a. 3',5' phosphatases.
    b. purine and pyrimidine polymerases.
    c. DNA ligases.
    d. DNA polymerases.
    e. DNA primases.

_____ 4. The scientists who demonstrated that DNA replication is semiconservative
    a. Watson and Crick.
    b. Meselson and Stahl.
    c. Hershey and Chase.
    d. Avery, MacLeod, and McCarty.
    e. Franklin and Wilkins.

_____ 5. In double-stranded DNA, adenine is joined to thymine and cytosine is joined to guanine by
    a. ionic bonds
    b. phosphates and sugars.
    c. covalent bonds.
    d. hydrogen bonds.
    e. purines and pyrimidins.

_____ 6. In most bacteria, such as _E. coli_, DNA is
    a. a single strand.
    b. circular.
    c. a single double-stranded molecule.
    d. absent.
    e. associated with structural proteins.

_____ 7. Which of the following base pairs is/are correct?
    a. A – T
    b. C – A
    c. G – C
    d. T – A
    e. T – C

_____ 8. The molecule(s) responsible for preventing supercoiling and preventing supercoiling & knot formation in replicating DNA is/are
    a. histones.
    b. chromatin.
    c. scaffolding proteins.
    d. topoisomerases.
    e. protosomes.

_____ 9. A mutation is
    a. a change in an enzyme.
    b. a change in a gene.
    c. a change in DNA.
    d. an alteration of mRNA.
    e. sometimes inheritable.

_____ 10. The 3' end of one DNA fragment is linked to the 5' end of another fragment by means of
    a. DNA ligase.
    b. helicase enzymes.
    c. a DNA polymerase.
    d. hydrogen bonds.
    e. complementary base pairing.

_____11. The direction for synthesis of DNA is
  a. 5' → 3'.
  b. 3' → 5'.
  c. 5' → 5'.
  d. 3' → 3'.
  e. variable.

_____12. The genetic code is carried in the
  a. DNA backbone.
  b. sequence of bases.
  c. arrangement of 5', 3' phosphodiester bonds.
  d. Okazaki fragments.
  e. histones.

_____13. The double helix structure of DNA was suggested as a result of X-ray diffraction data collected by
  a. Hershey and Chase.
  b. Griffith.
  c. Avery, MacLeod, and McCarty.
  d. Watson and Crick.
  e. Franklin and Wilkens.

_____14. In one molecule of DNA one would expect the composition of the two strands to be
  a. both either old or new.
  b. both all new.
  c. both partly new fragments and partly old parental fragments.
  d. one old, one new.
  e. unpredictable.

_____15. Chargaff's rules state or infer that
  a. [A] = [T].
  b. [G] = [C].
  c. ratio of purines to pyrimidines = 1.
  d. ratio of T to A = 1.
  e. ratio of G to C = 1.

_____16. Deoxyribose and phosphate are joined in the DNA backbone by
  a. one of four bases.
  b. purines.
  c. pyrimidines.
  d. phosphodiester bonds.
  e. the 1' carbon of the sugar.

_____17. The investigators credited with elucidating the structure of DNA are
  a. Hershey and Chase.
  b. Messelson and Stahl.
  c. Avery, MacLeod, and McCarty.
  d. Watson and Crick.
  e. Franklin and Wilkens.

VISUAL FOUNDATIONS

Color the parts of the illustration below as indicated.

RED ☐ thymine

GREEN ☐ adenine

YELLOW ☐ cytosine

BLUE ☐ guanine

ORANGE ☐ sugar

VIOLET ☐ phosphate

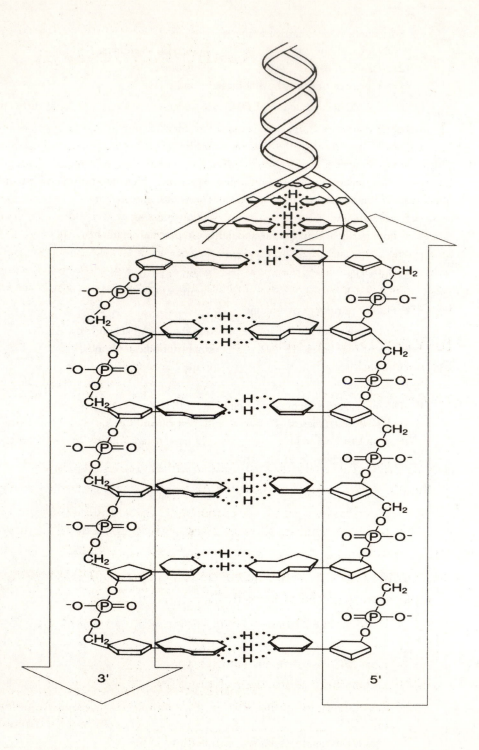

CHAPTER 13

Gene Expression

This chapter continues your examination of the transfer of genetic information from one cell generation to the next. You are now familiar with the behavior of chromosomes during sexual reproduction, how genes are arranged in chromosomes, how genes are transmitted from generation to generation, the chemical nature of genes, and how genes function. This chapter examines how cells convert the information in genes into proteins, thereby determining the phenotype of the organism. The manufacture of proteins by cells involves two major steps: the transcription of base sequences in DNA into the base sequences in RNA, and the translation of the base sequences in RNA into amino acid sequences in proteins. Of 64 possible codons, 61 code for amino acids and three serve as signals that specify the end of the coding sequence for a polypeptide chain. The genetic code is nearly universal, suggesting that it was established early in evolutionary history. A gene is a transcribable DNA nucleotide sequence that yields a product with a specific cellular function. A mutation is a change in a DNA nucleotide sequence. Genes can be altered by mutation in a number of ways.

REVIEWING CONCEPTS

Fill in the blanks.

INTRODUCTION

1. Gene expression involves a series of steps in which the information in the sequence of _____ specifies the makeup of the cell's proteins.

DISCOVERY OF THE GENE-PROTEIN RELATIONSHIP

Beadle and Tatum proposed the one-gene, one-enzyme hypothesis

2. Neurospora is an ideal experimental organism because it grows primarily as a haploid organism, allowing researchers to immediately identify a

   _____.

INFORMATION FLOW FROM DNA TO PROTEIN: AN OVERVIEW

DNA is transcribed to form RNA

3. DNA is transcribed to form three specific kinds of RNA. These are

   _____.

4. Each tRNA is specific for only one _____.

RNA is translated to form a polypeptide

5. Sequencing of amino acids in a protein involves two major steps: first is (a)_____, wherein DNA codes are read into a special messenger RNA, and second is (b)_____, involving the actual construction of polypeptide chains.

6. Each _____ in mRNA consists of three-bases that specify one amino acid in a polypeptide chain.

Biologists cracked the genetic code in the 1960s

7. _____ are the only amino acids that are specified by single codons.

8. There are (a)_____(#?) possible codons. Of these, (b)_____ code for amino acids, and (c)_____(#?) function to terminate protein synthesis.

9. The genetic code is read one triplet at a time from a fixed starting point that establishes the _____.

TRANSCRIPTION

10. mRNA synthesis is catalyzed by DNA-dependent_____.

11. Any point closer to the 3' end of transcribed DNA (and therefore toward the 5' end of mRNA) is said to be (a)_____ relative to a given reference point. Conversely, areas toward the 5' end of DNA and the 3' end of mRNA are (b)_____.

The synthesis of mRNA includes initiation, elongation, and termination

12. Transcription is initiated at sites containing specific base sequences called the _____ regions of DNA.

Messenger RNA contains base sequences that do not directly code for protein

13. Well before the protein-coding sequences at the 5' end of mRNA, a segment called the _____ contains recognition signals for ribosome binding.

14. Stop codons are followed by noncoding 3' _____.

TRANSLATION

15. The covalent bonds that join amino acids together are known as _____.

An amino acid is attached to t RNA before incorporation into a polypeptide

16. The specific enzymes that catalyze the formation of covalent bonds between amino acids and their respective tRNAs are the _____ _____.

The components of the translational machinery come together at the ribosomes

17. Translation begins with the formation of an _____ complex.

18. The codon _____ initiates the formation of the protein-synthesizing complex.

19. The addition of amino acids to a growing polypeptide chain is called _____.

20. Peptide bond formation requires the enzyme _____.

21. Newly formed polypeptides are released and ribosome subunits dissociate as the result of sequences in mRNA called _____.

VARIATIONS IN GENE EXPRESSION IN DIFFERENT ORGANISMS

Transcription and translation are coupled in prokaryotes

22. A _____ consists of an mRNA molecule that is bound to clusters of ribosomes.

Eukaryotic mRNA is modified after transcription and before translation

23. The original transcript in eukaryotes is known as _____.

Both noncoding and coding sequences are transcribed from eukaryotic genes

24. Intron is an abbreviation for (a)_____, and exon stands for (b)_____.

25. _____ are regions of protein tertiary structure that may have specific functions.

Several kinds of eukaryotic RNA have a role in gene expression

26. SiRNAs silence genes at the _____ level by selectively cleaving mRNA molecules with base sequences complementary to siRNA.

The definition of a gene has evolved as biologists have learned more about genes

27. A gene carries the information needed to produce

_____.

The usual direction of information flow has exceptions

28. Retroviruses are viruses that require _____.

MUTATIONS

29. A mutation is a change in the _____ in DNA.

Base substitution mutations result from the replacement of one base pair by another

30. (a)_____ mutations involve a change in only one pair of nucleotides. These can lead to the substitution of one amino acid for another, a so called (b)_____ mutation or a (c)_____ mutation involving the conversion of an amino acid-specifying codon to a termination codon.

Frameshift mutations result from the insertion or deletion of base pairs

31. In frameshift mutations, one of two nucleotide pairs are _____ from the molecule, altering the reading frame.

Some mutations involve larger DNA segments

32. Mobile genetic elements can disrupt the functions of some genes because they are

_____.

Mutations have various causes

33. _____ are regions of DNA that are particularly susceptible to mutations.

34. Agents that cause mutations are called _____.

BUILDING WORDS

Use combinations of prefixes and suffixes to build words for the definitions that follow.

Prefixes	The Meaning	Suffixes	The Meaning
anti-	against, opposite of	-gen	production of
poly-	much, many	-some	body

Prefix	Suffix	Definition
_____	-codon	1. A sequence of three nucleotides in tRNA that is complementary to ("opposite of"), and combines with, the three-nucleotide codon on mRNA.
ribo-	_____	2. An organelle (microbody) composed of RNA and proteins that functions in protein synthesis.
_____	_____	3. A submicroscopic complex consisting of many ribosomes attached to an mRNA molecule during translation.
muta-	_____	4. A substance capable of producing mutations.
carcino-	_____	5. A substance capable of producing cancer.

MATCHING

Terms:

a. Adenine
b. Codon
c. Elongation
d. Exon
e. Intron
f. mRNA
g. Promoter
h. Retrovirus
i. rRNA
j. Transcription
k. tRNA
l. Translocation
m. Uracil

For each of these definitions, select the correct matching term from the list above.

_____ 1. Site on DNA to which RNA polymerase attaches to begin transcription.

_____ 2. The making of mRNA from a DNA template.

_____ 3. An RNA virus that produces a DNA intermediate in its host cell.

_____ 4. A pyrimidine found in RNA.

_____ 5. A triplet of mRNA bases that specifies an amino acid or a signal to terminate the polypeptide.

_____ 6. The stage of translation during which amino acid chains are built.

_____ 7. The RNA responsible for base pairing during protein synthesis.

_____ 8. In eukaryotes, the coding region of DNA.

_____ 9. RNA that has been transcribed from DNA that specifies the amino acid sequence of a protein.

MAKING COMPARISONS

Fill in the blanks.

mRNA Codon	DNA Base Sequence	tRNA Anticodon	Amino Acid Specified
5' — AAA — 3'	3'—TTT—5'	3'—UUU—5'	Lysine (Lys)
#1	#2	3'— ACC —5'	Tryptophan (Trp)
5' — UUU — 3'	#3	#4	Phenylalanine (Phe)
5' — GGA — 3'	#5	3' — CCU — 5'	Glycine (Gly)
#6	3' — ATT — 5', 3' — ATC — 5', 3' — ACT — 5'	#7	None (stop codon)
5'—AUG—3'	#8	#9	Methionine (Met) (start codon)

MAKING CHOICES

Place your answer(s) in the space provided. Some questions may have more than one correct answer.

_____ 1. The following part of a spliceosome particle is involved in intron removal and regulation of transcription:

    a. SRP RNA   d. snRNA

    b. snoRNA   e. miRNA

    c. siRNA

_____ 2. The _____ selectively suppresses gene expression.

    a. SRP RNA   d. snRNA

    b. snoRNA   e. miRNA

    c. siRNA

_____ 3. Introns are

    a. intervening sequencess.   d. organized codes for specific amino acids.

    b. special integrated coding sequences.   e. HIV resistant sequences.

    c. noncoding sequences within a gene.

_____ 4. Energy for transfer of a peptide chain from the A site to the P site is provided by

    a. ATP.   d. ADP.

    b. enzymes.   e. guanosine triphosphate.

    c. GTP.

_____ 5. Transcription involves

    a. mRNA synthesis.   d. copying codes into codons.

    b. decoding of codons.   e. peptide bonding.

    c. copying DNA information.

_____ 6. Ribosomes attach to

    a. 5' end of mRNA.   d. tRNA.

    b. 3' end of mRNA.   e. RNA polymerase.

    c. an mRNA recognition sequence.

_____ 7. Translation involves
   a. hnRNA.
   b. decoding of codons.
   c. adapter molecules.
   d. copying codes into codons.
   e. copying codons into codes.

_____ 8. A large quantity of homogentisic acid in urine
   a. is associated with alkaptonuria.
   b. indicates a kidney disease.
   c. is caused by a block in the metabolic reactions that break down tyrosine.
   d. is due to absence of an oxidizing enzyme.
   e. indicates that a diuretic has been consumed.

_____ 9. Which of the following is/are termination codons?
   a. UAG
   b. UUA
   c. UAA
   d. GUA
   e. UGA

_____10. A recognition sequence for ribosome binding to mRNA is/are
   a. aminoacyl-tRNA.
   b. coding sequence.
   c. initiator.
   d. upstream from the coding sequence.
   e. leader sequence.

_____11. An "adapter" molecule containing an anticodon and a region to which an amino acid is bonded is a/an
   a. aminoacyl-tRNA.
   b. coding sequence.
   c. initiator.
   d. peptidyl transferase.
   e. leader sequence.

_____12. Beadle and Tatum concluded that
   a. one gene affects one polypeptide.
   b. one gene affects one enzyme.
   c. a mutation directly affects an enzyme.
   d. mutations have no effect on enzymes.
   e. proteins mutate when exposed to radiations.

_____13. _Neurospora_ was a good choice for the Beadle and Tatum studies because
   a. it is diploid.
   b. its asexual spores enable genetic analysis.
   c. it can be cultivated on minimal medium.
   d. it had no known mutant forms.
   e. the control strain could not use arginine.

_____14. Post-transcriptional modification and processing of mRNA takes place in
   a. cytoplasm.
   b. the nucleus.
   c. prokaryotes only.
   d. eukaryotes only.
   e. both prokaryotes and eukaryotes.

_____15. Initiation of protein synthesis involves
   a. the code AUG.
   b. initiation tRNA.
   c. loading initiation tRNA on the large ribosomal subunit.
   d. adding amino acids to a growing polypeptide chain.
   e. peptidyl transferase.

VISUAL FOUNDATIONS

Color the parts of the illustration below as indicated. Also label the 5' end of mRNA, the start codon, the leader sequence and circle the initiation complex.

RED	☐	formylated methionine tRNA
GREEN	☐	small ribosome subunit
YELLOW	☐	large ribosome subunit
PINK	☐	initiation factors
BLUE	☐	E site
ORANGE	☐	P site
VIOLET	☐	A site

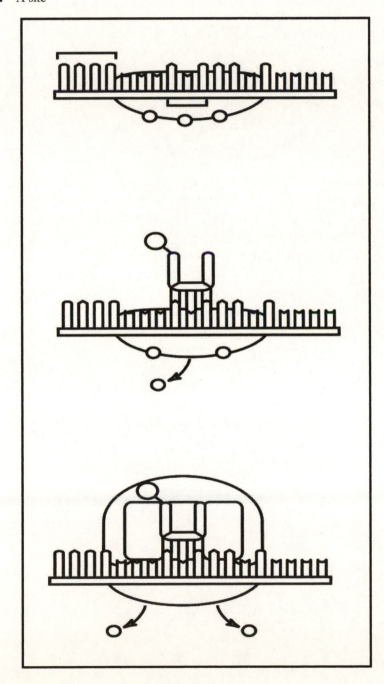

CHAPTER 14

Gene Regulation

Clearly, all of the genes of a multicellular organism are not expressed all at the same time. If they were, separate and distinct cells, tissues, and organs could not exist. Each type of cell in a multicellular organism has a characteristic shape, carries out very specific activities, and makes a distinct set of proteins. This is because only part of the genetic information is ever expressed in any given cell at any given time. This chapter looks at the mechanisms that control gene expression. The control mechanisms require information in the form of signals that originate either within the cell, other cells, or the environment and that interact with DNA, RNA, or protein. The mechanisms include controlling the amount of mRNA available, the rate of translation of mRNA, and the activity of the protein product. In prokaryotes, most gene regulation involves transcriptional level control of genes whose products are involved in resource utilization. In eukaryotes, gene regulation occurs at many levels, consistent with the greater complexity of these cells. Some genes whose products are required in large amounts exist as multiple copies in the chromosome. Some genes are inactivated by changes in chromosome structure.

REVIEWING CONCEPTS

Fill in the blanks.

INTRODUCTION

1. Mechanisms that regulate gene expression include:

   _____

   _____.

GENE REGULATION IN BACTERIA AND EUKARYOTES: AN OVERVIEW

2. _____ is the most efficient mechanism of gene regulation in bacteria.

GENE REGULATION IN BACTERIA

3. _____ are genes that encode essential proteins that are in constant use.

Operons in bacteria facilitate the coordinated control of functionally related genes

4. A gene complex consisting of a group of structural genes with related functions and the DNA sequences responsible for controlling them is known as an

   _____. .

5. A group of functionally related genes may be controlled by one _____ that is located upstream from the coding sequences.

6. The _____ is a sequence of bases that switches mRNA synthesis "on" or "off."

7. An _____ inactivates a repressor in order to turn "on" a gene or operon.

8. Repressible operons are turned "off" when a repressor is activated by binding to a _____.

Some posttranscriptional regulation occurs in bacteria

9. Protein levels can be controlled by transcriptional level control. Other regulatory mechanisms that occur *after* transcription are referred to as _____.

10. _____ is when a product blocks its own production by binding to an enzyme that is required to generate that product.

GENE REGULATION IN EUKARYOTIC CELLS

Eukaryotic transcription is controlled at many sites and by many different regulatory molecules

11. RNA polymerase in multicellular eukaryotes binds to a portion of the promoter known as the _____.

12. The efficiency of a eukaryotic promoter depends largely on the number and type of _____.

13. Highly coiled and compacted chromatin containing inactive genes is called (a)_____, whereas loosely coiled chromatin containing genes capable of transcription is referred to as (b)_____.

The mRNAs of eukaryotes have many types of posttranscriptional control

14. Through _____, the cells in each tissue produce their own version of mRNA corresponding to the particular gene.

Posttranslational chemical modifications may alter the activity of eukaryotic proteins

15. The addition or removal of phosphate groups is an example of _____, a mechanism used by eukaryotic cells to regulate protein activity.

16. Enzymes that modify chemicals by adding phosphate groups are called _____.

17. Enzymes that modify chemicals by removing phosphate groups are called _____.

BUILDING WORDS

Use combinations of prefixes and suffixes to build words for the definitions that follow.

Prefixes	The Meaning
co-	with, together, in association
eu-	good, well, "true"
hetero-	different, other
homo-	same

Prefix	Suffix	Definition
_____	-chromatin	1. Chromatin that appears loosely coiled. Because it is the only chromatin capable of transcription, it is considered to be the "good" or "true" chromatin.
_____	-chromatin	2. The inactive chromatin that appears highly coiled and compacted, and is generally not capable of transcription. (Chromatin that is "different from" or "other than" the euchromatin.)
_____	-repressor	3. A substance that, together with a repressor, represses protein synthesis in a specific gene.
_____	-dimer	4. A dimer in which the two component polypeptides are different.

MATCHING

Terms:

a. Allosteric binding site
b. Constitutive gene
c. Enhancer
d. Gene amplification
e. Inducible system

f. Operator
g. Operon
h. Repressible system
i. Repressor
j. Transcriptional control

k. Translational control

For each of these definitions, select the correct matching term from the list above.

_____ 1. One of the control regions of an operon.

_____ 2. A site located on an enzyme that enables a substance other than the normal substrate to bind to the molecule, and to change the shape of the molecule and the activity of the enzyme.

_____ 3. A regulatory protein that represses expression of a specific gene.

_____ 4. A regulatory mechanism that controls the rate at which a particular mRNA molecule is translated.

_____ 5. Process by which multiple copies of a gene are produced by selective replication, thus allowing for increased synthesis of the gene product.

_____ 6. Genes that are constantly transcribed.

_____ 7. In prokaryotes, a group of structural genes that are coordinately controlled and transcribed as a single message, plus their adjacent regulatory elements.

_____ 8. Regulatory elements that can be located long distances away from the actual coding regions of a gene.

_____ 9. Control in which the presence of a substrate induces the synthesis of an enzyme.

MAKING COMPARISONS

Fill in the blanks.

Regulatory Event	Name
Protein dimmers held together by leucine	Leucine zipper proteins
Genes are inducible only during specific phases of life cycle	#1
#2	Molecular chaperones
Phenotypic expression is determined by whether an allele is inherited from the male or the female parent	#3
#4	Posttranslational controls

Regulatory Event	Name
Adjusts the rate of synthesis in a metabolic pathway	#5
Help form an active transcription initiation complex thereby increasing the rate of RNA synthesis	#6
Regulate the rate of mRNA translation	#7

MAKING CHOICES

Place your answer(s) in the space provided. Some questions may have more than one correct answer.

_____ 1. Genes that code for proteins that are constantly needed for survival of an organism are called
     a. promoters.
     b. constitutive genes.
     c. operons.
     d. repressor genes.
     e. always "on".

_____ 2. A repressor usually controls an inducible gene by
     a. keeping it "turned off".
     b. suppressing mRNA production.
     c. reducing product resulting from an active inducer.
     d. inducing an antagonist.
     e. blocking DNA replication at specific sites.

_____ 3. The basic way(s) that cells control their metabolic activities is/are by
     a. regulating enzyme activity.
     b. mutations.
     c. developing different genes for different purposes.
     d. controlling the number of enzyme molecules.
     e. transduction.

_____ 4. Feedback inhibition is an example of
     a. transcriptional control.
     b. pretranscriptional control.
     c. a control mechanism affecting events after translation.
     d. an inducible system.
     e. a repressible system.

_____ 5. The lactose repressor
     a. can convert to an operator.
     b. is several bases upstream from the operator.
     c. is downstream from RNA polymerase coding sequences.
     d. becomes an activator when lactose is present.
     e. is continuously "on".

_____ 6. The genes and genetic information in different cells of multicellular organisms are
     a. all slightly different.
     b. distinctly different.
     c. identical in the same tissues, different in different tissues.
     d. identical.
     e. almost all identical.

_____ 7. Replication of specific genes in cells that need more of them is an example of
     a. magnification.
     b. amplification.
     c. induction.
     d. a positive control system.
     e. a regulon.

_____ 8. One would expect a gene involved in the manufacture of ATP to be
     a. "off" usually.
     b. always "on".
     c. facultative.
     d. a constitutive gene.
     e. sometimes "off," sometimes "on".

_____ 9. The tryptophan operon in *E. coli* is

    a. usually on.

    b. on in the absence of corepressor.

    c. repressed only when tryptophan binds to repressor.

    d. an inducible system.

    e. a repressible system.

_____10. UPEs

    a. are in DNA near promoters.

    b. are amino acids.

    c. seem to determine the strength of a repressor.

    d. are sequences of DNA bases.

    e. affect promoter activity.

_____11. The lactose operon in *E. coli*

    a. contains three structural genes.

    b. is turned on by a repressor.

    c. is a repressor.

    d. is transcribed as part of a single RNA molecule.

    e. is an inducible system.

_____12. "Zinc fingers" seems to be involved in

    a. unwinding DNA.

    b. activating transcription.

    c. insertion of a regulator domain into grooves of DNA.

    d. binding a regulatory protein to DNA.

    e. blocking posttranscriptional events.

_____13. A tightly coiled portion of DNA that contains inactive genes is

    a. called heterochromatin.

    b. called euchromatin.

    c. found in most eukaryotes and few prokaryotes.

    d. found only in prokaryotes.

    e. found only in eukaryotes.

VISUAL FOUNDATIONS

Color the parts of the illustration below as indicated. Also label the lactose operon, ribosomes and identify the enzymes produced in the presence of the inducer.

RED ☐ operator region

GREEN ☐ promoter region

YELLOW ☐ repressor gene

BLUE ☐ mRNA

ORANGE ☐ inducer

BROWN ☐ inactivated repressor protein

VIOLET ☐ active repressor protein

TAN ☐ structural gene

PINK ☐ RNA polymerase

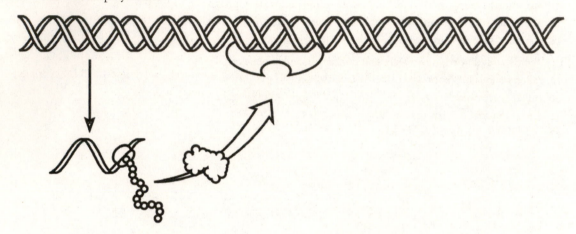

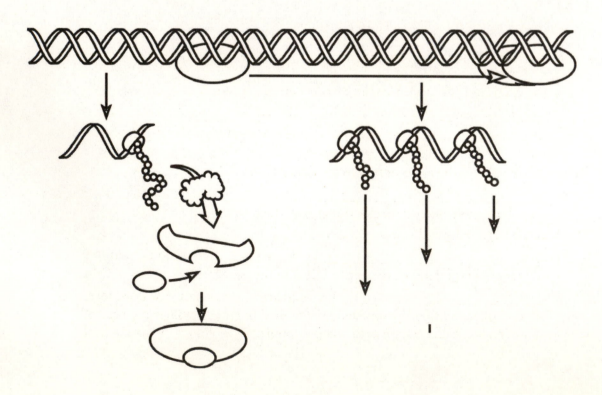

CHAPTER 15

DNA Technology and Genomics

Recombinant DNA technology involves introducing foreign DNA into cells of microorganisms where it replicates and is transmitted to daughter cells. In this way, a particular DNA sequence can be amplified to provide millions of identical copies that can be isolated in pure form. Once a piece of DNA is cloned, its functions can be studied and possibly engineered for a particular application. DNA can also be amplified in vitro. Most of our knowledge of the complex structure and control of eukaryotic genes, as discussed in the prior two chapters, and of the roles of genes in development, which will be taken up in Chapter 17, is derived from the application of these methods. Recombinant DNA technology has led to genetic engineering — the modification of the DNA of an organism to produce new genes with new characteristics. DNA technology has increasingly important applications in many fields of study and commerce, including evolution, medicine, pharmacology, and the food and forensic sciences. DNA technology has raised important safety concerns that scientists, following stringent guidelines, are addressing..

REVIEWING CONCEPTS
Fill in the blanks.

INTRODUCTION

1. The modification of an organism's DNA to produce new genes with new traits is called _____.

DNA CLONING
Restriction enzymes are "molecular scissors"

2. _____ cut DNA molecules at specific loci, producing precise fragments that can be incorporated into vector molecules.

3. Recombinant DNA vectors are usually constructed from
   _____.

4. A _____ sequence reads the same as its complement when both are read in the 5' to 3' direction.

Recombinant DNA forms when DNA is spliced into a vector

5. _____ are combination vectors with features from both bacteriophages and plasmids.

DNA can be cloned inside cells

6. A _____ is a collection of thousands of DNA fragments that represent the entire DNA in the genome.

7. _____ are radioactive strands of RNA or DNA that are complementary to specific targeted cloned fragments. They are used to identify a particular fragment in a large population of clones.

8. The enzyme _____ is used to synthesize single-stranded complementary DNA (cDNA) on a mRNA template.

The polymerase chain reaction is a technique for amplifying DNA in vitro

9. The polymerase chain reaction (PCR) technique for amplifying DNA is advantageous because it side steps the cumbersome, time-consuming process of cloning DNA. PCR is an *in vitro* process that alternately uses
   (a)_____ to replicate DNA with the use of
   (b)_____ to dissociate the replicated DNA strands for further replications.

10. The PCR technique must use a heat stable DNA polymerase named _____ polymerase after the organism from which it is obtained.

DNA ANALYSIS

Gel electrophoresis is used for separating macromolecules

11. In gel electrophoresis, nucleic acids migrate through the gel toward the _____ of the electric field because they are negatively charged.

DNA, RNA, and protein blots detect specific fragments

12. The method of detecting DNA fragments by separating them by gel electrophoresis and then transferring them to a nitrocellulose or nylon membrane is called a _____.

Restriction fragment length polymorphisms are a measure of genetic relationships

13. In molecular biology, a _____ is the presence of detectable variation in the genomes of different individuals in a population.

One way to characterize DNA is to determine its sequence of nucleotides

14. Automated DNA sequencing relies on the
    _____.

GENOMICS

Identifying protein-coding genes is useful for research and for medical applications

15. cDNA sequences that can identify protein-coding agents are known as
    _____.

One way to study gene function is to silence genes one at a time

16. Gene function is revealed in a procedure called _____,
    in which the researcher chooses and "knocks out" a single gene in an organism.

DNA microarrays are a powerful tool for studying how genes interact

17. _____ enable researchers to compare the activities of thousands of genes in normal and diseased cells from tissue samples.

The Human Genome Project stimulated studies on the genome sequences of other species

18. Several scientific fields, including
    _____ emerged in the wake of the Human Genome Project.

APPLICATIONS OF DNA TECHNOLOGIES

DNA technology has revolutionized medicine and pharmacology

19. _____, the use of specific DNA to treat a genetic disease by correcting the genetic problem, is another application of DNA technology.

DNA fingerprinting has numerous applications

20. STRs are highly _____ because they vary in length from one individual to another, making them useful in identifying individuals.

Transgenic organisms have incorporated foreign DNA into their cells

21. Transgenic organisms (plants or animals) are generally produced by injecting the DNA of a gene into a recipient's _____.

22. _____ is a research procedure in which the researcher chooses and in activates a single gene in an organism.

23. Transgenic plants that are resistant to insect pests, viral diseases, drought, heat, cold, herbicides, and salty or acidic soil are known as _____.

DNA TECHNOLOGY HAS RAISED SAFETY CONCERNS

24. The science of _____ uses statistical methods to quantify risks so they can be compared and contrasted.

BUILDING WORDS

Use combinations of prefixes and suffixes to build words for the definitions that follow.

Prefixes	The Meaning	Suffixes	The Meaning
bacterio-	bacteria	-phage	eat, devour
retro-	backward		
trans-	across		

Prefix	Suffix	Definition
_____	-genic	1. Pertains to organisms that have had their genome altered by recombinant DNA technology. (Genes are taken from one organism and transferred "across" into another organism.)
_____	-virus	2. An RNA virus that makes DNA copies of itself by reverse transcription.
_____ _____		3. A virus that infects and destroys bacteria.

MATCHING

Terms:

a.   Colony
b.   DNA ligase
c.   Genetic probe
d.   Kilobase
e.   Plasmid

f.   PCR
g.   Recombinant DNA
h.   Restriction enzyme
i.   Reverse transcriptase
j.   Vector

For each of these definitions, select the correct matching term from the list above.

_____ 1. Enzyme produced by retroviruses to enable the transcription of DNA from the viral RNA in the host cell.

_____ 2. Small, circular DNA molecules that carry genes separate from the main bacterial chromosome.

_____ 3. A cluster of genetically-identical cells.

_____ 4. Bacterial enzymes used to cut DNA molecules at specific base sequences.

_____ 5. Method used to produce large amounts of DNA from tiny amounts.

_____ 6. An agent, such as a plasmid or virus, that transfers genetic information.

_____ 7. Any DNA molecule made by combining genes from different organisms.

_____ 8. A radioactively-labeled segment of RNA or single-stranded DNA that is complementary to a target gene.

_____ 9. A unit of measurement applied to the size of DNA fragments.

MAKING COMPARISONS

Fill in the blanks.

Technique	Definition/Description
genetic engineering	Modifying the DNA of an organism to produce new genes with new traits
#1	Microbial proteins used to cut DNA in specific locations
#2	A segment of DNA that is homologous to a part of a DNA sequence that is being studied
#3	Method for replicating and isolating many copies of a specific recombinant DNA molecule.
#4	Amplifying DNA by alternate heat denaturization and use of heat-resistant DNA polymerase to replicate DNA strands *in vitro*
plasmids	#5
genomic DNA library	#6
#7	The study of all proteins encoded by the human genome

MAKING CHOICES

Place your answer(s) in the space provided. Some questions may have more than one correct answer.

_____ 1. Genes in eukaryotic organisms that do not code for proteins are

    a. introns.

    b. small PCRs.

    c. short tandem repeats (STRs).

    d. highly polymorphic.

    e. frequently used to diagnose genetic diseases.

_____ 2. Copies of DNA made from isolated mRNA by using reverse transcriptase are called

    a. genomic DNA.

    b. recombinant DNA.

    c. PCRs.

    d. complementary DNA (cDNA).

    e. probes.

_____ 3. Analysis of an individual's DNA based on his or her STRs is known as

    a. proteomics.

    b. DNA blotting analysis.

    c. DNA fingerprinting.

    d. cDNA library.

    e. genomic DNA library.

_____ 4. Detection of DNA fragments transferred to a nitrocellulose or nylon membrane after separating them by gel electrophoresis is known as a

    a. Northern blot.

    b. Southern blot.

    c. Easter blot.

    d. Western blot.

    e. fingerprinting blot.

_____ 5. The vector(s) commonly used to incorporate a DNA fragment into a carrier is/are

    a. prophages.

    b. *E. coli.*

    c. plasmids.

    d. translocation microphages.

    e. bacteriophages.

_____ 6. After "sticky ends" pair with other DNA molecules cut by the same enzyme, the two fragments can be covalently linked to form recombinant DNA by treating them with

    a. plasmozymes.

    b. restriction enzymes.

    c. DNA polymerase.

    d. DNA ligase.

    e. vector enzymes.

_____ 7. Restriction enzymes are normally used by bacteria to

    a. defend against viruses.

    b. remove extraneous introns.

    c. denature antibiotics.

    d. cut plasmids from the large chromosome.

    e. insert operators in active DNA.

_____ 8. Palindromic sequences are base sequences from one DNA strand that

    a. produce RNA polymerase.

    b. form RNA complementary to DNA.

    c. can be cut by restriction enzymes.

    d. cut specific fragments from DNA.

    e. read in the opposite direction as its complement.

_____ 9. cDNA libraries are compiled

    a. from introns.

    b. DNA copies of mRNA.

    c. from multiple copies of DNA fragments.

    d. from introns and exons.

    e. with the use of reverse transcriptase.

_____ 10. A base sequence that is palindromic to 3'-AGCTTAA-5' would read

    a. 3'-AGCTTAA-5'.

    b. 5'-TCGAATT-3'.

    c. 3'-TCGAATT-5'.

    d. 5'-UCGAAUU-3'.

    e. 3'-UCGAAUU-5'.

_____11. The Ti used to insert genes into plant cells is

    a. an RNA fragment.   d. a plasmid.

    b. not effective in dicots.   e. a vector.

    c. useful to increase yields of grain foods.

_____12. Restriction fragment length polymorphisms are

    a. complementary.   d. possibly due to mutations.

    b. RNA variants.   e. variants paired in order to obtain different clones.

    c. the result of changes in DNA.

_____13. DNA fragments for recombination are produced by cutting DNA

    a. and introducing a cleaved plasmid.   d. with bacteriophages.

    b. with restriction enzymes.   e. with DNA ligase.

    c. at specific base sequences.

_____14. Transgenic organisms are

    a. dead animals.   d. animals with foreign genes.

    b. vectors for retroviruses.   e. plants with foreign genes.

    c. used to produce recombinant proteins.

_____15. A radioactive strand of RNA or DNA that is used to locate a cloned DNA fragment is

    a. a genetic probe.   d. a restriction enzyme.

    b. the genome.   e. complementary to the ligase used.

    c. complementary to the targeted gene.

_____16. RNA viruses that make DNA copies of themselves by reverse transcription are

    a. plasmids.   d. used as genetic probes.

    b. retroviruses.   e. bacteriophages.

    c. double stranded.

VISUAL FOUNDATIONS

Color the parts of the illustration below as indicated. Also label DNA, pre-mRNA, mature mRNA, transcription, RNA processing, and double-stranded cDNA.

RED	☐	exon
GREEN	☐	intron
YELLOW	☐	reverse transcriptase
BLUE	☐	cDNA copy of mRNA
VIOLET	☐	mRNA
PINK	☐	DNA polymerase

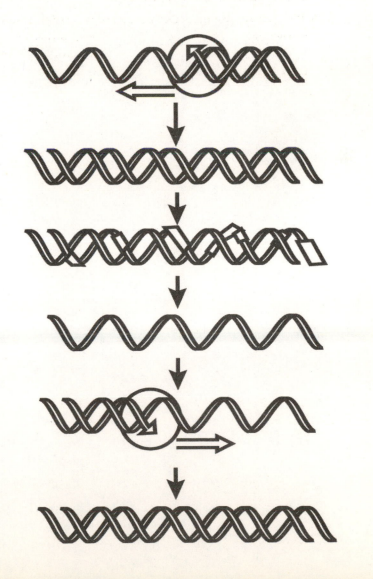

❑

The Human Genome

The principles of genetics apply to all organisms, including humans. Human geneticists, however, cannot make specific crosses of pure human strains as geneticists do with other organisms. Instead, they must rely on studies of other organisms and population studies of large extended families. Also, knowledge of human genetics has been greatly advanced in recent years by the mapping and sequencing of the human genome and by the study of genetic diseases in humans. New tools have been developed for genetic testing, screening, counseling, and therapy. Increasingly, work in the field of human genetics raises many ethical issues.

REVIEWING CONCEPTS

Fill in the blanks.

INTRODUCTION

1. In DNA sequencing, researchers identify the order of _____
   _____ to understand the genetic basis of human similarities and differences

STUDYING HUMAN GENETICS

Human chromosomes are studied by karyotyping

2. _____ is the study of chromosomes and their role in inheritance.

3. A _____ is an individual's chromosome composition.

Family pedigrees help identify certain inherited conditions

4. A _____ shows the transmission of genetic traits within a family over several generations and helps determine the exact interrelationships of the DNA molecules from related individuals.

The Human Genome Project sequenced the DNA on all human chromosomes

5. The Human Genome Project sequenced _____(#?) base pairs.

Comparative genomics has revealed several hundred DNA segments that are identical in both mouse and human genomes

6. About _____ (#?) DNA segments longer than 200 base pairs are known to be 100% identical between the mouse and human genomes.

Researchers use mouse models to study human genetic diseases

7. Researchers have used _____ to produce strains of mice that are homozygous or heterozygous to genetic diseases.

ABNORMALITIES IN CHROMOSOME NUMBER AND STRUCTURE

8. _____, the presence of multiple sets of chromosomes, is usually lethal in humans.

9. The abnormal presence or absence of one chromosome in a set is called (a)_____. For the affected chromosome, the normal condition, called disomy, becomes a (b)_____ condition with the addition of an extra chromosome, and a (c)_____ condition when one member of a pair is missing.

Down syndrome is usually caused by trisomy 21

10. _____ is a condition that results when chromosome 21 does not separate at anaphase.

Most sex chromosome aneuploidies are less severe than autosomal aneuploidies

11. Persons with _____ are nearly normal males, except for an extra X chromosome.

12. Persons with _____ are sterile females without Barr bodies.

Abnormalities in chromosome structure cause certain disorders

13. _____ is the attachment of a fragment of one chromosome to a nonhomologous chromosome.

14. A _____ is loss of part of a chromosome.

15. _____ are weak points where part of a chromatid appears to be attached to the rest of the chromosome by a thin thread of DNA.

GENETIC DISEASES CAUSED BY SINGLE-GENE MUTATIONS

16. PKU and akaptonuria are examples of disorders involving enzyme defects, collectively known as _____.

Most genetic diseases are inherited as autosomal recessive traits

17. Phenylketonuria results from an _____ deficiency.

18. Sickle cell anemia results from a _____ defect.

19. Cystic fibrosis results from defective _____.

20. In cystic fibrosis, a mutant protein causes the production of a very heavy _____ that eventually causes severe tissue damage.

21. Tay-Sachs disease results from abnormal (a)_____ metabolism in the (b)_____.

Some genetic diseases are inherited as autosomal dominant traits

22. Huntington's disease is caused by a rare _____ that affects the central nervous system.

Some genetic diseases are inherited as X-linked recessive traits

23. Hemophilia A is caused by the absence of a blood-clotting protein called _____.

GENE THERAPY

Gene therapy programs are carefully scrutinized

24. Gene therapy is a strategy that aims to replace a (a)_____ allele with a (b)_____ allele to treat some serious genetic diseases.

GENETIC TESTING AND COUNSELING

Prenatal diagnosis detects chromosome abnormalities and gene defects

25. _____ may be used to diagnose prenatal genetic diseases. It analyzes cells in amniotic fluid withdrawn from the uterus of a pregnant woman.

26. One technique designed to detect prenatal genetic defects is _____, which involves inspecting fetal cells involved in forming the placenta.

Genetic screening searches for genotypes or karyotypes

27. Newborn genetic screening is used primarily as the first step in _____ _____.

Genetic counselors educate people about genetic diseases

28. Genetic counselors use family histories of each partner to acquire the _____ that any given offspring will inherit a particular condition.

HUMAN GENETICS, SOCIETY, AND ETHICS

Genetic discrimination provokes heated debate

29. Genetic discrimination is discrimination against an individual or family members because of differences from the _____ genome in that individual.

Many ethical issues related to human genetics must be addressed

BUILDING WORDS

Use combinations of prefixes and suffixes to build words for the definitions that follow.

Prefixes	The Meaning
cyto-	cell
iso-	equal, "same"
poly-	much, many
trans-	across, beyond, through
tri-	three

Prefix	Suffix	Definition
_____	-genetics	1. The branch of biology that uses the methods of cytology to study genetics; the study of chromosomes and their role in inheritance.
_____	-ploidy	2. The presence of multiples of complete chromosome sets.
_____	-somy	3. Condition in which a chromosome is present in triplicate instead of duplicate (i.e., the normal pair).
_____	-location	4. An abnormality in which a part of a chromosome breaks off and attaches to another chromosome.

MATCHING

Terms:

a. Amniocentesis
b. Aneuploidy
c. Consanguineous mating
d. Disomic
e. Down Syndrome

f. Huntington's Disease
g. Karyotype
h. Klinefelter Syndrome
i. Monosomic
j. Nondisjunction

k. Nuclear sexing
l. Gene therapy

For each of these definitions, select the correct matching term from the list above.

_____ 1. A technique used to detect birth defects by sampling the fluid surrounding the fetus.

_____ 2. Abnormal separation of homologous chromosomes or sister chromatids caused by their failure to disjoin properly during cell division.

_____ 3. Genetic disease that causes mental and physical deterioration, uncontrollable muscle spasms, personality changes, and ultimately death.

_____ 4. Syndrome in which afflicted individuals have only 47 chromosomes.

_____ 5. Techniques that involve introducing normal copies of a gene into some of the cells of the body of a person afflicted with a genetic disorder.

_____ 6. Birth defect caused by an extra copy of human chromosome 21.

_____ 7. General term for the abnormal situation in which one member of a pair of chromosomes is missing.

_____ 8. Matings of close relatives.

_____ 9. Abnormality involving the presence of an extra chromosome or the absence of a chromosome.

_____10. The number and kinds of chromosomes present in the nucleus.

MAKING COMPARISONS

Fill in the blanks.

Genetic Disease	Abnormality	Method of Transmission	Clinical Description
Cystic fibrosis	Membrane proteins that transport chloride ion malfunction	Autosomal recessive trait	Characterized by abnormal secretions in respiratory and digestive systems
#1	Valine substitutes for glutamic acid in hemoglobin	#2	Abnormal RBCs block small blood vessels.
#3	Abnormal sex chromosome expression	#4	Small testes, sterile, abnormal breast development
Phenylketonuria (PKU)	#5	Autosomal recessive trait	Retardation caused by toxic phenylketones that damage the developing nervous system
#6	Lack of blood clotting Factor VIII	#7	Severe bleeding from even slight wounds
#8	Lack enzyme that breaks down a brain membrane lipid	#9	Blindness, severe retardation

Genetic Disease	Abnormality	Method of Transmission	Clinical Description
#10	#11	Autosomal dominant allele	Severe mental and physical deterioration, uncontrollable muscle spasms, personality changes, and ultimately death
#12	#13	Autosomal trisomy 21	Retardation, abnormalities

MAKING CHOICES

Place your answer(s) in the space provided. Some questions may have more than one correct answer.

_____ 1. The inactive X-chromosome seen as a region of darkly stained chromatin is called a
    a. translocated anomalie.
    b. Barr body.
    c. artifact.
    d. chromatic body.
    e. Klinefelter chromosome.

_____ 2. An individual with Turner syndrome would likely have a sex chromosome composition of
    a. XXX.
    b. XXY.
    c. XO.
    d. YO.
    e. XYY.

_____ 3. An individual with an XYY karyotype is
    a. male.
    b. .female
    c. fertile.
    d. sterile.
    e. likely to be tall.

_____ 4. IVF is a popular abbreviation for
    a. intravenous filtration.
    b. *in vitro* fertilization.
    c. a branch of stem cell research.
    d. a technique for fertilization in laboratory glassware.
    e. insufficient placental development leading to miscarriage.

_____ 5. A person with Klinefelter syndrome will
    a. have an enlarged head.
    b. be female.
    c. have 45 chromosomes.
    d. have two X chromosomes.
    e. have one Barr body per cell.

_____ 6. Amniocentesis involves
    a. examination of maternal blood.
    b. insertion of a needle into the uterus.
    c. examination of fetal cells.
    d. examination of karyotypes.
    e. appraisal of paternal abnormalities.

_____ 7. Translocations may involve
    a. loss of genes.
    b. acquisition of extra genes.
    c. duplication of genes.
    d. loss of a chromosome.
    e. acquisition of an extra chromosome.

_____ 8. A trisomic individual
    a. is missing one of a pair of chromosomes.
    b. may result from nondisjunction.
    c. has at least three X chromosomes.
    d. is a male.
    e. has an extra chromosome.

_____ 9. A person with Down syndrome will likely
a. have a father with Down syndrome.   d. be born of a mother in her teens.
b. have trisomy 21.   e. have 47 chromosomes.
c. be mentally retarded.

_____10. The most accurate statement about the frequency of abnormal genes is that they are found in
a. certain ethnic groups.   d. college students.
b. certain cultural groups.   e. everyone.
c. college professors.

_____11. An ideal organism for genetic studies would be one that
a. is polygenic.   d. produces one offspring per mating.
b. is heterozygous.   e. can be controlled.
c. has a short generation time.

_____12. The study of human genetics relies mainly on
a. pedigrees.   d. distribution of a trait in a population.
b. Punnett squares.   e. nondisjunction.
c. fetal karyotypes.

VISUAL FOUNDATIONS

Questions 1-3 show pedigrees of the type that are used to determine patterns of inheritance.
Indicate carriers by placing a dot in the appropriate squares ([•]) and circles (⊙).
Use the following list to answer questions 1-3:

a. x-linked recessive
b. x-linked dominant
c. autosomal dominant
d. autosomal recessive
e. cannot determine the pattern of inheritance

1. What is the pattern of inheritance for the following pedigree?

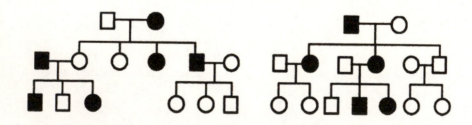

2. What is the pattern of inheritance for the following pedigree?

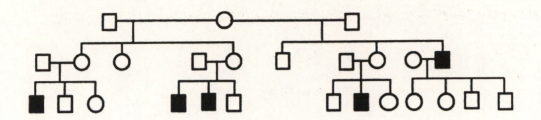

3. What is the pattern of inheritance for the following pedigree?

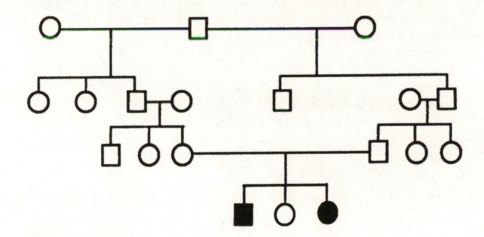

CHAPTER 17

❑

Developmental Genetics

Development encompasses all the changes that occur in the life of an individual organism. Of particular interest is the process by which cells specialize and organize into a complex organism. Groups of cells become gradually committed to specific patterns of gene activity and progressively organized into recognizable structures. Cellular specialization is not due to the loss of genes during development, but rather to differential gene activity. A variety of organisms and methodologies are used today to identify genes that control development and to determine how those genes work. Biotechnology, the subject matter of the previous two chapters, is making it possible for scientists to study differential gene expression in increasingly sophisticated ways. Many genes important in development are quite similar in a wide range of organisms, including humans.

REVIEWING CONCEPTS
Fill in the blanks.

INTRODUCTION
1. The technique to localize proteins in which a fluorescent dye is joined to an antibody that binds to a specific protein is called _____.

CELL DIFFERENTIATION AND NUCLEAR EQUIVALENCE
2. Development involves cell specialization, a developmental process involving a gradual commitment by each cell to a specific pattern of gene activity. This process is called (a)_____, and the final step in the process is referred to as (b)_____.

3. The distinctive organizational pattern, or form, that characterizes an organism is acquired through a process known as (a)_____. It proceeds through a series of steps known as (b)_____, by which groups of cells organize into identifiable structures.

4. A body of data supports the assertion that, in at least some organisms, almost all nuclei of differentiated cells are identical to each other and to the nucleus of the single cell from which they are descended, which has given rise to the concept of _____.

Most cell differences are due to differential gene expression

5. Much of the regulation that is important in development occurs at the _____ level.

A totipotent nucleus contains all the instructions for development

6. _____ cells are fully differentiated, yet, since they can still develop, they apparently still contain all of the original genetic equipment of their common single ancestral cell.

The first cloned mammal was a sheep

7. The main focus of cloning research is the production of _____ _____, in which foreign genes have been incorporated.

Stem cells divide and give rise to differentiated cells

8. Embryonic and adult stem cells are known as _____ stem cells because they can give rise to many, but not all, of the types of cells in an organism.

THE GENETIC CONTROL OF DEVELOPMENT

A variety of model organisms provide insights into basic biological processes

9. A model organism is a species chosen for biological studies because it has

_____

_____.

Many examples of genes that control development have been identified in *Drosophila*

10. The *Drosophila* life cycle includes _____, _____, _____, and _____ stages

11. The initial stages of development in *Drosophila* are under the control of _____ genes.

12. Once maternal genes have initiated pattern formation in the embryonic fruit fly, additional genes begin to act on development. Among these are the (a)_____ that begin to organize body segments and the (b)_____ that specify the developmental plan for each segment.

13. The _____ is a short DNA sequence that is consistently found in association with known developmental genes. It is useful as a probe for locating and cloning developmental mutants.

C. ELEGANS HAS A RELATIVELY RIGID DEVELOPMENTAL PATTERN

14. The fates of cells in adult roundworms are predetermined by a few _____ _____ that form in the early embryo.

15. Development in *Caenorhabditis elegans* is rigidly fixed in a predetermined pattern; that is, specific cells throughout the early embryo are already irrevocably "programmed" to develop in a particular way. This pattern of development is said to be _____.

THE MOUSE IS A MODEL FOR MAMMALIAN DEVELOPMENT

16. Early mice embryos can be fused to produce a _____, a term for offspring that exhibit a diversity of characteristics derived from two or more kinds of genetically dissimilar cells from different zygotes.

17. A self-regulating embryo exhibits a form of development referred to as _____. Such an embryo can still develop normally when it has extra cells or missing cells.

ARABIDOPSIS IS A MODEL FOR STUDYING PLANT DEVELOPMENT, INCLUDING TRANSCRIPTION FACTORS

18. In addition to the ABC genes, another class of genes designated _____ interacts with the B and C genes to specify the development of petals, stamens, and carpels.

CANCER AND CELL DEVELOPMENT

19. Tumors with cells that can metastasize are _____ tumors.

20. Cancer-causing genes are known as _____.

21. A gene that interacts with growth-inhibiting factors to block cell division is a _____ gene.

BUILDING WORDS

Use combinations of prefixes and suffixes to build words for the definitions that follow.

Prefixes	The Meaning	Suffixes	The Meaning
morpho-	form	-gen(esis)	production of
onco-	mass, tumor		
toti-	all		

Prefix	Suffix	Definition
_____	_____	1. The development and differentiation (production) of the form and structures of the body.
_____	-gene	2. A cancer-causing gene.
_____	-potent	3. The ability of some nuclei from differentiated plant and animal cells to provide information for the development of an entire organism; such cells contain all of the genetic material that would be present in the nucleus of a zygote.

MATCHING

Terms:

a. Determination
b. Differentiation
c. Homeobox
d. Homeotic gene
e. Induction

f. Oncogene
g. Somatic cell
h. Stem cell
i. Totipotent
j. Transgenic

k. Tumor
l. Zygotic gene

For each of these definitions, select the correct matching term from the list above.

_____ 1. An organism that has incorporated foreign DNA into its genome.

_____ 2. Mass of tissue that grows in an uncontrolled manner.

_____ 3. Development toward a more mature state; a process changing a young, relatively unspecialized cell to a more specialized cell.

_____ 4. The ability of a cell (or nucleus) to provide the information necessary for the development of an entire organism.

_____ 5. A phenomenon in which differentiation of a cell is influenced by interactions with particular neighboring cells.

_____ 6. A gene that controls the formation of specific structures during development.

_____ 7. Specific DNA sequence found in many genes that are involved in controlling the development of the body plan.

_____ 8. The progressive limitation of a cell line's potential fate during development.

_____ 9. Any of a number of genes that usually play an essential role in cell growth or division, and that cause the formation of a cancer cell when mutated.

_____10. A body cell not involved in reproduction.

MAKING COMPARISONS

Fill in the blanks.

Germ Layer	Organ, Organ System or Tissue	Specialized Cell
Mesoderm	Skeletal muscles	Striated muscle cell
#1	#2	Skin cell
#3	#4	Neuron
#5	Germ line cells	Egg and sperm (gametes)
Endoderm	#6	Tracheal cell
#7	Urinary bladder	Epithelial cell
#8	Blood	Red blood cell

MAKING CHOICES

Place your answer(s) in the space provided. Some questions may have more than one correct answer.

_____ 1. Gap genes in _Drosophila_

    a. determine segment polarity.   d. are complementary to maternal genes.

    b. act on all embryonic segments.   e. are a type of segmentation gene.

    c. are the first segmentation genes to act.

_____ 2. The concept of nuclear equivalence holds that

    a. adult cells are identical to one another.   d. adult cells are identical to egg cells.

    b. embryonic cells are identical to one another.   e. all adult somatic nuclei are identical.

    c. adult somatic nuclei are identical to the nucleus of the fertilized egg.

_____ 3. If destruction of a single cell early in development results in the absence of a structure in the adult, the embryo of that organism

    a. is mosaic.   d. contained founder cells.

    b. is a fruit fly.   e. is controlled to some extent by homeotic genes.

    c. contains predetermined cells.

_____ 4. Embryos that act as a self-regulating whole and can accommodate missing parts

    a. are transgenic.

    b. are mosaics.

    c. have regulative development patterns.

    d. have founder cells that give rise to adult parts.

    e. contain many genomic rearrangements.

_____ 5. Morphogens are

    a. chemical agents.

    b. derived from a homeobox.

    c. determined by embryonic segmentation genes.

    d. thought to exist in *Drosophila* eggs.

    e. unraveled portions of active DNA.

_____ 6. A group of cells grown in liquid medium from a single root cell is

    a. formed by a predetermined pattern.

    b. a mosaic.

    c. totipotent.

    d. predetermined.

    e. an embryoid body.

_____ 7. An embryo in which the fate of cells is determined early in development is said to be

    a. induced.

    b. mosaic.

    c. homeotic.

    d. transgenic.

    e. differentiated.

_____ 8. If differentiated cells can be induced to act like embryonic cells, the cells are

    a. formed by a predetermined pattern.

    b. mosaics.

    c. totipotent.

    d. predetermined.

    e. embryoids.

_____ 9. Pattern formation is a series of steps leading to

    a. determination.

    b. differentiation.

    c. predetermination.

    d. morphogenesis.

    e. the final step in cell specialization.

_____10. Differences in the molecular composition of different cells in a multicellular organism are due to

    a. loss of genes during development.

    b. transformation.

    c. regulation of the activities of different genes.

    d. nuclear equivalency.

    e. differential gene activity.

_____11. The process by which developing cells become committed to a specialization is

    a. determination.

    b. differentiation.

    c. transgenesis.

    d. morphogenesis.

    e. the first step in cell specialization.

_____12. The process by which the differentiation of an embryonic cell is influenced by neighboring embryonic cells is

    a. transformation.

    b. induction.

    c. mosaic development.

    d. an example of programmed cell death.

    e. determinate morphogenesis.

The Continuity of Life: Evolution

Introduction to Darwinian Evolution

The concept of evolution that all organisms that exist today evolved from earlier organisms links all fields of the life sciences into a unified body of knowledge. Evolution is firmly based on the genetic principles you learned in the previous section of the book. Charles Darwin discovered natural selection, the actual mechanism of evolution. According to this mechanism, each species produces more offspring than will survive to maturity, genetic variation exists among the offspring, organisms compete for resources, and individuals with the most favorable characteristics are most likely to survive and reproduce. Favorable characteristics are thus passed to succeeding generations. Over time, changes occur in the gene pools of geographically separated populations that may lead to the evolution of new species. A subsequent modification of Darwin's theory, the synthetic theory of evolution, explains Darwin's observation of variation among offspring in terms of mutation and recombination. An enormous body of scientific observations and experiments supports the concept of evolution. The testing of evolutionary hypotheses has demonstrated that not only is evolution a reality, but it occurs daily and hourly all around us.

REVIEWING CONCEPTS

Fill in the blanks.

INTRODUCTION

1. The genetic changes that bring about evolution do not occur in individual organisms, but rather in _____.

WHAT IS EVOLUTION?

2. The concept of evolution is the cornerstone of biology, because it _____ _____.

PRE-DARWINIAN IDEAS ABOUT EVOLUTION

3. _____ was probably the first to recognize that fossils are the remains of extinct organisms.

4. _____ proposed a theory of evolution in 1809, coincidentally in the year of Darwin's birth. His theory held that traits *acquired* during an organism's lifetime could be passed along to their offspring.

DARWIN AND EVOLUTION

5. Charles Darwin served as the naturalist on the ship (a)_____ during its five year cruise around the world. Central to the theory Darwin would propose were data about the similarities and differences among organisms in the (b)_____, a chain of islands west of mainland Ecuador.

6. In the mid-1800s, most believed that organisms did not change, but some understood the significance of variations among domestic animals that were intentionally induced by breeders, a process called _____.

7. _____ made the important mathematical observation that populations increase in size geometrically until checked by factors in the environment.

Darwin proposed that evolution occurs by natural selection

8. In 1858, both Darwin and _____ arrived at the conclusion that evolution occurred by natural selection.

9. Darwin published his monumental book, _____ _____ in 1859.

10. Darwin's mechanism of natural selection consists of four premises based on observations about the natural world. In summary, these are:

_____

_____.

The modern synthesis combines Darwin's theory with genetics

11. The "synthetic theory of evolution" is sometimes known as _____.

12. The synthetic theory of evolution explains variation in terms of _____ _____, which are inheritable and therefore potential explanations for *how* traits are passed from one generation to another.

Biologists study the effect of chance on evolution

13. Although it cannot be proven, it appears that _____ is a more important agent of evolutionary change than chance.

EVIDENCE FOR EVOLUTION

The fossil record provides strong evidence for evolution

14. The earliest fossils of *Homo sapiens* with anatomically modern features appear in the fossil record about _____ years ago.

15. _____ are the remains or traces of organisms that were preserved in large numbers over a relatively short geological time. They are used to identify specific sedimentary layers and to arrange strata in chronological order.

Comparative anatomy of related species demonstrates similarities in their structures

16. _____ features are those derived from the same structure in a common ancestor.

17. _____ have similar functions, whether or not they are structurally similar. They do not indicate close evolutionary ties.

18. The occasional presence of "remnant organs" in organisms is to be expected as ancestral species evolve and adapt to different modes of life. Such organs or parts of organs, called _____, are degenerate and have no apparent function.

The distribution of plants and animals supports evolution

19. The study of the past and present geographic distribution of plants and animals is called _____.

20. In 1915 Alfred Wegener proposed that a single land mass called Pangaea had broken apart in a process known as _____.

Developmental biology helps unravel evolutionary patterns

21. The evolution of new features depends on modification in _____ that already exist.

Molecular comparisons among organisms provide evidence for evolution

22. Codons code for a particular _____ in a polypeptide chain.

23. Determining the order of nucleotide bases in DNA is known as _____ _____.

Evolutionary hypotheses are tested experimentally

BUILDING WORDS

Use combinations of prefixes and suffixes to build words for the definitions that follow.

Prefixes	The Meaning
bio-	life
homo-	same

Prefix	Suffix	Definition
_____	-logous	1. Pertains to having a similarity of form due to having the same evolutionary origin.
_____	-geography	2. The study of the geographical distribution of living things.
_____	-plastic	3. Pertains to features with the same or similar functions in distantly related organisms.

MATCHING

Terms:

a. Adaptation
b. homoplastic
c. Artificial selection
d. Convergent evolution
e. Evolution
f. Fossil
g. Gene pool
h. Half-life
i. Index fossil
j. Synthetic theory of evolution
k. Range
l. Vestigial

For each of these definitions, select the correct matching term from the list above.

_____ 1. The modifications that improve the ability of an organism to adjust to its environment.

_____ 2. The independent evolution of structural or functional similarity in two or more organisms of widely-different, unrelated ancestry.

_____ 3. Parts or traces of an ancient organism preserved in rock.

_____ 4. The period of time required for one-half of the atoms of a radioisotope to change into a different atom.

_____ 5. A unified explanation of evolution combining Darwin's theory and Mendelian genetics.

_____ 6. Rudimentary; an evolutionary remnant of a formerly functional structure.

_____ 7. A fossil that is restricted to a narrow unit of time and is found in the same sedimentary layers in different geographical areas.

_____ 8. The genetic change in a population of organisms that occurs over time.

_____ 9. Similar in function or appearance, but not in origin or development.

_____10. The selection of traits that are desirable in plants or animals and breeding only those individuals that possess the desired traits.

MAKING COMPARISONS

Fill in the blanks.

Evidence for Evolution	How the Evidence Supports Evolution
Fossils	Direct evidence in the form of remains of ancient organisms
#1	Indicate evolutionary ties between organisms that have basic structural similarities, even though the structures may be used in different ways
#2	Indicate that organisms with different ancestries can adapt in similar ways to similar environmental demands
Vestigial organs	#3
Biogeography of plants and animals	#4
All organisms use a genetic code that is virtually identical	#5

MAKING CHOICES

Place your answer(s) in the space provided. Some questions may have more than one correct answer.

_____ 1. Which of the following statements is/are consistent with Darwin's theory of evolution?
   a. Organisms evolve by a process of natural selection.   d. Better adapted individuals are more likely to survive.
   b. Evolution progresses toward a more perfect state.   e. New species result from a need to survive.
   c. Accumulation of chance adaptations in a population can lead to new species.

_____ 2. The fossil record
   a. supports contemporary theories of evolution.   d. is consistent with Darwin's theories.
   b. is largely fabricated.   e. supports Lamarck's theory of evolution.
   c. supports evolution of plants and animals, but does not support evolution of humans.

_____ 3. The earliest fossils of _Homo sapiens_ with modern anatomical features appear no earlier than
   a. 1,000 years ago.   d. 5,000 BC.
   b. 10,000 years ago.   e. recorded human history.
   c. 100,000 years ago.

_____ 4. The first comprehensive theory of evolution, which proposed that the mechanism was an inner drive for improvement, was proposed by
   a. Aristotle.   d. Wallace.
   b. Lamarck.   e. da Vinci.
   c. Darwin.

_____5. The wings of a butterfly and a bat are
  a. homoplastic but not homologous.   d. both homoplastic and homologous.
  b. homologous but not homoplastic.   e. probably derived from a common ancestor.
  c. neither homoplastic nor homologous.

_____ 6. The humerus in a bird and human are
  a. homoplastic but not homologous.   d. both homoplastic and homologous.
  b. homologous but not homoplastic.   e. probably derived from a common ancestor.
  c. neither homoplastic nor homologous.

_____ 7. The person(s) who suggested that populations increase geometrically and that their numbers must eventually be checked was
  a. Malthus.   d. Wallace.
  b. Lamarck.   e. Lyell.
  c. Darwin.

_____ 8. That Earth's physical features were formed over long periods of time by a series of gradual changes was proposed by
  a. Aristotle.   d. Wallace.
  b. Lamarck.   e. Lyell.
  c. Darwin.

_____9. The first to propose that the mechanism for the evolution of organisms is by natural selection was
  a. Malthus.   d. Wallace.
  b. Lamarck.   e. Lyell.
  c. Darwin.

_____10. Structures of different organisms that have a similar form due to a common origin are
  a. homoplastic or analogous.   d. usually vestigial.
  b. artifacts.   e. found in fossils but never in contemporary organisms.
  c. homologous.

_____11. If Darwin had known and employed current biochemical techniques, it most likely would have resulted in
  a. a different theory of evolution.   d. confirmation of his theories.
  b. no theory of evolution.   e. an argument with Wallace.
  c. his conviction of immutable species.

_____ 12. The individuals who simultaneously concluded that organisms evolve as nature selects the fittest were
  a. Aristotle.   d. Wallace.
  b. Lamarck.   e. Lyell.
  c. Darwin.

_____13. Using microorganisms to clean up waste sites is known as
  a. biological control.   d. hazardous waste removal.
  b. reclamation.   e. bioincineration.
  c. bioremediation.

VISUAL FOUNDATIONS

Color the parts of the illustration below as indicated.

RED ☐ metacarpels

GREEN ☐ ulna

YELLOW ☐ carpels

BLUE ☐ radius

BROWN ☐ phalanges

VIOLET ☐ humerus

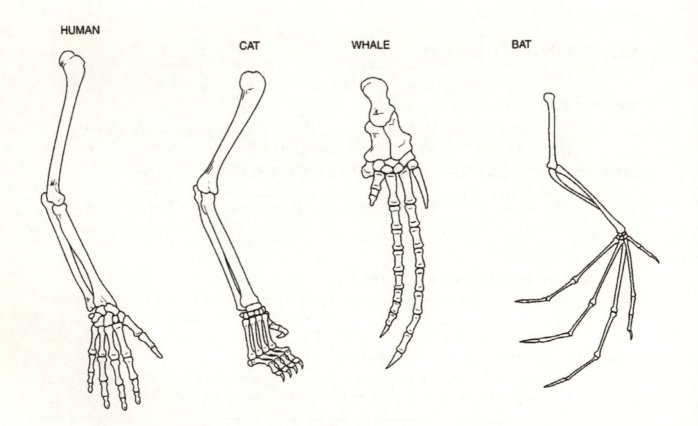

HUMAN CAT WHALE BAT

CHAPTER 19

Evolutionary Change in Populations

You learned in the previous chapter that evolution occurs in populations, not individuals, that natural selection results from differential survival and reproduction of individuals, that evolutionary change is inherited from one generation to the next, and that the principles of inheritance you learned in Chapter 11 underlie Darwinian evolution. This chapter examines the genetic variability within a population and of the forces that act on it. The factors responsible for evolutionary change are nonrandom mating, mutation, genetic drift, gene flow, and natural selection. Genetic variation in a gene pool is the raw material for evolutionary change. The gene pools of most populations contain a large reservoir of variability.

REVIEWING CONCEPTS
Fill in the blanks.

INTRODUCTION
1. Population genetics is the study of _____ within the population and the evolutionary forces that act on it.

GENOTYPE, PHENOTYPE, AND ALLELE FREQUENCIES
2. All of the alleles for all of the genes in a population are referred to as the _____ _____ for that population.

3. _____ is the proportion of a particular genotype in a population.

THE HARDY-WEINBERG PRINCIPLE
4. When the distribution of genotypes in a population conforms to the binomial equation (a)_____, the population is in a genetic equilibrium and is not evolving. This equilibrium is called the (b)_____.

Genetic equilibrium occurs if certain conditions are met
5. The proportion of alleles in successive generations does not change in a population when certain conditions are met. Among them, briefly stated, are the following five:
   _____
   _____.

Human MN blood groups are a valuable illustration of the Hardy-Weinberg principle
6. The MN blood group is of interest to population geneticists because the alleles for the MN blood group are _____.

MICROEVOLUTION
Nonrandom mating changes genotype frequencies
7. Mating among genetically similar individuals within a population is called _____.
8. The average number of survivors among offspring from a given genetic type compared to

other genetic types is a measure of _____.

9. Selection of mates based on a phenotype is known as _____.

MUTATION INCREASES VARIATION WITHIN A POPULATION

10. A _____ is an unpredictable change in DNA.

IN GENETIC DRIFT, RANDOM EVENTS CHANGE ALLELE FREQUENCIES

11. Production of random evolutionary changes in small breeding populations is known as _____.

12. Genetic drift may occur when the size of a population is suddenly reduced as a function of such temporary causes as a depleted food supply or disease. Such an event is called a _____.

GENE FLOW GENERALLY INCREASES VARIATION WITHIN A POPULATION

13. The migration of breeding individuals between populations causes a corresponding movement of alleles, or _____, that has significant evolutionary consequences.

NATURAL SELECTION CHANGES ALLELE FREQUENCIES IN A WAY THAT INCREASES ADAPTATION

14. Selection against phenotypic extremes, thereby favoring intermediate phenotypes, is called _____.

15. Selection that favors one particular phenotype over another is known as _____.

16. Selection that favors phenotypic extremes is called _____.

GENETIC VARIATION IN POPULATIONS

Genetic polymorphism exists among alleles and the proteins for which they code

17. Much of genetic polymorphism is not evident, because it doesn't produce distinct _____.

Balanced polymorphism exists for long periods

18. _____ occurs when a genotype such as Aa has a higher degree of fitness than either AA or aa.

19. Selection that acts to decrease the frequency of the more common phenotypes and increase the frequency of the less common types is called _____.

Neutral variation may give no selective advantage or disadvantage

20. Variation that does not alter the ability of an individual to survive and reproduce, and is therefore not adaptive, is called _____.

Populations in different geographical areas often exhibit genetic adaptations to local environments

21. Along with population variation, genetic differences often exist among different populations within the same species, a phenomenon known as _____.

BUILDING WORDS

Use combinations of prefixes and suffixes to build words for the definitions that follow.

Prefixes	The Meaning		Suffixes	The Meaning
micro-	small		-typ(e)	form
pheno-	visible			

Prefix	Suffix	Definition
_____	-evolution	1. Changes in allele frequencies over successive generations; involves small changes within a population.
_____ _____		2.. An organism's visible form or characteristics.

MATCHING

Terms:

a. Balanced polymorphism
b. Directional selection
c. Disruptive selection
d. Founder effect
e. Gene flow

f. Gene pool
g. Genetic drift
h. Hardy-Weinberg principle
i. Heterozygote advantage
j. Natural selection

k. Population
l. Stabilizing selection

For each of these definitions, select the correct matching term from the list above.

_____ 1. Natural selection that acts against extreme phenotypes and favors intermediate variants.

_____ 2. A phenomenon in which the heterozygous condition confers some special advantage on an individual that either homozygous condition does not.

_____ 3. A random change in gene frequency in a small, isolated population.

_____ 4. The movement of alleles between local populations, or demes, due to migration and subsequent interbreeding.

_____ 5. The gradual replacement of one phenotype with another due to environmental change.

_____ 6. The presence in a population of two or more genetic variants that are maintained in a stable frequency over several generations.

_____ 7. Genetic drift that results from a small number of individuals colonizing a new area.

_____ 8. All the alleles for all genes present in a population.

_____ 9. The principle that in a randomly-mating large population, regardless of dominance or recessiveness, the relative frequencies of allelic genes do not change from generation to generation.

_____10. A group of organisms of the same species that live in the same geographical area at the same time.

MAKING COMPARISONS
Fill in the blanks.

Category	Contributes to Evolutionary Change (yes or no)
Natural selection	#1
Gene flow	#2
Random mating	#3
Genetic drift	#4
Mutation	#5
Small population size	#6
Geographic stability	#7
Genetic polymorphism	#8
Neutral variation	#9

MAKING CHOICES
Place your answer(s) in the space provided. Some questions may have more than one correct answer.

_____ 1. The proportion of alleles in a population does not change if there is/are
a. only 10% of the alleles mutating in each generation.
b. random mating.
c. a large number of individuals in the population.
d. mating with individuals in a similar population.
e. natural selection of advantageous alleles.

_____ 2. When gradual environmental changes favor phenotypes at the extreme of the normal distribution curve, the result in time is likely to be
a. stabilizing selection.
b. no selection.
c. a shift in allele frequencies.
d. disruptive selection.
e. directional selection.

_____ 3. Changes in allele frequencies within a population are referred to as
a. evolution.
b. macroevolution.
c. microevolution.
d. p^2 shifts.
e. q^2 shifts.

_____ 4. A relatively quick, extreme environmental change that favors several phenotypes at the expense of the mean phenotype will likely result in
a. stabilizing selection.
b. no selection.
c. a shift in allele frequencies.
d. disruptive selection.
e. directional selection.

_____ 5. Random evolutionary changes in a small breeding population resulting from random changes in gene frequencies are referred to as
a. gene flow.
b. natural selection.
c. mutations.
d. genetic drift.
e. the heterozygote advantage.

_____ 6. The ultimate source of all new alleles is
     a. gene flow.   d. genetic drift.
     b. natural selection.   e. the heterozygote advantage.
     c. mutations.

_____ 7. If undisturbed by other forces, random sexual reproduction among members of a large population in nature leads to
     a. new species.   d. new traits.
     b. changes in gene frequencies.   e. a change in the frequency of 2pq, but not p^2 or q^2.
     c. generations of unchanged allele frequencies.

Use the following information, the Hardy-Weinberg equation, and the list below to answer questions 8-13. The phenotype coded for by the genotype "tt" is found among 400 individuals in a population of 10,000 randomly mating individuals.

a. 0.01	f. 0.16	k. 0.80	p. 4.00	u. 20.0
b. 0.02	g. 0.20	l. 1.00	q. 6.40	v. 32.0
c. 0.04	h. 0.32	m. 1.60	r. 9.60	w. 40.0
d. 0.08	i. 0.40	n. 2.00	s. 10.0	x. 64.0
e. 0.10	j. 0.64	o. 3.20	t. 16.0	y. 96.0

_____ 8. Frequency of homozygously recessive individuals.

_____ 9. Frequency of homozygously dominant individuals.

_____10. Frequency of q.

_____11. Percent of individuals containing one or more of the dominant alleles.

_____12. Frequency of p.

_____13. Frequency of heterozygous individuals.

VISUAL FOUNDATIONS

Color the parts of the illustration below as indicated. Also label directional selection, disruptive selection, and stabilizing selection.

RED   ☐   axis indicating number of individuals

GREEN   ☐   axis indicating phenotype variation

YELLOW   ☐   normal distribution

BLUE   ☐   less favored phenotypes

ORANGE   ☐   favored phenotypes

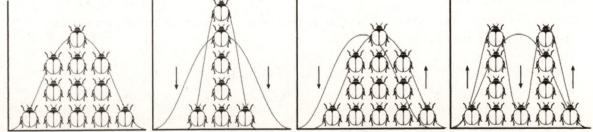

Speciation and Macroevolution

In the previous chapter, you learned how populations evolve. This chapter examines the evolution of species and macroevolution – the dramatic evolutionary changes that occur over long time spans. A species is generally viewed as a group of organisms with a common gene pool that is reproductively isolated from other organisms. Most species have two or more mechanisms that serve to preserve the integrity of the gene pool. Some mechanisms prevent fertilization from occurring in the first place. Others ensure reproductive failure should fertilization occur. Speciation, or the development of a new species, most commonly occurs when one population becomes geographically separated from the rest of the species and subsequently evolves. A new species can also evolve, however, within the same geographical region as its parent species. There are two theories about the pace of evolution. One says that there is a slow, steady change in species over time, and the other says there are long periods of little evolutionary change followed by short periods of rapid speciation. Macroevolution includes the origin of unusual features (e.g., wings with feathers), the evolution of many species from a single species, and the extinction of species.

REVIEWING CONCEPTS

Fill in the blanks.

INTRODUCTION

1. Although more than 99% of all species that ever existed are extinct, there are still an estimated _____(#?) living species today.

WHAT IS A SPECIES?

2. The 18th century biologist, _____, is generally considered the founder of modern taxonomy. His system for separating plants into different species based on their morphological characteristics formed the basis for modern systematics.

3. Although still very important in describing species, morphological features alone are not enough to delineate species. Biologists now define a species as a group of
   (a)_____ isolated organisms with a common
   (b)_____.

REPRODUCTIVE ISOLATION

Prezygotic barriers interfere with fertilization

4. Prezygotic isolating mechanisms prevent _____.

5. Prezygotic isolating mechanisms include (a)_____,
   which occurs because two groups reproduce at different times;
   (b)_____, involving different, incompatible
   courtship patterns; (c)_____, is due to
   anatomical differences that thwart successful matings; and

(d)_____, in which chemical differences between gametes prevents interspecific fertilization.

Postzygotic barriers prevent gene flow when fertilization occurs

6. Embryos are aborted due to _____.

7. _____ occurs when the gametes produced by interspecific hybrids are abnormal and nonfunctional.

Biologists are discovering the genetic basis of isolating mechanisms

8. _____ is a sperm protein in abalone that attaches to a lysine receptor protein located on the egg envelope, producing a hole in the egg envelope that permits the sperm to penetrate the egg.

SPECIATION

9. _____ is the evolution of a new species.

Long physical isolation and different selective pressures result in allopatric speciation

10. Allopatric speciation occurs when one population becomes _____ _____ from the rest of the species and subsequently evolves.

Two populations diverge in the same physical location by sympatric speciation

11. Sympatric speciation usually occurs *within* a geographical region as a result of _____.

12. (a)_____ is the possession of more than two sets of chromosomes, a common phenomenon in plants. When it results from the pairing of chromosomes from different species, it is known as (b)_____.

Reproductive isolation breaks down in hybrid zones

13. The _____ is an area of overlap in which populations, subspecies, or species come into contact and can interbreed.

THE RATE OF EVOLUTIONARY CHANGE

14. According to proponents of the theory of _____, evolution proceeds in spurts; that is, short periods of active evolution are followed by long periods of inactivity or stasis.

15. According to proponents of the theory of _____, populations slowly and steadily diverge from one another by the accumulation of adaptive characteristics within a population.

MACROEVOLUTION

Evolutionary novelties originate through modifications of pre-existing structures

16. Evolutionary "novelties" originate from mutations that alter developmental pathways. For example, (a)_____ occurs when developing body parts grow at different rates, and (b)_____ results from differences in the *timing* of development.

Adaptive radiation is the diversification of an ancestral species into many species

17. (a)_____ are new ecological roles made possible by an adaptive advancement. When an organism with a newly acquired evolutionary advancement assumes a new ecological role made possible by its advancement(s), its diversification is known as (b)_____.

Extinction is an important aspect of evolution

18. Extinction that is continuous, ongoing, and relatively low frequency is called _____.

19. _____ extinction is a relatively rapid and widespread loss of numerous species.

Is microevolution related to speciation and macroevolution?

20. Biologists hypothesize that (a)_____ processes explain (b)_____ patterns.

BUILDING WORDS

Use combinations of prefixes and suffixes to build words for the definitions that follow.

Prefixes	The Meaning
allo-	other
macro-	large, long, great, excessive
poly-	much, many

Prefix	Suffix	Definition
_____	-ploidy	1. The presence of multiples of complete chromosome sets.
_____	-patric	2. Originating in or occupying different (other) geographical areas.
_____	-polyploidy	3. Polyploidy following the joining of chromosomes from two different species (half of the chromosomes come from one parent and half from the other parent).
_____	-evolution	4. Large-scale evolutionary change.
_____	-metric	5. Varied rates of growth for different parts of the body during development (some parts grow at rates different from other parts).

MATCHING

Terms:

a. Adaptive radiation
b. Behavioral isolation
c. Extinction
d. Gametic isolation
e. Gradualism

f. Hybrid breakdown
g. Hybrid inviability
h. Hybrid sterility
i. Hybridization
j. Mechanical isolation

k. Punctuated equilibrium
l. Reproductive isolation
m. Speciation
n. Sympatric speciation

For each of these definitions, select the correct matching term from the list above.

_____ 1. A postzygotic isolating mechanism in which the embryonic development of an interspecific hybrid is aborted.

_____ 2. The concept that evolution proceeds with periods of inactivity followed by very active phases, so that major adaptations or clusters of adaptations appear suddenly in the fossil record.

_____ 3. The evolution of a new species within the same geographical region as the parent species.

_____ 4. The evolution of several to many related species from one or a few ancestral species in a relatively short period of time.

_____ 5. A prezygotic isolating mechanism in which sexual reproduction between two individuals cannot occur because of chemical differences between the gametes of the two species.

_____ 6. A model of evolution in which the evolutionary change of a species is due to a slow, steady transformation over time.

_____ 7. Sexual reproduction between individuals from closely related species.

_____ 8. Evolution of a new species.

_____ 9. An isolating mechanism in which gamete exchange between two groups is prevented because each group possesses its own characteristic courtship behavior.

_____10. The end of a lineage, occurring when the last individual of a species dies.

MAKING COMPARISONS

Fill in the blanks.

Reproductive Isolating Mechanism	How It Works
Behavioral isolation	Similar species have distinctive courtship behaviors
#1	Prevents the offspring of hybrids that are able to reproduce successfully from reproducing past one or a few generations
Gametic isolation	#2
#3	Interspecific hybrid survives to adulthood but is unable to reproduce successfully
Mechanical isolation	#4
#5	Interspecific hybrid dies at early stage of embryonic development
#6	Similar species reproduce at different times

MAKING CHOICES

Place your answer(s) in the space provided. Some questions may have more than one correct answer.

_____ 1. When large-scale phenotypic changes in populations justify placing them in taxonomic groups at the species level or higher it is known as

    a. macroevolution.
    b. paedomorphosis.
    c. polymorphic speciation.
    d. macromorphism.
    e. adaptive radiation.

_____ 2. The process by which an ancestral species evolves into many new species is known as

    a. macroevolution.
    b. paedomorphosis.
    c. polymorphic speciation.
    d. macromorphism.
    e. adaptive radiation.

_____ 3. A species typically has
   a. a common gene pool.
   b. an isolated gene pool.
   c. the capacity to reproduce with other species in the laboratory.
   d. members with different morphological characteristics.
   e. become reproductively isolated.

_____ 4. The fact that dogs come in many sizes, shapes, and colors supports the contention that
   a. dogs have one gene pool.
   b. dogs can produce hybrids.
   c. physical appearance alone is not enough to define a species.
   d. reproduction within a species produces sterile offspring.
   e. all dogs belong to one species.

_____ 5. A population that evolves within the same geographical region as its parent species is an example of
   a. cladogenesis.
   b. character displacement.
   c. anagenesis.
   d. sympatric speciation.
   e. allopatric speciation.

_____ 6. If two closely related species produce a fertile interspecific hybrid, the hybrid is the result of
   a. anagenesis.
   b. phyletic evolution.
   c. cladogenesis.
   d. diversifying evolution.
   e. allopolyploidy.

_____ 7. If two distinct populations of organisms occasionally interbreed in the wild, they are considered to be
   a. separate species.
   b. one species.
   c. reproductively isolated.
   d. one gene pool.
   e. evolving.

_____ 8. If flower-loving female botany majors were required to smell a flower before male botany majors would mate with them, and if male zoology majors would only mate with females that did not sniff flowers, zoology majors and botany majors would likely become
   a. separate species.
   b. one species.
   c. reproductively isolated.
   d. one gene pool.
   e. behaviorally isolated.

_____ 9. Evolution of organisms occurs by means of
   a. changes in individual organisms.
   b. uniformitarianism.
   c. changes in gene frequencies in the gene pool.
   d. extinctions.
   e. changes in populations.

_____10. A population that evolves as the result of its geographic separation from the rest of the species is an example of
   a. cladogenesis.
   b. character displacement.
   c. anagenesis.
   d. sympatric speciation.
   e. allopatric speciation.

_____11. If college-educated persons mated only at night and noncollege-educated persons mated only at the lunch hour, and no amount of laboratory experimentation could change these habits, these two populations would be considered
   a. separate species.
   b. one species.
   c. reproductively isolated.
   d. members of one gene pool.
   e. temporally isolated.

_____12. Hybrid breakdown affects the
   a. P_1 generation.
   b. F_1 generation.
   c. F_2 generation.
   d. F_3 generation.
   e. course of evolution.

_____13. The death of interspecific embryos during development is

a. called hybrid sterility.

b. one type of postzygotic barrier.

c. a form of anagenesis.

d. called hybrid inviability.

e. allopatric dissociation.

VISUAL FOUNDATIONS

Color the parts of the illustration below as indicated. Also label gradualism and punctuated equilibrium.

RED	☐	original species
GREEN	☐	last species to evolve
YELLOW	☐	time axis
BLUE	☐	first species to evolve
ORANGE	☐	structural change axis

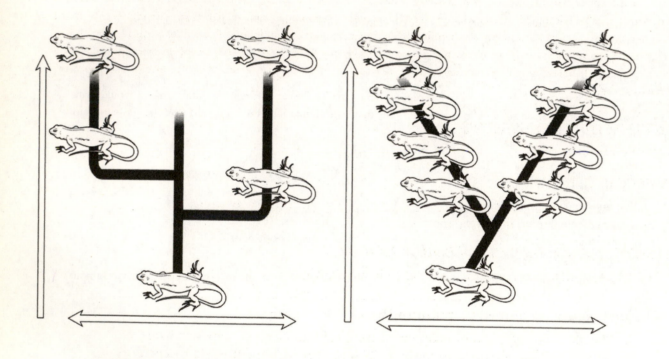

CHAPTER 21

The Origin and Evolutionary History of Life

The previous three chapters dealt with the evolution of organisms, but what about the evolution of life in the first place? This chapter addresses how life began and traces its long evolutionary history. Life on earth developed from nonliving matter. Exactly how this process occurred is not certain. Current models suggest that small organic molecules formed, accumulated, and became organized into complicated assemblages. Cells then evolved from these macromolecular assemblages. This process was enabled by the absence of oxygen and the presence of energy, the chemical building blocks of organic molecules, and sufficient time for the molecules to accumulate and react. The first cells to evolve were anaerobic and prokaryotic. They are believed to have obtained their energy from organic compounds in the environment. Later, cells evolved that could obtain energy from sunlight. Photosynthesis produced enough oxygen to change the atmosphere significantly and thereby alter the evolution of early life. Organisms evolved that had the ability to use oxygen in cell respiration. Eukaryotes are believed to have evolved from prokaryotes. Mitochondria, chloroplasts, and certain other organelles may have originated from symbiotic relationships between two prokaryotic organisms.

REVIEWING CONCEPTS

Fill in the blanks.

INTRODUCTION

1. _____ is the process by which life developed from nonliving matter

CHEMICAL EVOLUTION ON EARLY EARTH

2. The four basic requirements for the chemical evolution of life are: _____
_____.

Organic molecules formed on primitive Earth

3. The idea that organic molecules could form from nonliving components on primitive Earth originated in the early 20th century with the Russian biochemist (a)_____ and the Scottish biologist (b)_____.

4. In the 1950s, (a)_____ constructed conditions in the laboratory that were thought to prevail on primitive Earth. They started with an atmosphere of water (H_2O) and the three gases (b)_____
_____. These and other experiments have produced a variety of organic molecules.

THE FIRST CELLS

5. _____ are assemblages of organic polymers that form spontaneously, are organized, and to some extent resemble living cells.

Molecular reproduction was a crucial step in the origin of cells

6. (a)_____ may have been the first polynucleotide to carry "hereditary" information. Interestingly, some forms of this molecule, called (b)_____, can catalyze the formation of more forms.

Biological evolution began with the first cells

7. (a)_____, or ancient remains of microscopic life, suggest that cells may have been thriving as long as (b)_____(#?) billion years ago.

The first cells were probably heterotrophic

8. The first cells were (aerobic or anaerobic?) (a)_____ and (prokaryotic or eukaryotic?) (b)_____.

9. Organisms like ourselves that require preformed organic molecules (food), because they cannot synthesize them *de novo*, are called (a)_____. On the other hand, (b)_____ possess the chemical apparatus required to synthesize food.

Aerobes appeared after oxygen increased in the atmosphere

10. The evolution of photosynthesis ultimately changed early life and afforded vastly new opportunities for diversity because it generated _____.

Eukaryotic cells descended from prokaryotic cells

11. The _____ theory proposes that mitochondria, chloroplasts, and possibly centrioles and flagella evolved from prokaryotes engulfed by other cells.

THE HISTORY OF LIFE

Rocks from the Ediacaran period contain fossils of cells and simple animals

12. The oldest known fossils of multicellular animals are called _____.

A diversity of organisms evolved during the Paleozoic era

13. The Paleozoic era began about _____ years ago with a profound burst of evolution.

14. Later, in the (a)_____ period, the first vertebrate, the jawless, bony-armored fish called (b)_____ appeared.

Dinosaurs and other reptiles dominated the Mesozoic era

15. The Mesozoic era began about _____ years ago.

16. The Mesozoic era consisted of the three periods: (a)_____ _____. Mammals appeared in the (b)_____ period and birds appeared in the (c)_____ period.

The Cenozoic era is the Age of Mammals

17. The Cenozoic era can be divided into two periods: The (a)_____ _____ containing three epochs and the (b)_____ _____ containing four epochs.

BUILDING WORDS

Use combinations of prefixes and suffixes to build words for the definitions that follow.

Prefixes	**The Meaning**		**Suffixes**	**The Meaning**
aero-	air		-be (bios)	life
an-	without, not, lacking		-bio(nt)	life
auto-	self, same		-troph	nutrition, growth, "eat"
Pre-	before, prior to			
hetero-	different, other			
proto-	first, earliest form of			

Prefix	**Suffix**	**Definition**
_____	_____	1. An organism that produces its own food.
_____	_____	2. Spontaneous assemblages of organic polymers that may have been involved in the evolution of the earliest forms of life.
_____	_____	3. An organism that is dependent upon other organisms for food, energy, and oxygen.
_____	-cambrian	4. The geologic time period prior to the Cambrian.
_____	_____	5. An organism that requires air or free oxygen to live.
_____	-aerobe	6. An organism that does not require air or free oxygen to live.

MATCHING

Terms:

a. Cenozoic era
b. Cyanobacteria
c. Endosymbiont
d. Epoch
e. Era
f. Mesozoic era
g. Microfossil
h. Microsphere
i. Paleozoic era
j. Period
k. Ribozyme
l. Stromatolite

For each of these definitions, select the correct matching term from the list above.

____ 1. A geological time period that is a subdivision of an era, and is divided into epochs.

____ 2. Term applied to divisions of geological time that are divided into periods.

____ 3. The "Age of Mammals."

____ 4. A rocklike column composed of many minute layers of prokaryotic cells; a type of fossil evidence.

____ 5. A type of protobiont formed by adding water to abiotically-formed polypeptides.

____ 6. A major interval of geological time; a subdivision of a period.

____ 7. An organism that lives symbiotically inside a host cell.

____ 8. The remains of a microscopic organism.

____ 9. A molecule of RNA that has catalytic ability.

MAKING COMPARISONS
Fill in the blanks.

Biological Event	Period	Era
First fishes appear	#1	#2
The first mammals and first dinosaurs	#3	#4
Dinosaurs peak and become extinct	#5	#6
Age of Marine Invertebrates	#7	#8
Age of *Homo sapiens*	#9	#10
Amphibians and wingless insects appear	#11	#12
Apes appear	#13	#14
Reptiles appear	#15	#16

MAKING CHOICES

Place your answer(s) in the space provided. Some questions may have more than one correct answer.

_____ 1. The earth's surface is protected from the mutagenic effects of ultraviolet radiation by a layer of

    a. CO_2.
    b. N_2.
    c. H_2.
    d. O_3.
    e. H_2O.

_____ 2. It is thought that the mitochondra in eukaryotic cells evolved from

    a. anaerobic bacteria.
    b. aerobic bacteria.
    c. endosymbionts.
    d. primitive stem cells.
    e. intracytoplasmic protozoans.

_____ 3. The Earth is thought to be about how many years old?

    a. 10-20 billion.
    b. 4,000.
    c. 3.5 billion.
    d. 4.6 billion.
    e. 10-20 million.

_____ 4. About how many years passed between the Earth's formation and the appearance of the earliest life?

    a. 10-20 million
    b. 100-200 million
    c. 800-900 million
    d. five billion
    e. 10 billion

_____ 5. Earth's early atmosphere became oxygenated through the photosynthetic activity of

    a.. purple sulfur bacteria
    b. cyanobacteria.
    c. water-splitting autotrophs.
    d. green sulfur bacteria.
    e. hydrogen-sulfide bacteria.

_____ 6. Aerobes are generally better competitors than anaerobes because aerobic respiration

    a. is more efficient.
    b. splits water.
    c. is confined to symbiotic mitochondria.
    d. extracts more energy from a molecule.
    e. has had a longer period of evolution.

_____ 7. The early Earth's atmosphere included
    a. CO_2.
    b. N_2.
    c. H_2.
    d. CO.
    e. H_2O vapor.

_____ 8. The basic requirements for chemical evolution include
    a. DNA.
    b. oxygen.
    c. energy.
    d. water.
    e. time.

_____ 9. Experimenters that simulated conditions thought to have existed on early Earth found that these conditions could produce
    a. DNA.
    b. RNA.
    c. nucleotide bases.
    d. proteins.
    e. amino acids.

_____ 10. Strong support for the endosymbiotic theory of eukaryotic cell origins is derived from the fact that mitochondria and chloroplasts have
    a. unique DNA.
    b. two membranes.
    c. ribosomes.
    d. tRNA.
    e. endosymbionts.

_____ 11. The consensus among scientists is that the first cells were
    a. aerobic autotrophic prokaryotes.
    b. anaerobic autotrophic prokaryotes.
    c. aerobic heterotrophic prokaryotes.
    d. anaerobic heterotrophic prokaryotes.
    e. anaerobic heterotrophic eukaryotes.

_____ 12. Which of the following era/period combinations is/are mismatched?
    a. Cenozoic/Paleogene
    b. Paleozoic/Miocene
    c. Mesozoic/Jurassic
    d. Cenozoic/Cambrian
    e. Paleozoic/Devonian

_____ 13. The Paleozoic era
    a. began about 542 mya.
    b. lasted about 291 million years.
    c. consists of six periods.
    d. includes the Cambrian period.
    e. gave rise to many new species.

VISUAL FOUNDATIONS

Color the parts of the illustration below as indicated.

RED   ☐ DNA

GREEN   ☐ chloroplast

YELLOW   ☐ aerobic bacterium

BLUE   ☐ nuclear envelope

ORANGE   ☐ mitochondrion

BROWN   ☐ endoplasmic reticulum

TAN   ☐ photosynthetic bacterium

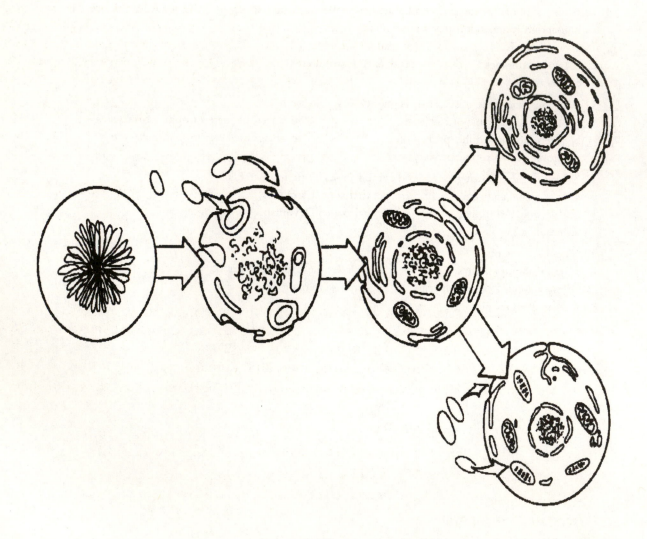

❑

CHAPTER 22

The Evolution of Primates

This is the last of five chapters that address the concept of evolution – the concept that links all fields of the life sciences into a unified body of knowledge. This chapter addresses the evolution of primates, with particular focus on humans and their ancestors. Primates evolved from small, tree-dwelling, shrewlike placental mammals. The first primates to evolve were the lemurs, galagos, and lorises. These were followed by the tarsiers and anthropoids. The early anthropoids branched into two groups, the New World Monkeys and the Old World Monkeys. The latter gave rise to apes that, in turn, gave rise to the hominids (humans and their ancestors). Hominid evolution began in Africa. The first hominid to have enough human features to be placed in the same genus as modern humans is *Homo habilis*. Anatomically modern *Homo sapiens* appeared approximately 195,000 years ago. The evolutionary increase in human brain size made cultural evolution possible.

REVIEWING CONCEPTS
Fill in the blanks.

INTRODUCTION
1. Paleontologists hypothesize that the first primates descended from small, shrewlike
   _____.

PRIMATE ADAPTATIONS
2. The first primates appeared on earth about _____ years ago.

PRIMATE CLASSIFICATION
3. There are three suborders in the order Primates: the (a)_____ include lemurs, lorises, and galagos; the (b)_____ includes tarsiers; and the (c)_____ includes monkeys, apes, and humans.

Suborder Anthropoidea includes monkeys, apes, and humans

4. A few monkeys have tails capable of wrapping around branches and serving as fifth limbs known as _____.

Apes are our closest living relatives

5. Apes and humans comprise a group called the _____.

6. The five living genera of hominoids are _____
   _____.

HOMINID EVOLUTION
The earliest hominids may have lived 6 mya to 7 mya

7. Earliest hominids most likely appeared about _____ years ago.

8. Hominid evolution probably began on the continent of _____.

Australopithecines are the immediate ancestors of genus *Homo*

9. The hominid species *Australopithecus anamensis* exhibits distinct phenotypic differences between the two sexes, a condition known as _____ _____.

10. "Lucy" was one of the most ancient hominids, a member of the species _____.

Homo habilis is the oldest member of genus *Homo*

11. *Homo habilis* was discovered in the country _____ in Africa.

12. *Homo habilis* appeared about _____ years ago. *H. habilis* was the first primate to consciously design useful tools.

Homo erectus apparently evolved from *Homo habilis*

13. *Homo erectus* evidently evolved from the species _____.

14. *Homo erectus* is about _____ years old. It had a larger brain than *H. habilis*, made more sophisticated tools, and discovered how to use fire.

Archaic *Homo sapiens* appeared between 400,000 and 200,000 years ago

15. Archaic *Homo sapiens* had a rich and varied culture, which included making _____.

Neandertals appeared approximately 230,000 years ago

16. Neandertal fossils were first discovered in _____.

Biologists debate the origin of modern *Homo sapiens*

17. The "_____" hypothesis holds that *Homo sapiens* arose in only in Africa.

18. The "_____" hypothesis asserts that modern humans arose independently in Asia, Africa, and Europe.

CULTURAL EVOLUTION

19. Cultural evolution is the progressive addition of _____ to the human experience. It is made possible by an evolutionary increase in brain size in humans.

20. Three of the most significant advances in cultural evolution were the

_____

_____

_____.

Development of agriculture resulted in a more dependable food supply

21. Evidence has shown that humans had begun to cultivate crops approximately _____ years ago.

Cultural evolution has had a profound impact on the biosphere

22. Cultural evolution has resulted in large-scale disruption and degradation of the _____.

BUILDING WORDS

Use combinations of prefixes and suffixes to build words for the definitions that follow.

Prefixes	The Meaning		Suffixes	The Meaning
anthrop(o)-	human		-oid	resembling, like
bi-	twice, two		-ped(al)	foot
pro-	"before"			
quadr(u)-	four			
supra-	above			

Prefix	Suffix	Definition
_____	_____	1. Member of the suborder Anthropoidea; animals that "resemble" humans.
_____	_____	2. Pertains to an animal that walks on four feet.
_____	_____	3. Pertains to an animal that walks on two feet.
homin-	_____	4. Member of the superfamily Hominoidea; a gibbon, orangutan, gorilla, chimpanzee, or human.
_____	-orbital	5. Situated above the eye socket.

MATCHING

Terms:

a. Arboreal
b. *Australopithecus*
c. Brachiate
d. Foramen magnum
e. Hominid
f. *Homo erectus*
g. *Homo habilis*
h. *Homo sapiens*
i. Neandertals
j. Paleoanthropologist
k. Prehensile

For each of these definitions, select the correct matching term from the list above.

____ 1. Any of a group of extinct and living humans.

____ 2. One of the earliest groups of *Homo sapiens*, according to some scientists.

____ 3. Tree-dwelling.

____ 4. The earliest hominid genus.

____ 5. Adapted for grasping by wrapping around an object.

____ 6. A scientist that studies human evolution.

____ 7. The first hominid to have enough human features to be placed in the same genus as modern humans.

____ 8. To swing arm to arm, from one branch to another.

____ 9. The opening in the base of the vertebrate skull through which the spinal cord passes.

MAKING COMPARISONS

Fill in the blanks.

Hominid	Time of Origin	Characteristic(s)
Cro-magnon	30,000 years ago	Possessed modern features, established a culture in what is now Europe
#1	About 4 million years ago (mya)	Small brain, walked upright, large canine teeth, jutting jaw, no evidence of tool use (famous 3.2 mya example was nicknamed "Lucy")
#2	230,000 years ago	Disappeared mysteriously approximately 28,000 years ago
#3	About 5.8 mya	Close to the "root" of the human family tree
#4	3 mya	Walked erect, humanlike hands and teeth, likely omnivorous, likely derived from *A. afarensis*
Homo habilis	#5	#6
Sahelanthropus sp.	#7	The earliest known hominid, close to the last common ancestor of hominids and chimpanzees
Homo erectus	#8	#9

MAKING CHOICES

Place your answer(s) in the space provided. Some questions may have more than one correct answer.

_____ 1. The suborders of primates are
   a. Hominoids.   d. Tarsiiformes.
   b. Hominids.   e. Anthropoidea.
   c. Prosimii.

_____ 2. The group(s) to which humans and their ancestors belong is/are
   a. Hominoids.   d. Tarsiiformes.
   b. Hominids.   e. Anthropoidea.
   c. Prosimii.

_____ 3. The group(s) to which apes belong is/are
   a. Hominoids.   d. Tarsiiformes.
   b. Hominids.   e. Anthropoidea.
   c. Prosimii.

_____ 4. Both apes and humans
   a. are hominoids.   d. lack tails.
   b. are hominids.   e. have opposable thumbs.
   c. can brachiate.

_____ 5. The first hominid that migrated to Europe and Asia is
    a. *H. habilis.*
    b. *H. erectus.*
    c. *H. sapiens.*
    d. australopithecines.
    e. "Lucy" and her descendants.

_____ 6. Archaic *Homo sapiens* appeared about how many years ago?
    a. 20,000-40,000
    b. 100,000-150,000
    c. 200,000-400,000
    d. 800,000-900,000
    e. one million

_____ 7. The Peking man and Java man are classified as
    a. *H. habilis.*
    b. *H. erectus.*
    c. *H. sapiens.*
    d. apes.
    e. hominids.

_____ 8. The "Neandertal man" appeared about _____ years ago.
    a. 800,000
    b. 230,000
    c. 30,000
    d. 1 million
    e. 2 million

_____ 9. The earliest hominid to be placed in the genus *Homo* is
    a. *H. habilis.*
    b. *H. erectus.*
    c. *H. sapiens.*
    d. australopithecines.
    e. "Lucy" and her descendants.

_____10. The earliest hominids
    a. had short canines.
    b. are in the genus *Homo.*
    c. are in the spcies *sapiens.*
    d. appeared about 6-7 mya.
    e. evolved in Africa.

_____11. Based on molecular similarities and other characteristics, it is thought that the nearest living relative of humans is the
    a. gorilla.
    b. monkey.
    c. gibbon.
    d. chimpanzee.
    e. ape.

_____ 12. Which of the following support(s) the "out-of-Africa hypothesis?"
    a. modern *H. sapiens* evolved from *H. erectus*
    b. newly evolved *H. sapiens* migrated to Europe
    c. results from recent molecular analyses
    d. racial differences.
    e. paternal mitochondrial DNA.

_____13. The immediate ancestor to the genus *Homo* is
    a. Prosimii.
    b. Tarsiiformes.
    c. Australopithecines.
    d. primitive apes.
    e. Therapsids.

_____14. The earliest Hominids may have lived as long as
    a. 3-4 mya.
    b. 4-5 mya.
    c. 5-6 mya.
    d. 6-7 mya.
    e. 7-8 mya.

VISUAL FOUNDATIONS

Color the parts of the illustration below as indicated.

VIOLET ☐ cerebrum

GREEN ☐ jaw (mandible bone)

BLUE ☐ supraorbital ridge

YELLOW ☐ canine teeth

ORANGE ☐ incisor teeth

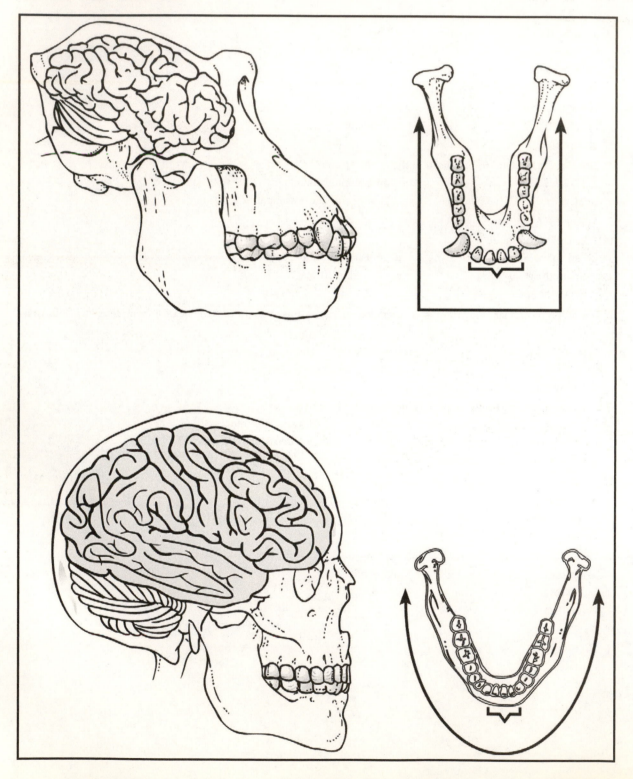

Color the parts of the illustration below as indicated. Label the anatomical difference between human and gorilla on each skeletal element.

RED   ❑   pelvis
GREEN   ❑   skull
YELLOW   ❑   first toe
BLUE   ❑   spine
ORANGE   ❑   arm
BROWN   ❑   leg

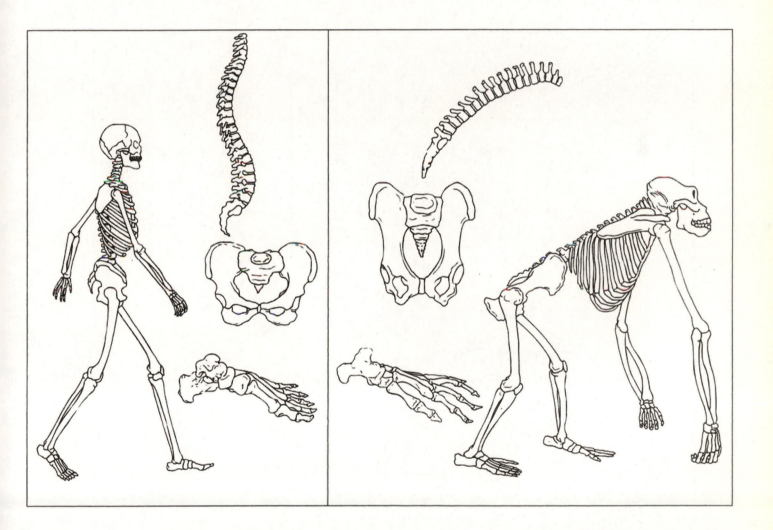

PART V

□

The Diversity of Life

CHAPTER 23

❑

Understanding Diversity: Systematics

Evolution, the subject matter of the previous five chapters, provides the framework for this section of the book -- an examination of The Diversity of Life. This chapter explores some of the approaches and methods used by biologists to classify organisms and to infer their evolutionary history and relationships. Systematics is the scientific study of the diversity of organisms and their evolutionary relationships. Taxonomy, an important aspect of systematics, is the science of naming, describing, and classifying organisms. Organisms are classified according to a system developed in the mid-18th century, a system in which each organism is given a two-part name; the first name is the genus and the second is the specific epithet. Together, the two names constitute the scientific species name of the organism. Classification is hierarchical, with one or more related genera constituting a family, and one or more related families constituting an order, and so on. The goal of systematics is to base classification on evolutionary change and relationships. Two or more different organisms are grouped together when evidence suggests that they have evolved from a common ancestor.

REVIEWING CONCEPTS
Fill in the blanks.

INTRODUCTION

1. The variety of organisms and ecosystems is referred to as
_____.

CLASSIFYING ORGANISMS

2. _____ is the science of naming, describing, and classifying organisms.

Organisms are named using a binomial system

3. The system developed by Carolus Linnaeus that assigns a two-part name to each kind of organism is logically referred to as the _____ of nomenclature.

4. The two-part name assigned to an organism consists first of the (a)_____, which is capitalized, followed by the decapitalized (b)_____.

Each taxonomic level is more general than the one below it

5. The classification system consists of several levels of organization. For example, a group of closely related species is placed in the same genus. In similar fashion, closely related genera are grouped together in the same (a)_____, which in turn are grouped and placed in orders. The next level in the hierarchy above orders is the (b)_____, followed in succession by the (c)_____.

6. In the system for classifying organisms, each grouping, or level, such as a particular species, or genus, or order, and so forth is called a _____.

Biologists are moving away from Linnaean categories

7. In the PhyloCode approach to classification, a group of organisms with a common ancestor is known as a _____.

DETERMINING THE MAJOR BRANCHES IN THE TREE OF LIFE

8. Gene swapping between organisms in one taxon and related organisms in another taxon is called _____.

RECONSTRUCTING PHYLOGENY

9. Modern classification is based on the reconstruction of _____, or phylogeny, as it is called.

Homologous structures are important in determining evolutionary relationships

10. A characteristic that superficially appears homologous but is actually independently acquired by convergent evolution or reversal is described as exhibiting _____.

Shared derived characters provide clues about phylogeny

11. Shared primitive characters suggest (ancient or recent?) _____ ancestry.

12. Shared derived characters indicate more (ancient or recent?) _____ common ancestry.

Biologists carefully choose taxonomic criteria

13. Organisms are typically classified on the basis of a _____ of traits rather than on any single trait.

Molecular homologies help clarify phylogeny

14. Comparisons of the (a)_____ of proteins, and the (b)_____of nucleotides are used to confirm evolutionary relationships.

15. The science of _____ focuses on molecular structure to identify evolutionary relationships.

Taxa are grouped based on their evolutionary relationships

16. When the organisms in a given taxon have evolved from different ancestors, the taxon is said to be a _____.

17. The taxonomic grouping that includes an ancestral species and all of its descendants is known as a _____.

CONSTRUCTING PHYLOGENETIC TREES

18. The phenetic system is a numerical taxonomy in which organisms are grouped according to the number of _____ they share.

19. The two major approaches to systematics are _____ _____.

Outgroup analysis is used in constructing and interpreting cladograms

20. An _____ is a taxon that is considered to have diverged earlier than the taxa under investigation.

A cladogram is constructed by considering shared derived characters

21. Membership in a clade cannot be established by shared _____ _____.

In a cladogram each branch point represents a major evolutionary step

22. A phylogram provides information about the relative number of _____ that have occurred within each lineage.

Systematists use the principle of parsimony to make decisions

23. _____ is based on the experience that the simplest explanation is probably the correct one.

BUILDING WORDS

Use combinations of prefixes and suffixes to build words for the definitions that follow.

Prefixes	The Meaning
mono-	alone, single, one
poly-	much, many
sub-	under, below

Prefix	Suffix	Definition
_____	-phylum	1. A distinctive subunit of a phylum.
_____	-phyletic	2. Refers to a taxon in which all of the subgroups have a common (single) ancestry.
_____	-phyletic	3. Refers to a taxon consisting of several evolutionary lines and not including a common ancestor.

MATCHING

Terms:

a. Ancestral character
b. Cladogram
c. Class
d. Derived character
e. Family
f. Order
g. Phenetics
h. Phylogeny
i. Species
j. Systematics
k. Taxon
l. Taxonomy

For each of these definitions, select the correct matching term from the list above.

_____ 1. A group of interbreeding individuals that live within the same population.

_____ 2. The systematic study of organisms based on similarities of many characters.

_____ 3. The science of naming, describing, and classifying organisms.

_____ 4. A characteristic that has remained unchanged in a species.

_____ 5. A branching diagram based on shared derived characters to show the evolutionary relationships between organisms.

_____ 6. A taxonomic group of any given level.

_____ 7. The scientific study of the diversity of organisms and their evolutionary relationships.

_____ 8. A characteristic that is not present in the ancestral lineage.

_____ 9. A taxon that comprises related genera.

_____10. The evolutionary history of organisms.

MAKING COMPARISONS

Fill in the blanks.

Domain	Kingdom	Characteristics
Bacteria	Bacteria	Unicellular prokaryotes, generally with cell walls composed of peptidoglycan
#1	#2	Unicellular prokaryotes lacking peptidoglycan in cell walls
#3	#4	Eukaryotic, generally unicellular protozoa, algae, slime molds, water molds
#5	#6	Heterotrophic eukaryotes with cell walls
#7	#8	Heterotrophic eukaryotes without cell walls
#9	#10	Autotrophic eukaryotes with cell walls

MAKING CHOICES

Place your answer(s) in the space provided. Some questions may have more than one correct answer.

_____ 1. The study of the diversity of organisms and their evolutionary relationships is
    a. taxonomy.   d. determinism.
    b. nomenclature.   e. classification.
    c. systematics.

_____ 2. The science of describing, naming, and classifying organisms is
    a. taxonomy.   d. determinism.
    b. nomenclature.   e. classification.
    c. systematics.

_____ 3. Arranging organisms into groups based on similarities that reflect their evolutionary histories is
    a. taxonomy.   d. determinism.
    b. nomenclature.   e. classification.
    c. systematics.

_____ 4. The systematist would likely be most interested in
    a. lumping taxa.   d. ontogeny.
    b. splitting taxa.   e. ancestries.
    c. evolutionary relationships.

_____ 5. Which of the following would the cladist emphasize when classifying organisms?
    a. phylogenetic trees.   d. number of shared characteristics.
    b. amino acid sequences.   e. antigen-antibody relationships.
    c. nucleotide sequences.

_____ 6. Which of the following is/are listed in order from taxa of greatest diversity to taxa of least diversity?
    a. Phylum, order, genus   d. Family, genus, species
    b. Family, class, order   e. Class, order, phylum
    c. Class, family, genus

_____ 7. Fungi are generally not classified as plants because they do not
   a. photosynthesize.
   b. have a true nucleus.
   c. have cell walls.
   d. have mitochondria.
   e. reproduce sexually.

_____ 8. Which of the following is/are used extensively by the molecular biologist to determine relationships?
   a. carbohydrates.
   b. nucleic acids.
   c. lipids.
   d. proteins.
   e. hormones.

_____ 9. The system of binomial nomenclature was developed by
   a. Charles Darwin.
   b. Carrolus Linnaeus.
   c. Paul Hebert.
   d. Carl Woese.
   e. Watson and Crick.

_____10. In a cladogram, a node represents
   a. the number of shared characteristics.
   b. the convergence of species.
   c. the basis for a new binomial.
   d. integration of higher taxa.
   e. a branch point for divergent species.

VISUAL FOUNDATIONS

Color the parts of the illustration below as indicated. Inside each large box, write the specific name of each category used in classifying the depicted organisms, and also indicate the specific epithet of the organism used to illustrate this hierachical organization in the species box.

RED ☐ species
TAN ☐ domain
GREEN ☐ phylum
YELLOW ☐ class
BLUE ☐ kingdom
ORANGE ☐ order
VIOLET ☐ family
PINK ☐ genus

CHAPTER 24

❑

Viruses and Prokaryotes

This and the following seven chapters examine the major groups of organisms that inhabit our planet. You will see how systematists have classified these organisms in domains and kingdoms based on current data from morphology, development, the fossil record, behavior, and molecular biology. This chapter examines the diversity and characteristics of viruses, viroids, prions, and prokaryotes. Viruses consist of a core of DNA or RNA surrounded by a protein coat. Some are surrounded by an outer envelope too. Whereas some viruses may or may not kill their hosts, others invariably do. Some integrate their DNA into the host DNA, conferring new properties on the host. Viroids consist of RNA with no protein coat. Prions appear to consist of only protein. Prokaryotes are mostly unicellular, but some form colonies or filaments. They have ribosomes, but lack membrane-bounded organelles. Their genetic material is a circular DNA molecule. Some have flagella. Whereas most prokaryotes get their nourishment either from dead organic matter or symbiotically from other organisms, some manufacture their own organic molecules. Although most prokaryotes reproduce asexually, some exchange genetic material. The former are thought to be the original prokaryotes from which all cellular life descended. Although most prokaryotes are harmless, a few are notorious for the diseases they cause. Prokaryotes have important medical and industrial uses.

REVIEWING CONCEPTS

Fill in the blanks.

INTRODUCTION

1. Pathogens are microorganisms such as bacteria, fungi, and protozoa that cause
_____.

VIRUSES

A virus consists of nucleic acid surrounded by a protein coat

2. The core of a virus is a _____.

3. Viruses are surrounded by a protein coat called a _____.

Viruses may have evolved from cells

4. According to the _____, some viruses may trace their origin to animal cells, others to plant cells or to bacterial cells.

The International Committee on Taxonomy of Viruses classifies viruses

5. Viruses can be classified based on the types of organisms they infect or on their
_____.

Bacteriophages are viruses that attack bacteria

6. The most common structure of a bacteriophage consists of a long nucleic acid molecule coiled within a _____.

Viruses reproduce only inside host cells

7. The two types of reproduction among viruses are _____
_____.

Lytic reproductive cycles destroy host cells

8. Viruses that have only a lytic cycle are said to be _____
_____.

9. The five steps that almost all lytic viral infections follow are

_____.

_____.

Temperate viruses integrate their DNA into the host DNA

10. Viruses that do not always destroy the host cell are referred to as

_____.

11. Viral DNA that is integrated into host DNA is called a
(a)_____, and bacterial cells carrying integrated viral DNA
are called (b)_____.

12. _____ occurs when bacteria exhibit new
properties as a result of integration of temperate viral DNA.

Many viruses infect vertebrates

13. RNA viruses called _____ transcribe their RNA genome
into a DNA strand that is then used as a template to produce more viral RNA.
HIV is one such virus.

14. The DNA polymerase used by retroviruses in transcription is called

_____.

Some viruses infect plant cells

15. The genome of most plant viruses consists of _____.

VIROIDS AND PRIONS

Viroids are the smallest known pathogens

16. Viroids are hardy and resist heat and ultraviolet radiation because of the
condensed folding of their _____.

Prions are protein particles

17. Prions are found in the brains of patients with fatal degenerative brain diseases
called _____.

PROKARYOTES

Prokaryotes have several common shapes

18. Spherical prokaryotes are called (a)_____. They may occur in pairs
called (b)_____, in chains called
(c)_____, or in clumps or bunches called
(d)_____.

19. Rod-shaped prokaryotes are known as _____.

20. A rigid spiral-shaped prokaryote is known as a (a)_____, and
a flexible spiral-shaped prokaryote is called a (b)_____.

21. A spirillum shaped like a comma is called a _____.

Prokaryotic cells lack membrane-enclosed organelles

22. In prokaryotes, there is a nuclear area that contains DNA and resembles the nucleus of an eukaryote; it is referred to as the _____.

A cell wall typically covers the cell surface

23. The Eubacterial cell wall includes _____, a polymer that consists of two unusual types of sugars linked with short polypeptides.

24. Bacteria that absorb and retain crystal violet stain are referred to as _____.

Many types of prokaryotes are motile

25. Prokaryotes have two main methods of motility: (a)_____, consisting of a basal body, a hook, and a single filament, and (b)_____, movement in response to chemicals in the environment.

Prokaryotes have a circular DNA molecule

26. In addition to their genomic DNA, many bacteria have small amounts of genetic information present as one or more _____, or circular fragments of DNA.

Most prokaryotes reproduce by binary fission

27. The three different types of reproduction in prokaryotes are _____ _____.

Bacteria transfer genetic information

28. Prokaryotes usually reproduce asexually by simple transverse binary fission, but sometimes genetic material is exchanged in one of several ways. In _____, fragments of DNA released by a cell are taken in by another bacterial cell.

29. In _____, a phage carries bacterial genes from one bacterial cell into another.

30. Prokaryotes of two different "mating types" may exchange genetic material during _____.

Evolution proceeds rapidly in bacterial populations

31. Because bacteria reproduce by binary fission, _____ are quickly passed on to new generations, and natural selection effects are quickly evident.

Some bacteria form endospores

32. _____ are dormant, extremely durable cells that can survive in very dry, hot, or frozen environments.

Many bacteria form biofilms

33. Dental plaque is an example of a _____, bacteria-formed dense film attached to a solid surface.

Metabolic diversity has evolved among prokaryotes

34. Most prokaryotes are free-living decomposers, sometimes called
_____, organisms that get their carbon and energy
from dead organic matter.

35. The purple nonsulfur bacteria are _____,
organisms that get their carbon from other organisms but use chlorophyll and
other photosynthetic pigments to trap energy from sunlight.

Most prokaryotes require oxygen

36. Most bacterial cells are (a)_____, or require oxygen for cellular
respiration. Some other bacteria, however, are (b)_____
_____ that use oxygen if it's available, or (c)_____
_____ that carry out anaerobic respiration.

THE TWO PROKARYOTE DOMAINS

Some Archaea survive in harsh environments

37. Based on their metabolism and ecology, we can identify three main types of
Archaea: _____.

Bacteria are the most familiar prokaryotes

38. Bergey's Manual divides bacteria into more than _____(#?) phyla.

IMPACT OF PROKARYOTES

Some prokaryotes cause disease

39. _____ are used to demonstrate that a specific
pathogen causes specific disease symptoms.

Prokaryotes are used in many commercial processes

40. More than 1000 different species of bacteria have been used in
_____, a process in which a contaminated site is exposed
to microorganisms that break down toxins and leave behind harmless by-products
such as carbon dioxide and chloride.

BUILDING WORDS

Use combinations of prefixes and suffixes to build words for the definitions that follow.

Prefixes	The Meaning	Suffixes	The Meaning
an-	without, lacking	-gen	production of
bacterio-	bacteria	-phage	eat, devour
endo-	within	-karyo(te)	nucleus
eu-	good, well, "true"		
exo-	outside, outer, external		
patho-	suffering, disease, feeling		

Prefix	Suffix	Definition
_____	_____	1. A virus that infects and destroys bacteria.
_____	-toxin	2. A poison that is secreted by bacterial cells into the external environment.

_____   -toxin

methano-   _____

_____   -spore

_____ _____

_____   -aerobe

pro-   _____

_____ _____

3. A poison that is a component "within" the cell wall of most gram-negative bacteria.

4. A bacterium that produces methane from carbon dioxide and water.

5. A thick-walled spore that forms within a bacterium in response to adverse conditions.

6. Any disease-producing organism.

7. An organism that metabolizes only in the absence of (without) molecular oxygen.

8. An organism that lacks a nuclear membrane.

9. An organism with a distinct nucleus surrounded by nuclear membranes.

MATCHING

Terms:

a. Bacillus
b. Binary fission
c. Capsid
d. Conjugation
e. Flagellum

f. Lysogenic conversion
g. Mycoplasma
h. Peptidoglycan
i. Pilus
j. Saprobe

k. Transduction
l. Transformation

For each of these definitions, select the correct matching term from the list above.

_____ 1. A genetic transfer mechanism that produces new DNA in bacteria when DNA from a new organism is combined with the DNA of the host cell.

_____ 2. A rod-shaped bacterium.

_____ 3. An eubacterium that lacks cell walls.

_____ 4. A structure used for locomotion in some bacteria.

_____ 5. A process whereby one cell divides into two similar cells.

_____ 6. The protein covering of a virus.

_____ 7. A component of eubacterial cell walls.

_____ 8. A method of reproduction in single-celled organisms in which two cells link and exchange nuclear material.

_____ 9. A type of genetic recombination resulting from transfer of genes from one organism to another by a virus.

_____10. A hair-like structure associated with prokaryotes.

MAKING COMPARISONS

Fill in the blanks.

Characteristic	Bacteria	Archaea	Eukarya
Peptidoglycan in cell wall	Present	Absent	#1
70S ribosomes	#2	#3	#4
Nuclear envelope	#5	#6	#7

Characteristic	Bacteria	Archaea	Eukarya
Simple RNA polymerase	#8	#9	#10
Membrane-bounded organelles	#11	#12	#13

MAKING CHOICES

Place your answer(s) in the space provided. Some questions may have more than one correct answer.

_____ 1. Viruses

a. are acellular.

b. do not carry on metabolic activities.

c. can only reproduce when occupying a cell.

d. do not produce rRNA.

e. are not included in any of the three domains.

_____ 2. A virulent virus

a. has a lytic cycle.

b. can reproduce outside of cells.

c. causes disease.

d. degrades its host cell's nucleic acids.

e. invade but do not destroy their host cell.

_____ 3. Herpes simplex virus type 2

a. causes cold sores.

b. causes genital herpes.

c. causes infectious mononucleosis.

d. is a DNA virus with an envelope.

e. causes chicken pox.

_____ 4. When a host bacterium exhibits new properties because of a prophage, the phenomenon is called

a. lysogenic conversion.

b. assembly.

c. transduction.

d. viroid induction.

e. reverse transcription.

_____ 5. The process by which a phage carries genes from one bacterium to another bacterium is called

a. lysogenic conversion.

b. assembly.

c. transduction.

d. viroid induction.

e. reverse transcription.

_____ 6. Viruses are usually grouped (classified) according to

a. size.

b. shape.

c. evolutionary relationships.

d. phylogeny.

e. type of nucleic acid.

_____ 7. The characteristic(s) of life that viruses do not exhibit is/are

a. presence of nucleic acids.

b. independent movement.

c. reproduction.

d. a cellular structure.

e. independent metabolism.

_____ 8. The majority of bacteria are

a. autotrophs.

b. heterotrophs.

c. pathogens.

d. saprobes.

e. photosynthesizers.

_____ 9. Exchange of genetic material between bacteria sometimes occurs by

a. fusion of gametes.

b. conjugation.

c. incorporation by a bacterium of DNA fragments from another bacterium.

d. transfer of genes in bacteriophages.

e. transduction.

_____10. A small, circular piece of DNA that is separate from the main chromosome is

a. called a plasmid.

b. a mesosome.

c. responsible for photosynthesis.

d. found in archaebacteria and cyanobacteria, but not eubacteria.

e. found in gram-negative, but not gram-positive bacteria.

_____11. Bacteria that retain crystal violet stain differ from those that do not retain the stain in having a

a. thick lipoprotein layer.

b. thick lipopolysaccharide layer.

c. thicker peptidoglycan layer.

d. cell wall.

e. mesosome.

_____12. Prokaryotic cells are distinguished from eukaryotic cells by

a. absence of a nuclear envelope.

b. absence of mitochondria.

c. absence of DNA in the genetic material.

d. absence of a plasma membrane.

e. bacteriorhodopsin in cell walls.

_____13. The coat surrounding the nucleic acid core of a virus is

a. a capsid.

b. made of protein.

c. a virion.

d. a viral envelope.

e. composed of capsomeres.

_____14. An important difference between virulent and temperate viruses is that only the temperate virus

a. destroys the host cell.

b. lyses the host cell.

c. does not always lyse host cells in the lysogenic cycle.

d. contains DNA.

e. infects eukaryotic cells.

_____15. The virus that causes AIDS and some types of cancer is a/an

a. paramyxovirus.

b. herpesvirus.

c. adenovirus.

d. retrovirus.

e. RNA virus.

VISUAL FOUNDATIONS

Color the parts of the illustration below as indicated. Also label cell wall, flagellum, and pili.

RED ☐ plasma membrane

BROWN ☐ outer membrane

YELLOW ☐ peptidoglycan layer

BLUE ☐ DNA

TAN ☐ capsule

GREEN ☐ plasmid

Color the parts of the illustration below as indicated. Also label helical shape, polyhedral shape, and combination (helical and polyhedral) shape.

RED ☐ DNA

GREEN ☐ RNA

VIOLET ☐ capsid

ORANGE ☐ fibers

BROWN ☐ tail

Color the parts of the illustration below as indicated. Also label Gram-positive cell wall and Gram-negative cell wall.

RED ☐ transport protein

YELLOW ☐ peptidoglycan layer

BLUE ☐ polysaccharides

ORANGE ☐ plasma membrane

BROWN ☐ outer membrane

VIOLET ☐ lipoprotein

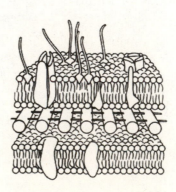

CHAPTER 25

Protists

The protist kingdom consists of a vast assortment of primarily aquatic eukaryotic organisms with diverse body forms, types or reproduction, modes of nutrition, and lifestyles. Biologists currently recognize as many as 60 major groups of protists. Protists range in size from microscopic single cells to meters-long multicellular organisms. They do not have specialized tissues. Protists obtain their nutrients autotrophically or heterotrophically. Whereas some protists are free-living, others form symbiotic associations ranging from mutualism to parasitism. Reproductive strategies vary considerably; most reproduce both sexually and asexually, others only asexually. Whereas some are nonmotile, most move by flagella, pseudopodia, or cilia. Protists are believed to be the first eukaryotes. Evidence suggests that mitochondria and chloroplasts may have originated as prokaryotes that became incorporated into larger cells.

REVIEWING CONCEPTS
Fill in the blanks.

INTRODUCTION

1. Protists are unicellular colonial or simple multicellular organisms that have a _____ cell organization.

INTRODUCTION TO THE PROTISTS

2. Organisms in the kingdom Protista do not comprise a naturally occurring phylogenetic group, but they are all (a)_____(eukaryotes or prokaryotes?) belonging to the domain (b)_____.

3. Possessing single multinucleated cells is a condition known as _____.

4. Most protists are aquatic, and many of them are floating microscopic forms called _____.

EVOLUTION OF THE EUKARYOTES
Mitochondria and chloroplasts probably originated from endosymbionts

5. In the serial endosymbiosis hypothesis, eukaryotic organelles such as mictochondria and chloroplasts arose from _____ _____ between larger cells and smaller prokaryotes that were incorporated and lived in them.

A consensus is emerging in eukaryote classification

6. Electron microscopy supports molecular data suggesting that certain protist taxa are _____.

REPRESENTATIVE PROTISTS

Excavates are anaerobic zooflagellates

7. Unlike virtually all other protests, excavates lack _____, or if they are present, they are atypical.

Discicristates include euglenoids and trypanosomes

8. Autotrophic euglenoids are single-celled, flagellated protists containing _____, the same pigments as those found in green algae and higher plants.

9. Autotrophic euglenoids store food in the form of the polysaccharide _____.

10. _____ is the genus name for the parasitic flagellate that causes African sleeping sickness.

Alveolates have flattened vesicles under the plasma membrane

11. Ciliates are distinct in having one or more small (a)_____ that function during the sexual process and a larger (b)_____ that regulates other cell functions.

12. "Cross fertilization" occurs among many ciliates during the sexual process known as _____.

13. The dinoflagellates are mostly unicellular, biflagellate, photosynthetic organisms; their alveoli contain interlocking _____ plates.

14. The _____ are photosynthetic endosymbionts that lack cellulose plates and flagella.

15. Apicomplexans are spore-forming parasites of _____.

16. Apicomplexans produce infective agents that are transmitted to the next host. These agents are called _____.

17. A sporozoan in the genus _____ can enter human RBCs and cause malaria.

Motile cells of heterokonts are biflagellate

18 The water molds, like the fungi, have a body called the _____.

19. Water molds reproduce asexually by biflagellated (a)_____ and sexually by means of (b)_____.

20. Diatoms have shells composed of _____(#?) parts.

21. Diatoms are mostly single-celled, with _____ impregnated in their cell walls.

22. When diatoms die, their shells accumulate over time, forming sediment called _____.

23. Most golden algae are flagellated, _____ organisms.

24. The largest and most complex of all algae are commonly called _____.

25. Most multicellular red algae attach to surfaces by means of a structure called the _____.

Red algae, green algae, and land plants are collectively classified as plants

26. The chloroplasts in red algae contain chlorophylls *a* and *d* and carotenoids, and in addition, the red pigment (a)_____ and the blue pigment (b)_____.

27. Green algae exhibit three kinds of sexual reproduction. In (a)_____ reproduction, the two gametes that fuse are apparently identical, whereas (b)_____ reproduction involves gametes of different sizes. Fusion of a nonmotile "egg" and a motile sperm-like gamete is called (c)_____ reproduction.

28. Green algae exhibit a wide diversity in size, complexity, and reproduction. They resemble plants, but some are endosymbionts, and others are symbiotic with fungi as "dual organisms" known as a _____.

Cercozoa are amoeboid cells enclosed in shells

29. Foraminiferans extend cytoplasmic projections through _____.

30. Actinopods project slender _____.

Amoebozoa have lobose pseudopodia

31. Amoebas glide along surfaces by flowing into cytoplasmic extensions called _____, meaning "false feet."

32. The parasitic species _____ causes amoebic dysentery.

33. Plasmodial slime molds form intricate stalked reproductive structures called _____, within which meiosis occurs.

34. When environmental conditions are suitable, spores open and one-celled haploid gametes of two types, one a flagellated form called a (a)_____, the other an ameboid (b)_____, emerge and fuse.

35. The cellular slime molds live vegetatively as individual, single-celled organisms, except during reproduction when they form a multicellular aggregate called a _____, which then differentiates and forms spores.

Opisthokonts include choanoflagellates, fungi, and animals

36. The single flagellum of the choanoflagellates is surrounded at the base by a delicate collar of _____.

BUILDING WORDS

Use combinations of prefixes and suffixes to build words for the definitions that follow.

Prefixes	The Meaning	Suffixes	The Meaning
cyto-	cell	-pod(ium)	foot, footed
iso-	equal, "same"	-zoa	animal
macro-	large, long, great, excessive		
micro-	small		
proto-	first, earliest form of		
pseudo-	false		

Prefix	Suffix	Definition
_____	_____	1. Single-celled, animal-like protists including amoebae, ciliates, flagellates, and sporozoans; members of this group are believed to have given rise to the earliest form of animal life.
_____	-pharynx	2. The "throat" of some zooflagellates, single-celled organisms.
_____	_____	3. A temporary protrusion of the cytoplasm of an ameboid cell that the cell uses for feeding and locomotion; a "false foot."
_____	-nucleus	4. A small nucleus found in ciliates.
_____	-nucleus	5. A large nucleus found in ciliates.
_____	-plasmodium	6. The aggregation of cells for reproduction in cellular slime molds; not a true plasmodium (multinucleate ameboid mass) as occurs in the plasmodial slime molds.
_____	_____	7. A type of sexual reproduction in which two gametes fuse (unite) that are identical ("the same") in size and structure.

MATCHING

Terms:

a. Actinopod
b. Axopod
c. Conjugation
d. Foraminiferan
e. Holdfast

f. Hypha
g. Mycelium
h. Oospore
i. Plankton
j. Plasmodium

k. Zoosporangium
l. Sporozoite

For each of these definitions, select the correct matching term from the list above.

_____ 1. One of the long, filamentous cytoplasmic projections that protrude through pores in the skeletons of actinopods.

_____ 2. A structure that produces tiny biflagellate zoospores.

_____ 3. The vegetative body of fungi; consists of a mass of hyphae.

_____ 4. Free-floating, mainly microscopic aquatic organisms found in the upper layers of the water.

_____ 5. Marine protozoans that produce chalky, many-chambered shells or tests.

_____ 6. A sexual process in which two individuals come together and exchange genetic material.

_____ 7. One of the filaments composing the mycelium of a fungus.

_____ 8. The basal root-like structure in multicellular algae that anchors the organism to a solid surface.

_____ 9. Multinucleate mass of cytoplasm that constitutes the feeding stage in the life cycle of slime molds.

_____10. Small infective agents produced by apicomplexans that are transmitted to the next host.

MAKING CHOICES

Place your answer(s) in the space provided. Some questions may have more than one correct answer.

_____ 1. Protists may be
a. unicellular.
b. colonial.
c. simple multicellular organisms.
d. eukayotic.
e. Archaea.

_____ 2. Mitochondria and chloroplasts are thought to have been derived from

    a. complex viruses.

    b. protozoa.

    c. endosymbionts.

    d. intrasymbiotic eukaryotes.

    e. aerobic bacteria.

_____ 3. Most zooflagellates are

    a. ciliated.

    b. colonial organisms.

    c. opportunistic autotrophs.

    d. heterotrophic.

    e. flagellated.

_____ 4. The *Paramecium*

    a. is a ciliate.

    b. is eukaryotic.

    c. is colonial, sometimes multicellular.

    d. possesses a pellicle.

    e. is a member of the taxon Cercozoa.

_____ 5. The parasite that causes malaria is in the phylum

    a. Ciliophora.

    b. Apicomplexa.

    c. Sarcomastigophora.

    d. Dinoflagellata.

    e. Euglenophyta.

_____ 6. A protozoan whose cells bear a striking resemblance to specialized cells in sponges is the

    a. foraminiferan.

    b. diatom.

    c. choanoflagellate.

    d. ameba.

    e. radiolarian.

_____ 7. One of the reasons that protists are placed in a separate kingdom is that many of them possess both plantlike and animal-like characteristics, which is particularly well-illustrated in the genus

    a. *Paramecium.*

    b. *Euglena.*

    c. *Plasmodium.*

    d. *Ameba.*

    e. *Didinium.*

_____ 8, Members of the genus *Euglena*

    a. are flagellated.

    b. do not possess a pellicle.

    c. possess chlorophyll.

    d. are heterotropic.

    e. are autotrophic.

_____ 9. A symbiotic relationship in which both "partners" benefit is known as

    a. commensalism.

    b. mutualism.

    c. parasitism.

    d. endosymbiosis.

    e. fraternism.

_____10. A symbiotic relationship in which one "partner" benefits and the other is unaffected is known as

    a. commensalism.

    b. mutualism.

    c. parasitism.

    d. endosymbiosis.

    e. fraternism.

VISUAL FOUNDATIONS

Color the parts of the illustration below as indicated. Also label cilia.

RED	☐	micronucleus
ORANGE	☐	oral groove
BLUE	☐	contractile vacuole
YELLOW	☐	food vacuole
TAN	☐	anal pore
VIOLET	☐	macronucleus

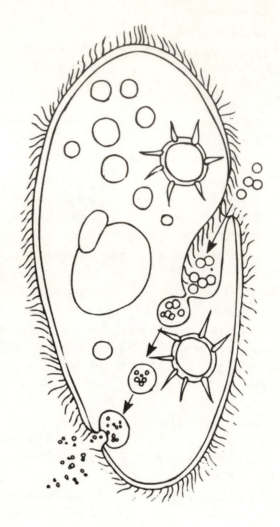

Color the parts of the illustration below as indicated. Label the portions of the illustration depicting sexual reproduction, and the portions depicting asexual reproduction. Also label diploid generation, haploid generation, mitosis, meiosis, and fertilization.

RED ☐ positive strain

GREEN ☐ negative strain

YELLOW ☐ zygote

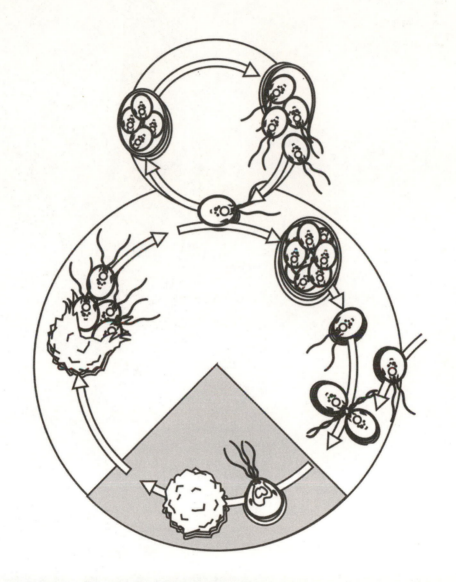

◻

Kingdom Fungi

Fungi are eukaryotes with cell walls. They digest their food outside their body and absorb the nutrients through the cell wall and cell membrane. Fungi are nonmotile and reproduce by means of spores that are formed either sexually or asexually. Some fungi are unicellular (the yeast form) and some are multicellular, having a filamentous body plan. Fungi are classified into five phyla. Most fungi are decomposers, and as such play a critical role in the cycle of life. Others, however, form symbiotic relationships with other organisms. Fungi are of both positive and negative economic importance. Some damage stored goods and building materials. Some provide food for humans. Some function in brewing beer and in baking. Some produce antibiotics and other drugs and chemicals of economic importance, and some cause serious diseases in humans and economically important animals and plants.

REVIEWING CONCEPTS
Fill in the blanks.

INTRODUCTION
1. Like prokaryotes, most fungi are _____ that obtain nutrients from dead organic matter.

CHARACTERISTICS OF FUNGI
2. Many fungi are _____(more or less?) sensitive to osmotic pressures than are bacteria.

Fungi absorb food from the environment

3. Fungi store excessive amounts of nutrients as _____
_____.

Fungi have cell walls that contain chitin

4. Cell walls of most fungi contain _____, which is highly resistant to microbial breakdown.

Most fungi have a filamentous body plan

5. The two main types of fungi based on body plan are the _____
_____.

Fungi reproduce by spores

6. Some form of sexual conjugation often occurs, sometimes producing hyphal cells with two distinctly different unfused nuclei. Such cells are said to be
(a)_____, whereas cells containing only one haploid nucleus are
(b)_____.

FUNGAL DIVERSITY

Fungi are assigned to the opisthokont clade

7. Based on structural characters and molecular data, the fungi, once considered a part of the (a)_____ kingdom, are now viewed as more closely related to (b)_____.

Diverse groups of fungi have evolved

8. Based on the characteristics of sexual spores, fruiting bodies, and molecular data, fungi are generally assigned to the five phyla _____

_____

_____.

9. When individuals within a given taxa do not share a common ancestor they are said to be _____.

10. Deuteromycetes are called "imperfect fungi" simply because no one has observed them to have a _____ in their life cycles.

Chytrids have flagellate spores

11. Most chytrids are unicellular or composed of a few cells that form a simple body called a (a)_____, which may have slender extensions called (b)_____.

Zygomycetes reproduce sexually by forming zygospores

12. The zygomycete species _____ is the well-known black bread mold.

13. Black bread mold is _____, meaning that an individual fungal hypha is self-sterile and mates only with a hypha of a different mating type.

Glomeromycetes are symbionts with plant roots

14. The symbiotic relationships between fungi and the roots of plants are called _____.

Ascomycetes reproduce sexually by forming ascospores

15. Ascomycetes produce sexual spores in sacs called (a)_____, and asexual spores called (b)_____ at the tips of (c)_____.

16. As the asci develop, they are surrounded by intertwining hyphae that develop into a fruiting body known as an _____.

Basidiomycetes reproduce sexually by forming basidiospores

17. Basidiomycetes develop an enlarged, club-shaped hyphal cell called a (a)_____, on the surface of which four spores called (b)_____ develop.

18. What we call a mushroom grows from a compact mass of hyphae along the mycelium called a (a)_____. A mushroom is more formally referred to as a (b)_____.

ECOLOGICAL IMPORTANCE OF FUNGI

19. Most fungi are free-living decomposers that absorb nutrients from organic wastes and dead organisms. In the process, they release (a)_____ to the atmosphere and return (b)_____ to the soil.

Fungi form symbiotic relationships with some animals

20. Symbiotic relationships have formed between certain fungi and animals; the fungi break down _____ that the animals ingest but are incapable of digesting.

Mycorrhizae are symbiotic relationships between fungi and plant roots

21. Some fungi are parasites. Others are mutualistic symbionts, notable among which are the _____ that live on the roots of plants.

Lichens are symbiotic relationships between a fungus and a photoautotroph

22. The phototroph portion of the lichen is usually a (a)_____. The fungus is sometimes a basidiomycete, but usually an (b)_____.

23. Lichens typically assume one of three forms: (a)_____, a low, flat growth; (b)_____, a flat growth with leaf-like lobes; or (c)_____, an erect and branching growth.

24. Lichens reproduce asexually by fragmentation, wherein pieces of lichen called _____ break off and begin growing after landing on a suitable substrate.

ECONOMIC, BIOLOGICAL, AND MEDICAL IMPACT OF FUNGI

Fungi provide beverages and food

25. (a)_____ are the fungi used to make wine and beer, and to produce baked goods. Wine is produced when the fungus ferments (b)_____, beer results from fermentation of (c)_____, and (d)_____ bubbles cause bread to rise.

26. Roquefort and Camembert cheeses are produced using the genus (a)_____, and the imperfect fungus (b)_____ is used to make soy sauce from soybeans.

27. Some of the most toxic mushrooms, such as the "destroying angel" and "death cap," belong to the genus _____.

28. The chemical _____ found in some mushrooms causes hallucinations and intoxication.

Fungi are important to modern biology and medicine

29. Alexander Fleming noticed that bacterial growth is inhibited by the mold (a)_____, and this discovery eventually led to the development of the most widely used of all antibiotics, (b)_____.

30. An ascomycete that infects cereal plant flowers produces a structure called an _____ where seeds would normally form. When livestock or humans eat grain or grain products contaminated with this fungus, they may be poisoned by the extremely toxic substances contained therein.

Some fungi cause animal diseases

31. Some species of Aspergillus produce potent mycotoxins called _____ that harm the liver and are known carcinogens.

Fungi cause many important plant diseases

32. All plants are susceptible to fungal diseases. Fungi enter plants through stomata, or wounds, or by dissolving a portion of cuticle with the enzyme _____.

33. Parasitic fungi often extend specialized hyphae called _____ into host cells in order to obtain nutrients from host cell cytoplasm.

BUILDING WORDS

Use combinations of prefixes and suffixes to build words for the definitions that follow.

Prefixes	The Meaning		Suffixes	The Meaning
coeno-	common		-cyt(ic)	cell
hetero-	different		-karyo(tic)	nucleus
homo-	same		-phore	bearer
mono-	alone, single, one		-troph	nutrition, growth, "eat"
sapro-	rotten			

Prefix	Suffix	Definition
_____	_____	1. Pertains to an organism made up of a multinucleate, continuous mass of cytoplasm enclosed by one cell wall (all nuclei share a common cell).
_____	_____	2. Pertains to hyphae that contain only one nucleus per cell.
_____	-thallic	3. Pertains to an organism that has two different mating types.
conidio-	_____	4. Hyphae in the ascomycetes that bear conidia.
_____	-thallic	5. Pertains to an organism that can mate with itself.

MATCHING

Terms:

a. Ascocarp
b. Ascospore
c. Basidiocarp
d. Basidium
e. Budding

f. Chitin
g. Conidium
h. Dikaryotic
i. Haustorium
j. Lichen

k. Mycelium
l. Mycorrhizae
m. Soredium
n. Zygospore

For each of these definitions, select the correct matching term from the list above.

_____ 1. Compound organism composed of a photosynthetic green alga or cyanobacterium and a fungus.

_____ 2. The vegetative body of fungi; consists of a mass of hyphae.

_____ 3. The sexual spores produced by an ascomycete.

_____ 4. The fruiting body of a basidiomycete.

_____ 5. Asexual reproductive body produced by lichens.

_____ 6. Mutualistic associations of fungi and plant roots that aid in the absorption of materials.

_____ 7. Asexual reproduction in which a small part of the parent's body separates from the rest and develops into a new individual.

_____ 8. The club-shaped spore-producing organ of certain fungi.

_____ 9. Cells having two nuclei.

_____ 10. A component of fungal cell walls; a polymer that consists of subunits of a nitrogen-containing-sugar.

MAKING COMPARISONS
Fill in the blanks.

Phylum	Mode of Sexual Reproduction	Mode of Asexual Reproduction	Representative(s) of the Group
#1	Zygospores	Haploid spores	Black bread mold
#2	Ascospores	Conidia	Yeasts, cup fungi, morels, truffles, and blue-green and pink molds
#3	Basidiospores	Uncommon	Mushrooms, puffballs, rusts, and smuts
#4	Unknown	Blastospores	Symbiotic fungi

MAKING CHOICES
Place your answer(s) in the space provided. Some questions may have more than one correct answer.

_____ 1. Fungi
     a. are heterotrophs.   d. possess cell walls.
     b. are photoautotrophs.   e. digest food outside their bodies.
     c. are eukaryotes.

_____ 2. Fungal hyphae that contain two genetically distinct nuclei within each cell are known as
     a. dihaploid.   d. *2n*.
     b. disomic.   e. *n + n*.
     c. dikaryotic.

_____ 3. A lichen can be composed of a/an
     a. alga and fungus.   d. alga and basidiomycete.
     b. photoautotroph and fungus.   e. alga and ascomycete.
     c. cyanobacterium and ascomycete.

_____ 4. A common fungal infection of the mucous membranes of the mouth, throat, or vagina is
     a. an autoimmune response.   d. ergotism.
     b. St. Anthony's fire.   e. candidiasis.
     c. histoplasmosis.

_____ 5. Fungi can reproduce
     a. sexually.   d. by simple division.
     b. asexually.   e. by budding.
     c. by spore formation.

_____ 6. The black fungus growing on a piece of bread
     a. is heterothallic.   d. has sporangia on the tips of stolons.
     b. has male and female strains.   e. is in the phylum Zygomycota.
     c. has coenocytic hyphae.

_____ 7. The genus _Penicillium_
    a. is an ascomycete.   d. produces the antibiotic penicillin.
    b. is a sac fungus.   e. produces the flavor in Brie cheese.
    c. produces the flavor in Roquefort cheese.

_____ 8. Lichens can be used as indicators of air pollution because they
    a. cannot excrete absorbed elements.   d. do not grow well in polluted areas.
    b. tolerate sulfur dioxide.   e. overgrow polluted areas.
    c. can endure large quantities of toxins.

_____ 9. A mass of filamentous hyphae is called a
    a. hypha.   d. thallus.
    b. mycelium.   e. ascocarp.
    c. conidium.

_____10. Yeast participates in the brewing of beer by
    a. adding vital amino acids.   d. producing ethyl alcohol.
    b. fermenting grain sugars.   e. converting barley to hops.
    c. fermenting fruit sugars.

_____11. If you eat just any mushroom that you find in the wild, there's a chance that you will
    a. die.   d. not become nauseated or die.
    b. become intoxicated.   e. ingest the hallucinogenic drug psilocybin.
    c. see colors that aren't really there.

_____12. A fungus infection throughout the body obtained by exposure to bird droppings is likely
    a. an autoimmune response.   d. ergotism.
    b. St. Anthony's fire.   e. candidiasis.
    c. histoplasmosis.

_____13. Chytrids (aka chytridiomycetes)
    a. are fungi.   d. have flagellated spores.
    b. are funguslike protists.   e. are among the latest in their kingdom to evolve.
    c. inhabit damp or wet environments.

VISUAL FOUNDATIONS

Color the parts of the illustration below as indicated.

RED	❑	haustoria
GREEN	❑	hypha
YELLOW	❑	epidermal cells
BROWN	❑	spore
VIOLET	❑	stoma

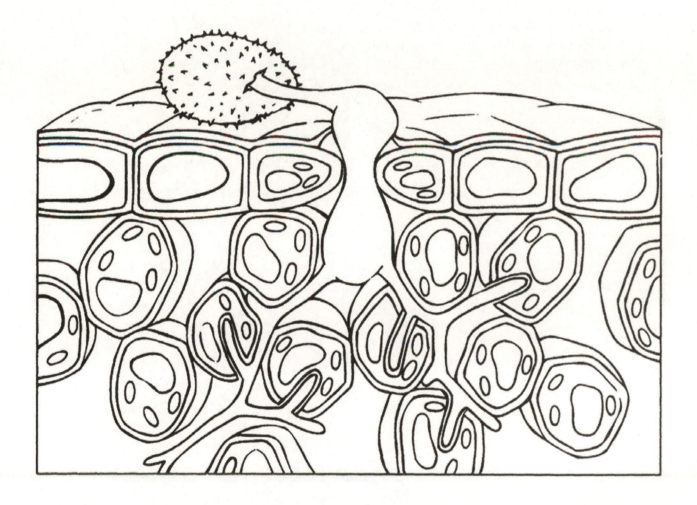

Color the parts of the illustration below as indicated. Also label asexual reproduction, sexual reproduction, and the diploid and haploid generations.

RED	☐	gamete type A
GREEN	☐	gamete type B
YELLOW	☐	haploid zoospore
BLUE	☐	diploid thallus
ORANGE	☐	haploid thallus
BROWN	☐	motile zygote
TAN	☐	resting sporangium
VIOLET	☐	diploid zoospore

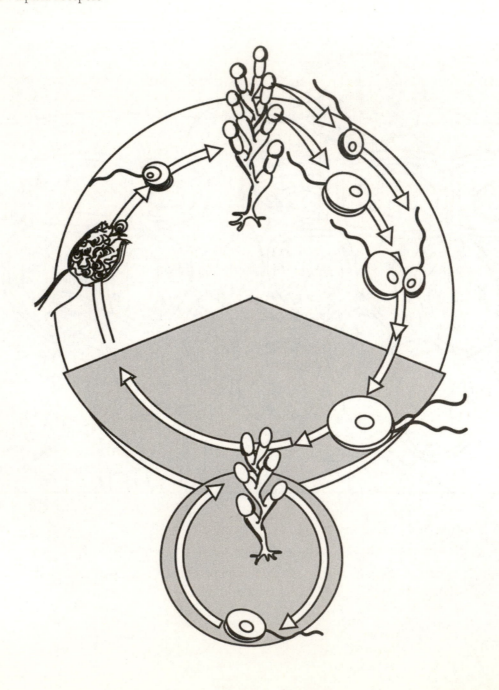

CHAPTER 27

The Plant Kingdom: Seedless Plants

Plants, believed to have evolved from green algae, are multicellular and exhibit great diversity in size, habitat, and form. They photosynthesize to obtain their energy. To survive on land, plants evolved 1) a waxy surface layer that protects against water loss, 2) openings in the surface layer that allow for gas exchange, 3) a strengthening polymer in the cell walls that enables them to grow tall, and 4) multicellular sex organs that protect the developing embryo. The plant kingdom consists of four major groups — bryophytes, seedless vascular plants, gymnosperms, and flowering plants. Except for the bryophytes, all plants have a vascular system. A vascular system allows plants to achieve larger sizes because water and nutrients can be transported over great distances to all parts of the plant. The focus of this chapter is on the seedless plants, i.e., the bryophytes and the ferns and their allies. Unlike the bryophytes, the ferns and their allies also have a larger and more dominant diploid stage, which is a trend in the evolution of land plants. Most of the seedless plants produce only one kind of spore as a result of meiosis. A few, however, produce two types of spores, an evolutionary development that led to the evolution of seed plants.

REVIEWING CONCEPTS
Fill in the blanks.

INTRODUCTION

1. Because of the numerous characteristics that they share, plants are thought to have evolved from ancient forms of _____.

2. Plants and the group from which they are thought to be derived have the same photosynthetic pigments, which are
   (a)_____ and they
   both store excess carbohydrates as (b)_____.

ADAPTATIONS OF PLANTS

3. The adaptation of plants to land environments involved a number of important adaptations. Among them was the (a)_____ that covers aerial parts, protecting the plant against water loss. Openings in this covering also developed. These (b)_____ facilitate the gas exchange that is necessary for photosynthesis.

4. The sex organs of plants, or _____, as they are called, are multicellular structures containing gametes.

The plant life cycle alternates haploid and diploid generations

5. In plants, the male gametes form in the (a)_____, and the female gametes in the (b)_____.

6. Plants have alternation of generations, spending part of their life cycle in the haploid (gametophyte) stage and part in the diploid (sporophyte) stage. The first stage in the sporophyte generation is the
(a)_____, and the haploid (b)_____ are the first stage of the gametophyte generation.

Four major groups of plants evolved

7. There are four major groups of plants. Of these, only the
(a)_____ lack a vascular (conducting) system. The two groups that reproduce and disperse primarily via spores are the (b)_____.
The (c)_____ have exposed seeds on a stem or in a cone, while the (d)_____ produce seeds within a fruit.

8. _____ is a polymer in the cell walls of large, vascular plants that strengthens and supports the plant and its conducting tissues.

BRYOPHYTES

9. Bryophytes can be separated into three groups, known in the vernacular as the
_____.

Moss gametophytes are differentiated into "leaves" and "stems"

10. Each moss plant has tiny, hairlike absorptive structures called
_____.

11. Among the most important mosses are those with large water storage vacuoles in their cells. These mosses are in the genus (a)_____, and are known by their common name, the (b)_____.

Liverwort gametophytes are either thalloid or leafy

12. Liverworts reproduce sexually and asexually. In sexual reproduction, the haploid gametangia are called _____.

13. In asexual reproduction, the thallus may simply branch and grow. Another form of asexual reproduction involves the formation of tiny balls of tissue called _____ on the thallus, which can grow into a new liverwort thallus.

Hornwort gametophytes are inconspicuous thalloid plants

14. Hornworts belong to the phylum _____ and live in disturbed habitats such as fallow fields and roadsides.

Bryophytes are used for experimental studies

15. Botanists use certain bryophytes as experimental models to study
_____ , i.e., plant responses to varying periods of night and day length.

Details of bryophyte evolution are based on fossils and on structural and molecular evidence

16. Fossil evidence indicates that _____ are probably the first group of plants to arise from the common plant ancestor.

SEEDLESS VASCULAR PLANTS

17. Ferns and fern allies have two basic forms of leaves: the small
(a)_____ with its single vascular strand and the larger
(b)_____ with multiple vascular strands.

Club mosses are small plants with rhizomes and short, erect branches

18. Club mosses were major contributors to our present-day _____
_____.

Ferns are a diverse group of spore-forming vascular plants

19. In ferns, the sporophyte is composed mainly of an underground stem called the
(a)_____, from which extend roots and leaves called (b)_____.

20. (a)_____, the spore cases on the leaves of ferns, often occur in
clusters called (b)_____. The mature fern gametophyte is a tiny, green, often
heart-shaped structure called a (c)_____.

21. The main organs of photosynthesis in Psilotum are the
(a)_____ that exhibit (b)_____ branching
(dividing into two equal halves).

22. The small haploid gametophyte of the whisk fern is a nonphotosynthetic
subterranean plant that apparently obtains nourishment through a symbiotic
relationship with a _____.

Some ferns and club mosses are heterosporous

23. Bryophytes, horsetails, whisk ferns, and most ferns and club mosses produce only
one type of spore, a condition known as (a)_____. Some ferns and
club mosses exhibit (b)_____, the production of two different
types of spores.

24. The strobilus of the club moss Selaginella has two kinds of sporangia. The small
(a)_____ produce "microspore mother cells" or
(b)_____ that undergo meiosis to form tiny, haploid
(c)_____ that develop into sperm-producing male gametophytes.
Egg-producing female gametophytes develop from
(d)_____.

Seedless vascular plants are used for experimental studies

25. The _____ is the area at the tip of the root or
shoot where growth occurs.

Seedless vascular plants arose more than 420 mya

26. Microscopic spores of early vascular plants appear in the fossil record earlier than
_____, suggesting that even older megafossils
of simple vascular plants may be discovered.

BUILDING WORDS

Use combinations of prefixes and suffixes to build words for the definitions that follow.

Prefixes	The Meaning		Suffixes	The Meaning
arch(e)-	primitive		-angi(o)(um)	vessel, container
bryo-	moss		-gon(o)(ium)	sexual, reproductive
gamet(o)-	sex cells, eggs and sperm		-phyte	plant
hetero-	different, other		-spor(e)(y)	spore
homo-	same			
mega-	large, great			
micro-	small			
spor(o)-	spore			
xantho-	yellow			

Prefix	Suffix	Definition
_____	-phyll	1. Yellow plant pigment.
_____	_____	2. Special structure (container) of plants, protists, and fungi in which gametes are formed.
_____	_____	3. The female reproductive organ in primitive land plants.
_____	_____	4. The gamete-producing stage in the life cycle of a plant.
_____	_____	5. The spore-producing stage in the life cycle of a plant.
_____	-phyll	6. A small leaf that contains one vascular strand.
_____	-phyll	7. A large leaf that contains multiple vascular strands.
_____	_____	8. Special structure (container) of certain plants and protists in which spores and sporelike bodies are produced.
_____	_____	9. Production of one type of spore in plants (all spores are the same type).
_____	_____	10. Production of two different types of spores in plants, microspores and megaspores.
_____	_____	11. Large spore formed in a megasporangium.
_____	_____	12. Small spore formed in a microsporangium.
_____	_____	13. Mosses, liverworts, and their relatives.

MATCHING

Terms:

a. Antheridium   f. Protonema   k. Strobilus
b. Dichotomous   g. Rhizoid   l. Thallus
c. Lignin   h. Rhizome   m. Xylem
d. Phloem   i. Sorus
e. Prothallus   j. Stoma

For each of these definitions, select the correct matching term from the list above.

_____ 1. Vascular tissue that conducts dissolved organic molecules in plants.

_____ 2. A strengthening polymer found in the walls of cells that function for support and conduction..

_____ 3. Hair-like absorptive structures similar in function to roots that extend from the base of the stem of mosses, liverworts, and fern prothallia.

_____ 4. A cluster of sporangia (in the ferns).

_____ 5. A type of branching in which the branches or veins always branch into two more or less equal parts.

_____ 6. The heart-shaped, haploid gametophyte plant found in ferns, whisk ferns, club mosses, and horsetails.

_____ 7. The male gametangium in certain plants.

_____ 8. A conelike structure that bears sporangia.

_____ 9. Vascular tissue that conducts water and dissolved minerals in plants.

_____10. A horizontal underground stem.

MAKING COMPARISONS

Fill in the blanks.

Plant	Nonvascular or Vascular	Method of Reproduction	Dominant Generation
Hornworts	Nonvascular	Seedless, reproduce by spores	Gametophyte
Angiosperms	#1	Seeds enclosed within a fruit	#2
Ferns	#3	Seedless, reproduce by spores	#4
Club mosses	#5	Seedless, reproduce by spores	#6
Gymnosperms	#7	#8	Sporophyte
Mosses	#9	#10	#11
Horsetails	Vascular	#12	Sporophyte
Whisk ferns	Vascular	#13	#14

MAKING CHOICES

Place your answer(s) in the space provided. Some questions may have more than one correct answer.

_____ 1. Spores grow
    a. into gametophyte plants.
    b. into sporophyte plants.
    c. into a haploid plant.
    d. to form a plant body by meiosis.
    e. to form a plant body by mitosis.

_____ 2. Plants all have
    a. chlorophyll a.
    b. chlorophyll b.
    c. xanthophyll.
    d. carotene.
    e. yellow pigments.

_____ 3. A strobilus is
    a. on a diploid plant.
    b. on a haploid plant.
    c. on a vascular plant.
    d. found on horsetails.
    e. found on plants that do not have true leaves.

_____ 4. The gametophyte generation of a plant
    a. is diploid.
    b. is haploid.
    c. produces haploid spores.
    d. produces haploid gametes by mitosis.
    e. produces haploid gametes by meiosis.

_____ 5. When a gamete produced by an archegonium fuses with a gamete produced by a male gametophyte plant, the result is
    a. an anomaly.
    b. a diploid zygote.
    c. the first stage in the sporophyte generation.
    d. the first stage in the gametophyte generation.
    e. called fertilization.

_____ 6. Plant sperm cells form in
    a. diploid gametophyte plants.
    b. haploid sporophyte plants.
    c. haploid gametophyte plants.
    d. antheridia.
    e. archegonia.

_____ 7. The leafy green part of a moss is the
    a. sporophyte generation.
    b. gametophyte generation.
    c. rhizoid.
    d. product of buds from a protonema.
    e. thallus.

_____ 8. The sporophyte generation of a plant
    a. is diploid.
    b. is haploid.
    c. produces haploid gametes.
    d. produces haploid spores by mitosis.
    e. produces haploid spores by meiosis.

_____ 9. Liverworts
    a. are bryophytes.
    b. are vascular plants.
    c. can produce archegonia and antheridia on a haploid gametophyte.
    d. contain a medicine that cures liver disease.
    e. are in the same class as hornworts.

_____ 10. The spore cases on a fern are
    a. usually on the fronds.
    b. formed by the haploid generation.
    c. called sporangia.
    d. often arranged in a sorus.
    e. precursors to the fiddlehead.

____11. The leaves of vascular plants that evolved from stem branches

    a. are megaphylls.

    b. are microphylls.

    c. evolutionarily derived from stem tissue.

    d. contain one vascular strand.

    e. contain more than one vascular strand.

____12. Land plants are thought to have evolved from

    a. fungi.

    b. green algae.

    c. bryophytes.

    d. charophytes.

    e. mosses.

VISUAL FOUNDATIONS

Color the parts of the illustration below as indicated. Also label haploid gametophyte generation, diploid sporophyte generation, fertilization, and meiosis.

RED	☐	zygote
GREEN	☐	archegonium
YELLOW	☐	egg
BLUE	☐	sperm
ORANGE	☐	antheridium

BROWN	☐	capsule
TAN	☐	gametophyte plants
PINK	☐	sporophyte
VIOLET	☐	spore

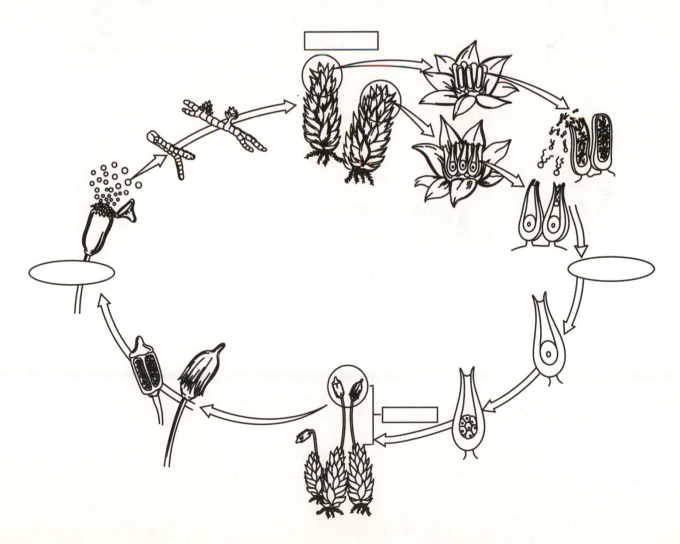

Color the parts of the illustration below as indicated.

RED ☐ zygote

GREEN ☐ archegonium

YELLOW ☐ egg

BLUE ☐ sperm

BROWN ☐ sporangium

TAN ☐ gametophyte

PINK ☐ mature sporophyte

VIOLET ☐ spores

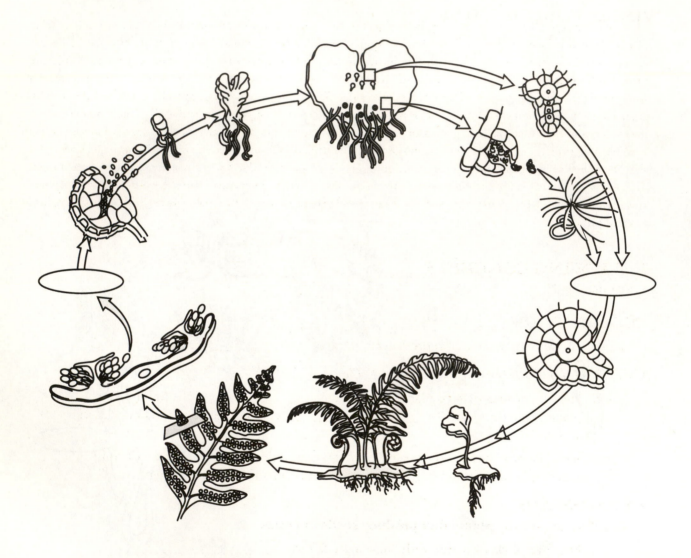

CHAPTER 28

The Plant Kingdom: Seed Plants

The most successful plants on earth are the seed plants, also known as the gymnosperms and angiosperms (flowering plants). Whereas gymnosperm seeds are totally exposed or borne on the scales of cones, those of angiosperms are encased within a fruit. All seeds are multicellular, consisting of an embryonic root, stem, and leaves. They are protected by a seed coat that allows them to survive until conditions are favorable for germination, and they contain a food supply that nourishes the seed until it becomes self-sufficient. Seed plants have vascular tissue, and they exhibit alternation of generations in which the gametophyte generation is significantly reduced. They produce both microspores and megaspores. There are four phyla of gymnosperms and one of angiosperms. Seeds and seed plants have been intimately connected with the development of human civilization. Human survival, in fact, is dependent on angiosperms, for they are a major source of human food. Angiosperm products play a major role in world economics. The organ of sexual reproduction in angiosperms is the flower. Flowering plants have several evolutionary advancements that account for their success. The fossil record indicates that seed plants evolved from seedless vascular plants.

REVIEWING CONCEPTS

Fill in the blanks.

INTRODUCTION

1. The primary means of reproduction in plants is by _____.

AN INTRODUCTION TO SEED PLANTS

2. The two groups of seed plants are the (a)_____, from the Greek "naked seed," and the (b)_____, meaning "seed enclosed in a vessel or case."

3. Seed plants all have two types of vascular tissues: (a)_____ for conducting water and minerals and (b)_____ for conducting dissolved sugar.

GYMNOSPERMS

Conifers are woody plants that produce seeds in cones

4. The four phyla of plants with "naked seeds" are: _____
   _____.

5. The conifers are the largest group of gymnosperms. Most of them are _____, which means that the male and female reproductive organs are in different places on the same plant.

6. A few examples of the most familiar conifers include _____
   _____.

7. The leaflike structures that bear sporangia on male cones of conifers are (a)_____, at the base of which are two microsporangia containing many "microspore mother cells," also called (b)_____. These cells develop into the (c)_____.

8. Female cones have (a)_____ on the upper surface of each cone scale, within which are megasporocytes. Each of these cells produces four (b)_____, three of which disintegrate, while the fourth develops into the egg producing (c)_____.

Cycads have seed cones and compound leaves

9. Cycads are in the phylum _____.

10. The cycads are palmlike or fernlike gymnosperms that reproduce in a manner similar to pines, except cycads are _____, meaning that male cones and female cones are on *separate* plants.

Ginkgo biloba is the only living species in its phylum

11. *Ginko biloba* is in the phylum _____.

Gnetophytes include three unusual genera

12. Gnetophytes belong to the phylum _____.

13. The gnetophytes share a number of advances over the rest of the gymnosperms, one of which is the presence of distinct _____ in their xylem tissues.

FLOWERING PLANTS

Monocots and eudicots are the two largest classes of flowering plants

14. A few examples of monocots include _____
_____.

15. A few examples of eudicots include _____
_____.

16. The monocots have floral parts in multiples of (a)_____, and their seeds contain (b)_____(#?) cotyledon(s). The nutritive tissue in their mature seeds is the (c)_____.

17. The eudicots have floral parts in multiples of (a)_____, and their seeds contain (b)_____(#?) cotyledons. The nutritive tissue in their mature seeds is usually in the (c)_____.

Flowers are involved in sexual reproduction

18. The four main organs in flowers are the (a)_____
_____, each of them arranged in whorls. The "male" organs are the (b)_____ and the "female" organs are the (c)_____. Flowers with both "male" and "female" parts are said to be (d)_____.

19. Pollen forms in the (a)_____, a saclike structure on the tip of a stamen, and ovules form within the (b)_____.

The life cycle of flowering plants includes double fertilization

20. The megasporocyte in an ovule produces four haploid (a)_____, three of which disintegrate while the remaining one develops into the female gametophyte or (b)_____, as it is also called.

21. Microsporocytes in the anther produce four haploid (a)_____, each of which develops into a (b)_____.

22. Double fertilization is a phenomenon that is unique to flowering plants. It results in the formation of two structures, namely, the _____ _____.

Seeds and fruits develop after fertilization

23. As a seed develops from an ovule following fertilization, the ovary wall surrounding it enlarges dramatically and develops into a _____.

Flowering plants have many adaptations that account for their success

24. Pollen transfer results in (a)_____, which mixes the genetic material and promotes (b)_____ among the offspring.

Studying how flowers evolved provides insights into the evolutionary process

25. Many botanists have concluded that stamens and carpels are probably derived from _____.

THE EVOLUTION OF SEED PLANTS

26. (a)_____ is an extinct group of plants descended from ancestral seedless vascular plants. This group probably gave rise to conifers and to another group of extinct plants called the (b)_____, which in turn are thought to be the ancestors of cycads and possibly ginkgo.

Our understanding of the evolution of flowering plants has made great progress in recent years

27. Flowering plants evolved from _____.

BUILDING WORDS

Use combinations of prefixes and suffixes to build words for the definitions that follow.

Prefixes	The Meaning
angio-	vessel, container
di-	two, twice, double
gymno-	naked, bare, exposed
mon(o)-	alone, single, one

Prefix	Suffix	Definition
_____	-sperm	1. A plant having its seeds enclosed in an ovary (a "vessel" or "container").
_____	-sperm	2. A plant having its seeds exposed or naked.

_____   -oecious   3. Pertains to plant species in which the male and female reproductive parts are in different locations on one plant.

_____   -oecious   4. Pertains to plant species in which the male and female reproductive parts are on two different plants.

MATCHING

Terms:

a. Anther
b. Calyx
c. Carpel
d. Corolla
e. Endosperm

f. Fertilization
g. Ovary
h. Ovule
i. Petals
j. Pistil

k. Pollination
l. Sepals
m. Style

For each of these definitions, select the correct matching term from the list above.

_____ 1. The part of the stamen in flowers that produces microspores and, ultimately, pollen.

_____ 2. The nutritive tissue that is found at some point in all flowering plant seeds.

_____ 3. The union of a male gamete and a female gamete to produce a zygote.

_____ 4. The colored cluster of modified leaves that constitute the next-to-outermost portion of a flower.

_____ 5. In seed plants, the transfer of pollen from the male to the female part of the plant.

_____ 6. The neck connecting the stigma to the ovary of a carpel.

_____ 7. The part of a plant that develops into seed after fertilization.

_____ 8. The outermost parts of a flower, usually leaflike in appearance, that protect the flower as a bud.

_____ 9. The collective term for the sepals of a flower.

_____10. The female reproductive unit of a flower that bears the ovules.

MAKING COMPARISONS

Fill in the blanks.

Plant Features	Monocot	Eudicot
Vascular bundles in stems	Scattered vascular bundles	Vascular bundles arranged in a circle
Growth tissue type	Herbaceous	#1
Nutritive material in mature seeds	#2	#3
Leaf shape	Long, narrow leaves	#4
Seeds	#5	2 cotyledons
Flowers	Floral parts in multiples of 3	#6
Leaf venation	#7	#8

MAKING CHOICES

Place your answer(s) in the space provided. Some questions may have more than one correct answer.

_____ 1. The major difference between gymnosperms and angiosperms is that in gymnosperms
a. self fertilization occurs.
b. seeds are exposed ("naked").
c. there is no ovary wall surrounding ovules.
d. flowers generally have 4 or 5 sepals and petals.
e. the mature ovary is a fruit.

_____ 2. The petals of a flower are collectively known as the
a. corolla.
b. calyx.
c. gynoecium.
d. androecium.
e. florus perfecti.

_____ 3. The sepals of a flower are collectively known as the
a. corolla.
b. calyx.
c. gynoecium.
d. androecium.
e. florus perfecti.

_____ 4. The triploid endosperm of angiosperms develops from fusion of
a. two sperm and one polar nucleus.
b. two polar nuclei and one sperm.
c. three polar nuclei.
d. one diploid ovule and one haploid sperm.
e. one haploid egg and one diploid sperm.

_____ 5. Which of the following is/are correct about a pine tree?
a. Sporophyte generation is dominant.
b. Male gametophyte produces an antheridium.
c. Gametophyte is dependent on sporophyte for nourishment.
d. Nutritive tissue in seed is gametophyte tissue.
e. Pollen grain is an immature male gametophyte.

_____ 6. A perfect flower has
a. stamens only.
b. carpels only.
c. both stamens and carpels.
d. anthers.
e. ovules.

_____ 7. In gymnosperms, the pollen grain develops from
a. microspore cells.
b. spores.
c. the male gametophyte.
d. the gametophyte generation.
e. meiosis of cells in the microsporangium.

_____ 8. Dioecious plants with naked seeds, motile sperm, and whose pollen is carried by air or insects are
a. ginkgoes.
b. cycads.
c. extinct.
d. gnetophytes.
e. gymnosperms.

_____ 9. A pine tree has
a. sporophylls.
b. megasporangia on the female cones.
c. separate male and female parts on the same tree.
d. two sizes of spores in separate cones.
e. female cones that are larger than male cones.

_____ 10. In angiosperms, of the four haploid megaspores that result from meiosis of the megaspore mother cell, one of them becomes the
a. male gametophyte generation.
b. female gametophyte generation.
c. embryo sac.
d. archegonium.
e. antheridium.

VISUAL FOUNDATIONS

Color the parts of the illustration below as indicated. Also label haploid gametophyte generation, diploid sporophyte generation, fertilization, and meiosis.

RED	☐	zygote
GREEN	☐	female cone
YELLOW	☐	megasporangium, megaspore,
BLUE	☐	pollen grain
ORANGE	☐	microsporangium, microspore

BROWN	☐	male cone
TAN	☐	sporophyte
PINK	☐	gametophyte
VIOLET	☐	embryo

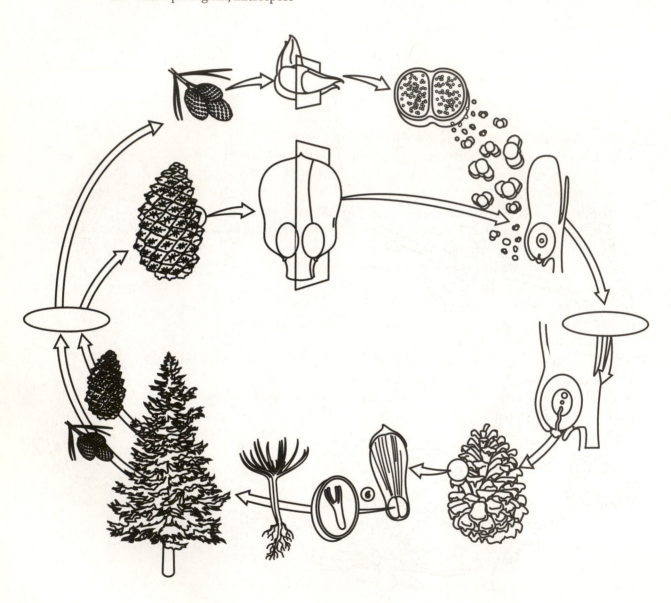

Color the parts of the illustration below as indicated. Also label haploid gametophyte generation, diploid sporophyte generation, double fertilization, and meiosis.

RED ☐ zygote

GREEN ☐ female floral part

YELLOW ☐ megasporangium, megaspore

BLUE ☐ pollen tube

ORANGE ☐ microsporangium, microspore

BROWN ☐ male floral part

TAN ☐ male and female sporophytes

PINK ☐ male and female gametophytes

VIOLET ☐ embryo

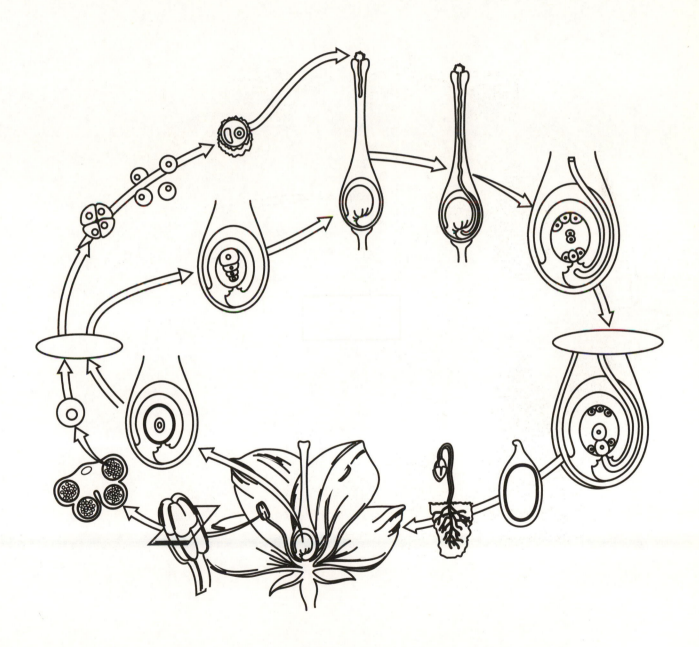

CHAPTER 29

The Animal Kingdom: An Introduction to Animal Diversity

This is the first of three chapters discussing the animal kingdom. Animals are eukaryotic, multicellular, heterotrophic organisms. They are made up of localized groupings of cells with specialized functions. One grouping, for example, might function in locomotion, another might respond to external stimuli, and another might function in sexual reproduction. Animals inhabit virtually every environment on earth. They may be classified in a variety of ways — as radially symmetrical or bilaterally symmetrical; as those with no body cavity (the acoelomates), a body cavity between mesoderm and endoderm (the pseudocoelomates), or a body cavity within mesoderm (the coelomates); or those in which the first opening that forms in the embryonic gut becomes either the mouth (the protostomes) or the anus (the deuterostomes). This chapter discusses the three phyla that diverged early in the evolutionary history of the animal kingdom: the Parazoa (sponges), the Cnidaria (hydras, jelly fish, sea anemones, and corals), and the Ctenophora (comb jellies).

REVIEWING CONCEPTS

Fill in the blanks.

INTRODUCTION

1. An estimated _____(%?) of all animal species that ever inhabited our planet are extinct.

ANIMAL CHARACTERS

2. Animals are _____, or consumers that depend on producers for their raw materials and energy.

ADAPTATIONS TO HABITATS

Marine habitats offer many advantages

3. Of the three major environments, the most hospitable to the majority of animals is _____.

Some animals are adapted to freshwater habitats

4. _____ environments are hypotonic to tissue fluids.

Terrestrial living requires major adaptations

5. Land environments tend to _____ animals.

ANIMAL ORIGINS

Molecular systematics helps biologists interpret the fossil record

6. The rapid appearance of an amazing variety of body plans in the fossil record is known as the _____.

Biologists develop hypotheses about the evolution of development

7. Mutations in _____ genes could have resulted in rapid changes in body plans.

RECONSTRUCTING ANIMAL PHYLOGENY

Animals exhibit two main types of body symmetry

8. There are many anatomical terms that are used to describe the locations of body parts. Among them are: (a)_____ for front and (b)_____ for rear; (c)_____ for a back surface and (d)_____ for the under or "belly" side.

9. Other locations of body parts include (a)_____ for parts closer to the midline and (b)_____ for parts farther from the midline, towards the sides; (c)_____ toward the head end; and (d)_____ when away from the head, toward the tail.

Animal body plans are linked to the level of tissue development

10. In the early development of all animals except sponges, cells form layers called _____ that give rise to linings and tissues.

Biologists group animals according to type of body cavity

11. Animals can also be grouped according to the presence or absence of a body cavity and, when present, the type of body cavity. For example, the "lowly" flatworms are called _____ because they do not have a body cavity.

12. Animals with body cavities are either (a)_____, possessing a body cavity of sorts between the mesoderm and endoderm, or (b)_____ with a true body cavity within the mesoderm.

Bilateral animals form two main groups based on differences in development

13. Coelomate animals can be grouped by shared developmental characteristics. In mollusks, annelids, and arthropods — all protostomes — the blastopore develops into the (a)_____, whereas in echinoderms and chordates — both deuterostomes — the blastopore becomes the (b)_____.

14. Early cell divisions in deuterostomes are either parallel to or at right angles to the polar axis, a pattern of division known as (a)_____; in protostomes, early cell divisions are diagonal to the polar axis, a type of division referred to as (b)_____.

Biologists have identified major animal groups based on structure

15. Animals that have true tissues are classified as _____.

Molecular data contribute to our understanding of animal relationships

16. One important innovation of body forms has been _____, a body plan in which certain structures are repeated producing body compartments.

THE PARAZOA: SPONGES
Collar cells characterize sponges

17. Because they have collar cells, sponges are thought to have evolved from
    _____.

18. The sponge is a cellular sac perforated by numerous "pores." In simple sponges,
    water enters the internal cavity or (a)_____ through these openings,
    then exits through the (b)_____.

19. Between the outer and inner cell layers of the sponge body is a gelatin-like layer
    supported by slender skeletal spikes, or _____.

THE RADIATA: ANIMALS WITH RADIAL SYMMETRY AND TWO CELL LAYERS
Cnidarians have unique stinging cells

20. The phylum Cnidaria is divided into three classes: Hydras and hydroids are in the
    class (a)_____, jellyfish are in (b)_____, and sea
    anemones and corals are in (c)_____.

21. The epidermis of hydra contains specialized "stinging cells" called
    _____.

22. Stinging "thread capsules" in the stinging cells of hydra are called
    _____. These capsules release a long thread that entraps prey.

23. Cnidarians are characterized by radial symmetry; stinging cells; and two definite
    tissue layers, the outer (a)_____ and the inner
    (b)_____ separated by a gelatinous (c)_____;
    and a nerve net.

24. Among the jellyfish, members of the Scyphozoa, the _____ stage is the
    dominant body form.

25. The _____ stage is small and inconspicuous, and may even be absent.

26. Anthozoan polyps produce eggs and sperm, and the fertilized egg develops into a
    small, ciliated larva called a _____.

Comb Jellies have adhesive glue cells that trap prey

27. Ctenophores are like Cnidarians in that they consist of two cell layers separated
    by a thick, jellylike _____.

BUILDING WORDS
Use combinations of prefixes and suffixes to build words for the definitions that follow.

Prefixes	The Meaning	Suffixes	The Meaning
deutero-	second	-coel(om)(y)	cavity
ecto-	outer, outside, external	-stome	mouth
entero-	intestine		
gastro-	stomach		
meso-	middle		
proto-	first, earliest form of		
pseudo-	false		
schizo-	split		

Prefix	Suffix	Definition
_____	_____	1. A body cavity between the mesoderm and endoderm; not a true coelom.
_____	-derm	2. The outermost germ layer.
_____	-derm	3. The middle layer of the three basic germ layers.
_____	_____	4. Major division of the animal kingdom in which the mouth forms from the first opening in the embryonic gut (the blastopore); the anus forms secondarily.
_____	_____	5. Major division of the animal kingdom in which the mouth forms from the second opening in the embryonic gut; the first opening (the blastopore) forms the anus.
_____	_____	6. The process of coelom formation in which the mesoderm splits into two layers.
_____	_____	7. The process of coelom formation in which the mesoderm forms as "outpocketings" of the developing intestine, eventually separating and forming pouches that become the coelom.
spongo-	_____	8. The central cavity in the body of a sponge.
_____	-dermis	9. The tissue lining the gut cavity ("stomach") that functions in digestion in certain phyla.

MATCHING

Terms:

a. Acoelomate
b. Auricle
c. Cephalization
d. Coelom
e. Collar cell

f. Cuticle
g. Dorsal
h. Hermaphroditic
i. Invertebrate
j. Nematocyst

k. Pseudocoelom
l. Sessile
m. Ventral

For each of these definitions, select the correct matching term from the list above.

_____ 1. A body cavity not completely lined with mesoderm.

_____ 2. Possessing sex organs of both the male and the female.

_____ 3. Without a body cavity (coelom).

_____ 4. The main body cavity of most animals.

_____ 5. A unique cell having a flagellum surrounded by a collar of microvilli.

_____ 6. The clustering of neural tissues at the anterior (leading) end of an animal.

_____ 7. Referring to the belly aspect of an animal's body.

_____ 8. A stinging structure in cnidarians.

_____ 9. Permanently attached to one location.

MAKING COMPARISONS
Fill in the blanks.

Phylum	Examples	Body Plan	Key Characteristics
Porifera	Sponges	Asymmetrical; sac-like body with pores, central cavity, and osculum	Choanocytes
Cnidaria	#1	Radial symmetry; diploblastic; gastrovascular cavity with one opening	Tentacles with cnidocytes (stinging cells); polyp and medusa body forms;
Ctenophora	Comb jellies	#2	Eight rows of cilia; tentacles with adhesive glue cells
#3	Flatworms, tape worms, planarians, flukes	#4	No body cavity; some cephalization
#5	Clams, snails, squids	Biradial symmetry; triploblastic; organ systems; complete digestive tube	#6
Annelida	#7	Biradial symmetry; triploblastic; organ systems; complete digestive tube	#8
#9	Qheel animals	Biradial symmetry; triploblastic; organ systems; complete digestive tube	#10
Nematoda	Roundworms: Ascaris, hookworms, trichina worms	Biradial symmetry; triploblastic; organ systems; complete digestive tube	#11
Arthropoda	#12	Biradial symmetry; triploblastic; organ systems; complete digestive tube	#13
#14	Starfish, sea urchins, sand dollars	#15	Endoskeleton; water vascular system; tube feet
Hemichordata	Acorn worms	Biradial symmetry; triploblastic; organ systems; complete digestive tube	#16
#17	Tunicates, lancelets, vertebrates	Biradial symmetry; triploblastic; organ systems; complete digestive tube	#18

MAKING CHOICES

Place your answer(s) in the space provided. Some questions may have more than one correct answer.

_____ 1. Marine carnivores with a proboscis that can be everted to capture prey are
    a. members of Nemertea.   d. known as ribbon worms.
    b. are protostomes.   e. members of Mollusca.
    c. are deuterostomes.

_____ 2. The taxon that has biradial symmetry, tentacles with adhesive glue cells, and eight rows of cilia is
    a. Scyphozoa.   d. Cnidaria.
    b. Hydrozoa.   e. Porifera.
    c. Ctenophora.

_____ 3. Examples of acoelomate eumetazoans include
    a. sponges.   d. those groups whose coelom developed from a blastocoel.
    b. insects.   e. cnidarians.
    c. flatworms.

_____ 4. The phylum/phyla that does/do not have specialized nerve cells is/are
    a. Echinodermata.   d. Nemertea.
    b. Cnidaria.   e. Porifera.
    c. Platyhelminthes.

_____ 5. When a group of cells moves inward to form a sac in early embryonic development, and that pore ultimately develops into a mouth, the group of animals is considered to be a
    a. protostome.   d. parazoan.
    b. deuterostome.   e. schizocoelomate.
    c. pseudocoelomate.

_____ 6. The one invertebrate phylum with radial symmetry and distinctly different tissues is
    a. Porifera.   d. Nemertea.
    b. Cnidaria.   e. Echinodermata.
    c. Platyhelminthes.

_____ 7. The roundworms
    a. are pseudocoelomates.   d. include hookworms.
    b. have a mastax.   e. are nematodes.
    c. are in the phylum Platyhelminthes.

_____ 8. Sessile, marine animals with no medusa stage and a partitioned gastrovascular cavity include
    a. hydrozoans.   d. jellyfish.
    b. scyphozoans.   e. corals, sea anemones, and their close relatives.
    c. anthozoans.

_____ 9. The taxon that has "thread capsules" in epidermal stinging cells is
    a. Scyphozoa.   d. Cnidaria.
    b. Hydrozoa.   e. Ctenophora.
    c. Porifera.

_____ 10. Deuterostomes characteristically have
    a. radial cleavage.   d. schizocoely.
    b. spiral cleavage.   e. enterocoely.
    c. indeterminate cleavage.

VISUAL FOUNDATIONS

Color the parts of the illustration below as indicated. Label the animals depicted as acoelomate, pseudocoelomate, or coelomate.

RED	☐	mesoderm
GREEN	☐	endoderm
YELLOW	☐	coelom
BROWN	☐	pseudocoelom
BLUE	☐	ectoderm
ORANGE	☐	mesenchyme

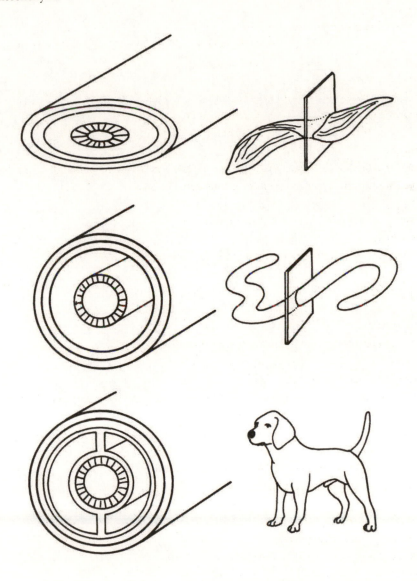

The Animal Kingdom: The Protostomes

This chapter discusses the protostomes, animals with coeloms in which the first opening that forms in the embryonic gut gives rise to the mouth. Protostomes have a complete digestive tract, and most have well-developed circulatory, excretory, and nervous systems. The coelom confers advantages to animals over those without a coelom. For one thing, it separates muscles of the body wall from those of the digestive tract, permitting movement of food independent of body movements. Also, it serves as a space in which organs develop and function, and the fluid it contains helps transport food, oxygen, and wastes. Additionally, it can be used as a hydrostatic skeleton, permitting a greater range of movement than in acoelomate animals and conferring shape to the body of soft animals. The coelomate protostomes include the Lophotrochozoa (platyhelminthes, nemerteans, mollusks, annelids, and the lophophorate phyla) and the Ecdysozoa (nematodes and arthropods).

REVIEWING CONCEPTS

Fill in the blanks.

INTRODUCTION

1. More than _____(%?) of animal species belong to the Bilateria.

IMPORTANCE OF THE COELOM

2. The coelom is completely lined with _____.

3. Among the many advantages derived from having a coelom are:

   _____

   _____

   _____.

THE LOPHOTROCHOZOA

Flatworms are bilateral acoelomates

4. Flatworms are characterized by (a)_____ symmetry, cephalization, (b)_____(#?) definite tissue layers, well-developed organs, a simple nervous system, and a brain consisting of two masses of nervous tissue called (c)_____.

5. The four classes of Platyhelminthes are (a)_____, the free-living flatworms; (b)_____, the two classes of flukes; and (c)_____, the tapeworms.

6. The organs of chemoreception on planarians that resemble "ears" are called _____.

7. Planarians trap prey in a mucous secretion, then extend the _____ outward to "vacuum" them into their digestive tube.

8. Most tapeworms have suckers and/or hooks on the _____ for attachment to their hosts.

9. In tapeworms, each segment, or _____, contains both male and female reproductive organs.

Phylum Nemertea is characterized by the proboscis

10. The proboscis organ is a long, hollow, muscular tube that can be rapidly _____.

Mollusks have a muscular foot, visceral mass, and mantle

11. Mollusks are soft-bodied animals usually covered by a shell; they possess a ventral foot for locomotion and a (a)_____ that covers the visceral mass. The phylum Mollusca is second only to phylum (b)_____ in number of fossil species.

12. Most mollusks have an open circulatory system, meaning the blood flows through a network of open sinuses called the blood cavity or _____.

13. Most marine mollusks have larval stages, the first a free-swimming, ciliated form called a (a)_____ larva, and the second, when present, a (b)_____ larva with a shell, foot, and mantle.

14. A distinctive feature of chitons is their shell, which is composed of _____(#?) overlapping plates.

15. Chitons are in the class _____.

16. Class Gastropoda, the largest and most successful group of mollusks, includes the snails, slugs, and their relatives. It is a class second in size only to the _____.

17. The gastropod shell (when present) is coiled, and the visceral mass is twisted, a phenomenon known as _____.

18. Foreign matter lodged between the shell and (a)_____ of bivalves may cause deposition of the compound (b)_____, forming a pearl.

19. The _____ is the part of a scallop that we humans eat.

20. Class Cephalopoda includes the squids and octopods, which are active predatory animals. The foot is divided into (a)_____ that surround the mouth, (b)_____(#?) of them in squids, (c)_____(#?) of them in octopods.

21. The term Cephalopoda literally means _____.

Annelids are segmented worms

22. In polychaetes and earthworms, body segments are separated from one another by partitions called (a)_____. Also, each body segment bears a pair of bristle-like structures called (b)_____ that function to anchor the body during locomotion.

23. Class Polychaeta consists of marine worms characterized by bristled _____, used for locomotion and gas exchange.

24. The term Polychaeta literally means _____.

25. The common earthworm is in the phylum (a)_____, the class (b)_____, and the genus and species (c)_____.

26. The two parts of the earthworm's "stomach" are the (a)_____, where food is stored, and the (b)_____, where ingested materials are ground into bits.

27. The respiratory pigment in earthworms is (a)_____; the excretory system consists of paired (b)_____ in most segments; and the gas exchange surface is the (c)_____.

The lophophorate phyla are distinguished by a ciliated ring of tentacles

28. There are three lophophorate phyla: (a)_____, or lampshells; (b)_____, wormlike sessile animals; and (c)_____, or microscopic aquatic animals.

Rotifers have a crown of cilia

29. Protonephridia with _____ remove excess water from the body and may also excrete metabolic wastes.

THE ECDYSOZOA

Roundworms are of great ecological importance

30. Members of the phylum Nematoda play key ecological roles as _____ _____.

Arthropods are characterized by jointed appendages and an exoskeleton of chitin

31. The word "Arthropoda" means (a)_____, which is one of the phylum's distinguishing characteristics. Another is the (b)_____, an armor-like body covering composed of (c)_____.

32. Gas exchange in terrestrial arthropods occurs in a system of thin, branching tubes called _____.

33. The members of the class (a)_____, or "centipedes," have (b)_____ pair(s) of legs per body segment.

34. The members of the class (a)_____, or "millipedes," have (b)_____(#?) pair(s) of legs per body segment.

35. The subphylum of arthropods that lacks antennae and includes horseshoe crabs and arachnids is _____.

36. Subphylum Chelicerata includes the (a)_____ (the horseshoe crabs) and the (b)_____ (spiders, mites, and their relatives).

37. The arachnid body consists of a (a)_____ and abdomen; there are (b)_____(#?) pairs of jointed appendages.

38. Gas exchange in arachnids occurs in tracheae and/or across thin, vascularized plates called _____.

39. Silk glands in spiders secrete an elastic protein that is spun into web fibers by organs called _____.

40. The subphylum _____ includes lobsters, crabs, shrimp, and barnacles.

41. Crustaceans are characterized by the (a)_____, the third pair of appendages used for biting food; (b)_____ appendages; and (c)_____(#?) pairs of antennae.

42. _____ are the only sessile crustaceans.

43. Lobsters, crayfish, crabs, and shrimp are members of the largest crustacean order, the _____.

44. Among the many highly specialized appendages in decapods are the (a)_____, two pairs of feeding appendages just behind the mandibles; the (b)_____ that are used to chop food and pass it to the mouth; the (c)_____ or pinching claws on the fourth thoracic segments; and the pairs of (d)_____ on the last four thoracic segments.

45. Appendages on the first abdominal segment of decapods are part of the (a)_____ system. (b)_____ on the next four abdominal segments are paddle-like structures adapted for swimming and holding eggs.

46. Insects are described as articulated, meaning (a)_____, and tracheated, meaning (b)_____, hexapods, which means (c)_____.

47. Adult insects typically have (a)_____(#?) pairs of legs, (b)_____(#?) pairs of wings, and (c)_____(#?) pair(s) of antennae. The excretory organs are called (d)_____.

BUILDING WORDS

Use combinations of prefixes and suffixes to build words for the definitions that follow.

Prefixes	The Meaning		Suffixes	The Meaning
arthro-	joint, jointed		-pod	foot, footed
bi-	twice, two			
cephalo-	head			
exo-	outside, outer, external			
hexa-	six			
tri-	three			
uni-	one			

Prefix	Suffix	Definition
_____	-skeleton	1. An external skeleton, such as the shell of arthropods.
_____	-valve	2. A mollusk that has two shells (valves) hinged together, as the oyster, clam, or mussel; a member of the class Bivalvia.
_____	-ramous	3. Consisting of or divided into two branches.
_____	-ramous	4. Unbranched; consisting of only one branch.
_____	-lobite	5. An extinct marine arthropod characterized by two dorsal grooves that divide the body into three longitudinal parts (or lobes).
_____	-thorax	6. Anterior part of the body in certain arthropods consisting of the fused head and thorax.

_____ _____   7. Having six feet; an insect.

_____ _____   8. An animal with paired, jointed legs.

MATCHING

Terms:

a. Chelicerae
b. Coelom
c. Hemocoel
d. Malpighian tubule

e. Mandible
f. Metamorphosis
g. Molting
h. Radula

i. Tracheae
j. Trochophore

For each of these definitions, select the correct matching term from the list above.

____ 1. Network of large spaces where body tissues are directly bathed in blood.

____ 2. A rasplike structure in the mouth region of certain mollusks.

____ 3. The first pair of appendages in arthropods.

____ 4. The shedding and replacement of the arthropod exoskeleton.

____ 5. Transition from one developmental stage to another, such as from a larva to an adult.

____ 6. A fluid-filled space lined by mesoderm that lies between the digestive tube and the outer body wall.

____ 7. Internal branching air tubes throughout the body of many terrestrial arthropods and some terrestrial mollusks.

____ 8. A larval form found in mollusks and many polychaetes.

____ 9. A mouth part in certain arthropods.

MAKING COMPARISONS

Fill in the blanks.

Category	Mollusks	Annelids	Arthropods
Representative animals	Clams, snails, squid	Earthworms, marine worms	Crustaceans, insects
Level of organization	#1	#2	Organ system
Nervous system	3 pairs of ganglia, sense organs	#3	#4
Circulation	Open system, closed in cephalopods	#5	#6
Digestion	#7	Complete digestive tract	#8
Reproduction	Sexual: separate sexes	#9	Sexual: separate sexes
Life style	Herbivores, carnivores, scavengers, suspension feeders	#10	Herbivores, carnivores, scavengers

MAKING CHOICES

Place your answer(s) in the space provided. Some questions may have more than one correct answer.

_____ 1. The excretory organ(s) in earthworms is/are the
   a. parapodia.   d. lophophores.
   b. kidneys.   e. crop.
   c. metanephridia.

_____ 2. The so-called "wheel animals"
   a. are unicellular.   d. move by spinning head over tail, hence the name "wheel animals".
   b. are generally aquatic.   e. are rotifers.
   c. have no cell mitosis after completing embryonic development.

_____ 3. Roundworms are
   a. nematodes.   d. annelids.
   b. mollusks.   e. cephalopods.
   c. chilopods.

_____ 4. Earthworms are
   a. nematodes.   d. annelids.
   b. mollusks.   e. cephalopods.
   c. chilopods.

_____ 5. The animal that has a shell consisting of eight separate, overlapping dorsal plates is
   a. an insect.   d. in the class Polyplacophora.
   b. a chiton.   e. a mollusk.
   c. in the same class as squids.

_____ 6. Earthworms are held together in copulation by mucous secretions from the
   a. clitellum.   d. epidermis.
   b. typhlosole.   e. prostomium.
   c. seminal receptacles.

_____ 7. Earthworms are
   a. annelids.   d. protostomes.
   b. oligochaets.   e. acoelomate animals.
   c. in the phylum Polychaeta.

_____ 8. Butterflies, termites, and black widows all
   a. are insects.   d. have closed circulatory systems.
   b. are arthropods.   e. have an abdomen.
   c. have paired, jointed appendages.

_____ 9. A marine animal with parapodia and a trochophore larva might very well be
   a. a tubeworm.   d. a leech.
   b. an oligochaet.   e. a polychaet.
   c. bilaterally symmetrical.

_____10. An octopus is a

    a. vertebrate.   d. protostome.

    b. mollusk.   e. cephalopod.

    c. chilopod.

_____11. Animals with most of the organs in a visceral mass that is covered by a mantle are

    a. annelids.   d. mollusks.

    b. oligochaets.   e. cheliceratans.

    c. onychophorans.

_____12. Clams, scallops, and oysters are

    a. nematodes.   d. annelids.

    b. mollusks.   e. cephalopods.

    c. chilopods.

_____13. Insects are

    a. hexapods.   d. in the same subphylum as lobsters and crabs.

    b. arthropods.   e. in the same subphylum as spiders.

    c. chelicerates.

_____14. Spiders are

    a. hexapods.   d. in the same subphylum as lobsters and crabs.

    b. arthropods.   e. in the same subphylum as insects.

    c. chelicerates.

VISUAL FOUNDATIONS

Color the parts of the illustration below as indicated.

RED ☐ digestive tract

GREEN ☐ shell

YELLOW ☐ foot

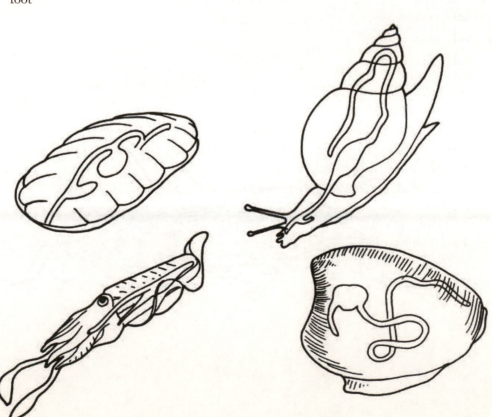

Color the parts of the illustration below as indicated.

RED ☐ dorsal and ventral blood vessels
GREEN ☐ metanephridium
YELLOW ☐ nerve cord
BLUE ☐ cerebral ganglia
ORANGE ☐ pharynx
BROWN ☐ intestine
TAN ☐ coelom

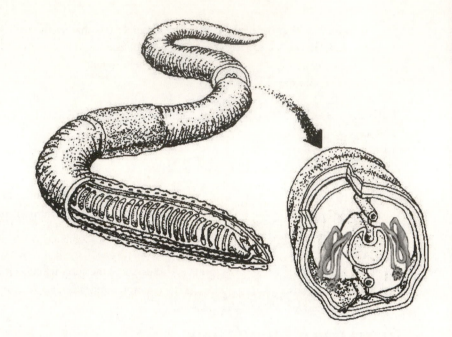

Color the parts of the illustration below as indicated.

RED ☐ heart
GREEN ☐ Malpighian tubules
YELLOW ☐ brain, nerve cord
BLUE ☐ ovary
ORANGE ☐ digestive gland
BROWN ☐ intestine

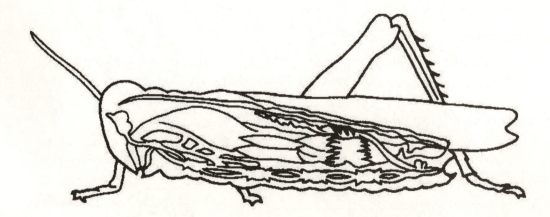

CHAPTER 31

The Animal Kingdom: The Deuterostomes

This chapter discusses the deuterostomes, animals with coeloms in which the second opening that develops in the embryo becomes the mouth, and the first becomes the anus. The deuterostomes include the echinoderms, hemichordates, and the chordates. The echinoderms and the chordates are by far the more ubiquitous and dominant deuterostomes. The echinoderms all live in the sea. They have spiny skins, a water vascular system, and tube feet. The chordates include animals that live both in the sea and on land. They have a notochord, a dorsal tubular nerve cord, and pharyngeal gill slits, at least at some time in their life cycle. There are three groups of chordates: the tunicates, the lancelets, and the vertebrates. The vertebrates are characterized by a vertebral column, a cranium, pronounced cephalization, a differentiated brain, muscles attached to an endoskeleton, and two pairs of appendages. There are 10 classes of living vertebrates: 6 of fishes and 4 of animals with four limbs. The fishes include the jawless, cartilaginous, and bony fishes, and, the four-limbed vertebrates include the amphibians, reptiles, birds, and mammals. Mammals are classified first on the basis of whether embryonic development occurs within an egg, a maternal pouch, or an organ of exchange, i.e., a placenta.

REVIEWING CONCEPTS

Fill in the blanks.

INTRODUCTION

1. The two major phyla assigned to the deuterostomes are the
   _____.

WHAT ARE DEUTEROSTOMES?

2. Deuterostomes are characterized by radial cleavage. The cleavage is
   _____, which means the fate of their cells is fixed
   later in development than is the case in protostomes

ECHINODERMS

3. Echinoderm larvae have (a)_____ symmetry; most adults have
   (b)_____ symmetry. Echinoderms also have a true
   (c)_____ that houses the internal organs.

Members of class Crinoidea are suspension feeders

4. Class Crinoidea includes sea lilies and feather stars. Unlike other echinoderms,
   the oral surface of crinoids is located on the _____ surface.

Many members of class Asteroidea capture prey

5. Sea stars have a _____ from which radiate five or more arms or
   rays.

6. The arms or rays on sea stars contain hundreds of small _____, each of
   which contains a canal that leads to a central canal.

Class Ophiuroidea is the largest class of echinoderms

7. Ophiuroids tube feet are used to (a)_____ and not for (b)_____.

Members of class Echinoidea have movable spines

8. Class Echinoidea includes the sea urchins and sand dollars, animals that lack arms; they have a solid shell called a _____, and their body is covered with spines.

9. Sea urchins use their tube feet for _____.

Members of class Holothuroidea are elongated, sluggish animals

10. The circulatory system of sea cucumbers functions to _____ _____.

CHORDATE CHARACTERS

11. Chordates are deuterostome coelomates with _____ symmetry.

12. The characteristics that distinguish chordates from all other groups are the dorsal, flexible rod called the (a)_____; the single, dorsal, hollow, and tubular (b)_____; the embryonic (c)_____ that develop into functional respiratory structures in aquatic chordates and entirely different structures in terrestrial chordates; and the postanal tail.

INVERTEBRATE CHORDATES

Tunicates are common marine animals

13. Larval tunicates have typical chordate characteristics and superficially resemble _____.

Lancelets may be closely related to vertebrates

14. Subphylum Cephalochordata consists of the lancelets. A common genus, Branchiostoma, also known as _____, is often selected as a representative, typical chordate.

15. The oral hood is a vestibule of the digestive tract just anterior to the mouth. Food particles swept into the mouth are trapped in mucus in the _____, then passed back to the intestine.

Systematists are making progress in understanding chordate phylogeny

16. Conodonts were simple fishlike chordates with fins, large eyes, and complex toothlike hooks that were probably used for _____.

INTRODUCING THE VERTEBRATES

The vertebral column is a key vertebrate character

17. Subphylum Vertebrata includes animals with a backbone or (a)_____ _____, a braincase or (b)_____, and concentration of nerve cells and sense organs in a definite head, a phenomenon known as (c)_____.

Vertebrate taxonomy is a work in progress

18. The extant vertebrates are currently assigned to (a)_____(#?) classes of fishes and (b)_____(#?) of tetrapods. The four classes of tetrapods are (c)_____.

JAWLESS FISHES

19. Some of the earliest known vertebrates, the _____, consisted of several groups of small, armored, jawless fishes that lived on the ocean floor and strained their food from the water.

20. Contemporary hagfishes and lampreys have neither jaws nor paired _____.

EVOLUTION OF JAWS AND LIMBS: JAWED FISHES AND AMPHIBIANS

Members of class Chondrichthyes are cartilaginous fishes

21. The skin of cartilaginous fish contains numerous _____, toothlike structures composed of layers of enamel and dentine.

22. (a)_____ are sensory grooves along the sides of all fish that contain cells capable of perceiving water movements. Similarly, the (b)_____ on the shark's head can sense very weak electrical currents, like those generated by another animal's muscle contractions.

23. Sharks may be (a)_____, that is, lay eggs; or (b)_____, which means their eggs hatch internally; or (c)_____, giving live birth to their young.

The ray-finned fishes gave rise to modern bony fishes

24. In the Devonian period, bony fishes diverged into two major groups: the (a)_____ fish, or actinopterygians, and the (b)_____ with their fleshy, lobed fins and lungs.

25. The actinopterygians gave rise to the modern bony fish in which lungs became modified as _____.

Descendants of the lungfishes moved onto the land?

26. A Tiktaalik was a transitional form between _____ _____ and lived 375 mya.

Amphibians were the first successful land vertebrates

27. The three orders of modern amphibians are: (a)_____, the amphibians with long tails, including salamanders, mud puppies, and newts; the (b)_____, tailless frogs and toads; and (c)_____, the wormlike caecilians.

28. Amphibians use lungs and their (a)_____ for gas exchange. They have a (b)_____(#?)-chambered heart with systemic and pulmonary circulations, and they have (c)_____ in the skin that help keep the body surface moist.

AMNIOTES

Our understanding of amniote phylogeny is changing

29. Diapsid amniotes have (a)_____ pairs of openings in the temporal bones whereas synapsid amniotes have (b)_____ pair.

Reptiles have many terrestrial adaptations

30. Reptiles are true terrestrial animals; fertilization is internal; most secrete a protective shell around the egg; the embryo develops an amnion and other extraembryonic membranes. Most reptiles have a (a)_____(#?)-chambered heart, and are (b)_____, which means they cannot regulate body temperature.

We can assign extant reptiles to four groups

31. Reptiles have been assigned to four orders. These are the _____ _____.

Are birds really dinosaurs?

32. One of the earliest known birds, _____, was about the size of a pigeon, had jawbones armed with reptilian-type teeth, and had a long reptilian tail covered with feathers.

Some dinosaurs had feathers

33. Insulating feathers have contributed to the evolution of _____, permitting animals to be more active.

Modern birds are adapted for flight

34. Adaptations for flight in birds include feathers, wings, and light hollow bones. Birds also have a (a)_____(#?)-chambered heart, very efficient lungs, a high metabolic rate, and they are (b)_____, meaning they can maintain a constant body temperature. They excrete wastes as semi-solid (c)_____. Birds also have a well-developed nervous system and excellent vision and hearing. Birds are the only animals that have (d)_____.

35. The saclike portion of a bird's digestive system that temporarily stores food is the (a)_____, whereas the (b)_____ portion of the stomach secretes gastric juices, and the (c)_____ grinds the food.

Mammals are characterized by hair and mammary glands

36. The distinguishing characteristics of mammals are _____ _____. They also maintain a constant body temperature, and they have a highly developed nervous system and a muscular diaphragm.

37. Mammals probably evolved from (a)_____, a reptilian group, about 200 million years ago in the (b)_____ period.

38. _____ are mammals that lay eggs. They include the duck-billed platypus and the spiny anteater.

39. (a)_____ are pouched mammals. The young are born immature and complete development in the (b)_____, where they are nourished from the mammary glands.

BUILDING WORDS

Use combinations of prefixes and suffixes to build words for the definitions that follow.

Prefixes	The Meaning		Suffixes	The Meaning
a-	without		-derm	skin
Chondr(o)-	cartilage		-gnath(an)	jaw
endo-	within		-ichthy(es)	fish
echino-	sea urchin, "spiny"		-pod(al)	foot, footed
tetra-	four		-ur(o)(a)	tail
Uro-	tail			

Prefix	Suffix	Definition
_____	_____	1. A fish without jaws (jawless fish); a member of the class of vertebrates including lampreys and hagfishes.
An-	_____	2. An order of amphibia with legs but no tail; tailless frogs or toads.
_____	_____	3. Pertains to those amphibians with no feet, i.e., the wormlike caecilians.
_____	_____	4. The class comprising the cartilaginous fishes.
_____	_____	5. A vertebrate with four legs; a member of the superclass Tetrapoda.
_____	-dela	6. The order that comprises the amphibians with long tails, e.g., the salamanders, mudpuppies, and newts.
_____	_____	7. A spiny-skinned animal.
_____	-skeleton	8. Bony and cartilaginous-supporting structures within the body that provide support from within.

MATCHING

Terms:

a. Amnion
b. Amphibian
c. Ctylosaur
d. Cotylosaur
e. Labyrinthodont

f. Marsupium
g. Neoteny
h. Notochord
i. Oviparous
j. Ovoviviparous

k. Placenta
l. Placoderm
m. Therapsid
n. Vertebrate

For each of these definitions, select the correct matching term from the list above.

_____ 1. The dorsal, longitudinal rod that serves as an internal skeleton in the embryos of all, and in the adults of some, chordates.

_____ 2. The pouch in which marsupial young develop.

_____ 3. Animals that are egg-layers.

_____ 4. An extra-embryonic membrane that forms a fluid-filled sac for the protection of the developing embryo.

_____ 5. An extinct jawed fish.

_____ 6. A group of mammal-like reptiles of the Permian period that gave rise to the mammals.

_____ 7. An organ of exchange between developing embryo and mother in eutherian mammals.

_____ 8. A chordate possessing a bony vertebral column.

MAKING COMPARISONS

Fill in the blanks.

Vertebrate Class	Representative Animals	Heart	Skeletal Material	Other Characteristics
Myxini	Hagfish	Two-chambered heart	Cartilage	Jawless, most primitive vertebrates, gills,
Chondrichthyes	#1	Two-chambered heart	#2	#3
#4	Salmon, tuna	Two-chambered heart	#5	Gills, swim bladder
Amphibia	#6	#7	Bone	Tetrapods, aquatic larva metamorphoses into terrestrial adult, moist skin and lungs for gas exchange, ectothermic
#8	Birds	#9	#10	Amniotes with feathers, many adaptations for flight, endothermic, high metabolic rate
#11	Monotremes, marsupials, placentals	Four-chambered heart	#12	#13

MAKING CHOICES

Place your answer(s) in the space provided. Some questions may have more than one correct answer.

_____ 1. Humans are members of the taxon(a)

    a. Lagomorpha.   d. that includes placental mammals.

    b. Rodentia.   e. Cetacea.

    c. that has chisel-like incisors that grow continually.

_____ 2. The phylum(a) that many biologists believe had a common ancestry with our own phylum is/are

    a. Mollusca.   d. Echinodermata.

    b. Annelida.   e. Chordata.

    c. Arthropoda.

_____ 3. The group(s) of animals that maintain a constant internal body temperature is/are

    a. Reptilia.   d. Amphibia.

    b. Mammalia.   e. Pisces.

    c. Aves.

_____ 4. The group(s) of animals with a part of the stomach that secretes gastric juices and a separate part of the stomach that grinds food is/are

a. Reptilia.   d. Amphibia.

b. Mammalia.   e. Pisces.

c. Aves.

_____ 5. A turtle is a member of the taxon(a)

a. Reptilia.   d. Amphibia.

b. Chelonia.   e. Anura.

c. Urodela.

_____ 6. Mammals probably evolved from

a. amphibians.   d. monotremes.

b. a type of fish.   e. therapsids.

c. reptiles.

_____ 7. The lateral line organ is found

a. only in sharks.   d. in all fishes.

b. only in bony fish.   e. in all fish except placoderms.

c. only in agnathans.

_____ 8. Members of the phylum Chordata all have

a. an embryonic notochord.   d. well-developed germ layers.

b. embryonic pharyngeal gill slits.   e. bilateral symmetry.

c. a dorsal, tubular nerve cord.

_____ 9. The "spiny-skinned" animals include

a. sea stars.   d. lancets.

b. sand dollars.   e. marine worms.

c. snails.

_____10. The acorn worm

a. has a notochord.   d. is in the phylum Chordata.

b. is a deuterostome.   e. is a hemichordate.

c. is in the same subphylum as *Amphioxus*.

_____11. The group(s) of animals with both four-chambered hearts and a double circuit of blood flow is/are the

a. earthworms.   d. class Aves.

b. birds.   e. amphibians.

c. mammals.

_____12. A hermaphroditic animal that traps food in mucus secreted by cells of the endostyle is a/an

a. holothuroidean.   d. sea cucumber.

b. enchinoderm.   e. tunicate.

c. urochordate.

_____13. The first successful land vertebrates were

a. tetrapods.   d. amphibians.

b. reptiles.   e. in the superclass Pisces.

c. chondrichthyes.

_____14. Cats are members of the taxon(a)

a. Lagomorpha.   d. that includes placental mammals.

b. Rodentia.   e. Cetacea.

c. that has chisel-like incisors that grow continually.

_____15. Unique features of the echinoderms include
    a. radial larvae.   d. gas exchange by diffusion.
    b. ciliated larvae.   e. water vascular system.
    c. endoskeleton composed of $CaCO_3$ plates.

_____16. A frog is a member of the taxon(a)
    a. Reptilia.   d. Amphibia.
    b. Chelonia.   e. Anura.
    c. Urodela.

VISUAL FOUNDATIONS

Color the parts of the illustration below as indicated. Also label postanal tail.

RED ☐ heart
GREEN ☐ mouth
YELLOW ☐ brain and dorsal, hollow nerve tube
BLUE ☐ notochord
ORANGE ☐ pharyngeal gill slits
PINK ☐ pharynx, intestine
TAN ☐ muscular segments

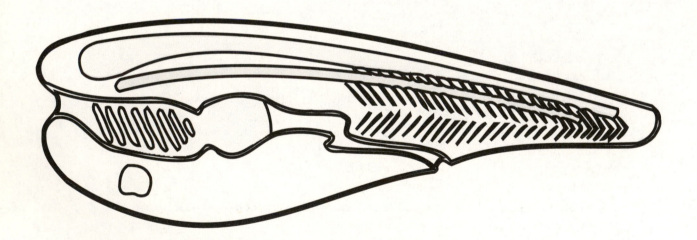

PART VI

Structure and Life Processes in Plants

CHAPTER 32

Plant Structure, Growth, and Differentiation

This is the first of six chapters addressing the integration of plant structure and function. This chapter examines the external structure of the flowering plant body; the organization of its cells, tissues, and tissue systems; and its basic growth patterns. The plant body is organized into a root system and a shoot system. Plants are composed of cells that are organized into tissues, and tissues that are organized into organs. Each organ performs a single function or group of functions, and is dependent on other organs for its survival. All vascular plants have three tissue systems: the dermal tissue system, the vascular tissue system, and the ground tissue system. The tissue systems are composed of various simple and complex tissues. Plants grow by increasing both in girth and length. Unlike growth in animals, plant growth is localized in areas of unspecialized cells, and involves cell proliferation, elongation, and differentiation.

REVIEWING CONCEPTS

Fill in the blanks.

INTRODUCTION

1. Vascular plants characterized by flowers, double fertilization, endosperm, and seeds enclosed within fruits are called _____.

PLANT STRUCTURE AND LIFE SPAN

2. Plants characteristically have similar body plans consisting of three main parts, namely _____.

3. Herbaceous plants that grow, reproduce, and die in one year are called _____.

Plants have different life history strategies

4. Characteristic features of an organism's life cycle, particularly as they relate to its survival and successful reproduction, are known as _____ _____.

THE PLANT BODY

The plant body consists of cells and tissues

5. A (a)_____ is a group of cells that form a structural and functional unit. (b)_____ are composed of only one kind of cell whereas (c)_____ have two or more kinds of cells.

6. Roots, stems, leaves, flower parts, and fruits are _____ because each is composed of all three tissue systems.

The ground tissue system is composed of three simple tissues

7. The three primary functions of parenchyma tissue include _____.

8. Collenchyma tissue is composed of living cells that function primarily to _____ the plant.

9. The two types of sclerenchyma cells are _____.

10. Cellulose microfibrils are cemented together by a matrix of _____.

The vascular tissue system consists of two complex tissues

11. Xylem conducts (a)_____ from roots to stems and leaves. Xylem also contains (b)_____ for storage and (c)_____ for support.

12. Phloem is a complex tissue that functions to conduct _____ throughout the plant.

13. Highly specialized cells in phloem, called (a)_____, conduct food in solution. The cells' end walls, called (b)_____, have a series of holes through which cytoplasm extends from one sieve tube element into the next.

The dermal tissue system consists of two complex tissues

14. The dermal tissue system, the (a) _____, provides a protective covering over plant parts. Epidermal cells secrete a waxy layer called the (b)_____ that restricts water loss. (c)_____ are openings in this layer through which gases diffuse.

15. The periderm replaces the epidermis in the stems and roots of older woody plants. Its primary function is for _____.

16. The epidermis contains special outgrowths, or hairs, called _____, which occur in many sizes and shapes and have a variety of functions.

PLANT MERISTEMS

17. Plant growth is localized in regions called (a)_____, and involves three cell activities, namely, (b)_____.

18. Plants have two kinds of meristematic growth: (a)_____ growth, an increase in length; and (b)_____ growth, an increase in the girth of the plant.

Primary growth takes place at apical meristems

19. The root apical meristem consists of three zones, which, in order from the root cap inward, are the a)_____,
(b)_____, and
(c)_____.

20. (a)_____(developing leaves) and
(b)_____ (developing buds) arise from the shoot apical meristem.

Secondary growth takes place at lateral meristems

21. There are two lateral meristems responsible for secondary growth (increase in girth), the (a)_____, a layer of meristematic cells that forms a long, thin, continuous cylinder within the stem and root, and the

(b)_____, a thin cylinder or irregular arrangement of meristematic cell in the outer bark.

22. The outermost covering over woody stems and roots is the _____.

BUILDING WORDS

Use combinations of prefixes and suffixes to build words for the definitions that follow.

Prefixes	The Meaning
bi-	two
epi-	on, upon
stom-	mouth
trich-	hair

Prefix	Suffix	Definition
_____	-ome	1. A hair or other special outgrowth growing out from the epidermis of plants.
_____	-ennial	2. A plant that takes two years to complete its life cycle.
_____	-dermis	3. Along with the periderm, provides a protective covering on the surface of plants.
_____	-a	4. A small pore ("mouth") in the epidermis of plants.

MATCHING

Terms:

a.	Apical meristem	f.	Meristem	k.	Sclerenchyma
b.	Companion cell	g.	Parenchyma	l.	Shoot system
c.	Dermal tissue	h.	Perennial	m.	Tracheid
d.	Ground tissue	i.	Periderm	n.	Vascular cambium
e.	Lateral meristem	j.	Root system	o.	Xylem

For each of these definitions, select the correct matching term from the list above.

_____ 1. Vascular tissue that conducts water and dissolved minerals through the plant.

_____ 2. The chief type of water-conducting cell in the xylem of gymnosperms.

_____ 3. An area of dividing tissue located at the tips of roots and shoots in plants.

_____ 4. Secondary meristem that produces the secondary xylem and secondary phloem.

_____ 5. The type of tissue system that gives rise to the epidermis.

_____ 6. Plant cell that has thick secondary walls, is dead at maturity and functions in support.

_____ 7. A nucleated cell in the phloem responsible for loading and unloading sugar into the sieve tube member.

_____ 8. A plant that lives longer than two years.

_____ 9. General term for all localized areas of mitosis and growth in the plant body.

_____10. Plant cells that are relatively unspecialized, are thin walled, and function in photosynthesis and in the storage of nutrients.

MAKING COMPARISONS

Fill in the blanks.

Tissue	Tissue System	Function	Location in Plant Body
Parenchyma	Ground tissue	Photosynthesis, storage, secretion	Throughout plant body
Collenchyma	#1	Flexible structural support	#2
#3	Ground tissue	#4	Throughout plant body; common in stems and certain leaves, some nuts and pits of stone fruit
#5	#6	Conducts water, dissolved minerals	Extends throughout plant body
Phloem	Vascular tissue	#7	#8
Epidermis	#9	#10	#11
#12	Dermal tissue	#13	Covers body of woody plants

MAKING CHOICES

Place your answer(s) in the space provided. Some questions may have more than one correct answer.

_____ 1. Plants with the potential for living more than two years are called
　　a. annuals.　　　　　　　　　d. perennials.
　　b. biennials.　　　　　　　　e. polyannuals.
　　c. triennials.

_____ 2. The most common type of cell and tissue found throughout the plant body is the
　　a. parenchyma.　　　　　　　d. vascular tissue.
　　b. sclerenchyma.　　　　　　e. xylem.
　　c. collenchyma.

_____ 3. The kind of growth that results in an increase in the girth of the plant is known as
　　a. differentiation.　　　　　d. primary growth.
　　b. elongation.　　　　　　　e. secondary growth.
　　c. apical meristem growth.

_____ 4. All plant cells have
　　a. cell walls.　　　　　　　　d. primary cell walls.
　　b. secondary growth.　　　　e. secondary cell walls.
　　c. the capacity to form a complete plant.

_____ 5. Hairlike outgrowths of plant epidermis are called
　　a. periderm.　　　　　　　　d. trichomes.
　　b. fiber elements.　　　　　e. companion cells.
　　c. lateral buds.

_____ 6. The types of cells in phloem include
  a. tracheids.
  b. vessel elements.
  c. parenchyma.
  d. sieve tube elements.
  e. fiber cells.

_____ 7. The cell type found throughout the plant body that often functions in photosynthesis, secretion, and storage is called
  a. sclerenchyma.
  b. collenchyma.
  c. parenchyma.
  d. companion cells.
  e. a tracheid.

_____ 8. Increase in the girth of a plant is due to growth of the
  a. vascular cambium.
  b. cork cambium.
  c. area of cell maturation.
  d. area of cell elongation.
  e. lateral meristems.

_____ 9. Localized areas of cell division resulting in plant growth are called
  a. primordia.
  b. mitotic zones.
  c. meristems.
  d. primary growth centers.
  e. secondary growth centers.

_____ 10. The types of cells in xylem include
  a. tracheids.
  b. vessel elements.
  c. parenchyma.
  d. sieve tube members.
  e. fiber cells.

VISUAL FOUNDATIONS

Color the parts of the illustration below as indicated. Also label area of cell division and area of cell maturation.

RED ☐ apical meristem

GREEN ☐ area of cell elongation

YELLOW ☐ root cap

ORANGE ☐ root hairs

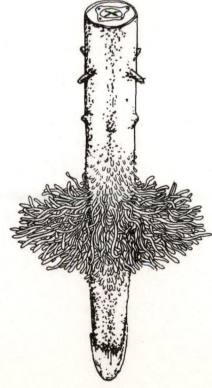

Color the parts of the illustration below as indicated. Also label shoot system, root system, rosette of basal leaves, and leaf.

RED	☐	flower
GREEN	☐	blade
YELLOW	☐	internode
BLUE	☐	petiole
ORANGE	☐	node
BROWN	☐	taproot
TAN	☐	stem
VIOLET	☐	axillary bud

CHAPTER 33

Leaf Structure and Function

This chapter, the second of six addressing the integration of plant structure and function, discusses the structural and physiological adaptations of leaves. Leaves, as the principal photosynthetic organs in plants, are highly adapted to collect radiant energy, convert it into the chemical bonds of carbohydrates, and transport the carbohydrates to the rest of the plant. They are also adapted to permit gas exchange, and to receive water and minerals transported into them by the plant vascular system. Water loss from leaves is controlled in part by the cell walls of the epidermal cells, by presence of a surface waxy layer secreted by epidermal cells, and by tiny pores in the leaf epidermis that open and close in response to various environmental factors. Most of the water absorbed by land plants is lost, either in the form of water vapor or as liquid water. Survival at low temperatures is facilitated in some plants by the loss of leaves. The leaves of many plants are modified for functions other than photosynthesis.

REVIEWING CONCEPTS

Fill in the blanks.

INTRODUCTION

LEAF FORM AND STRUCTURE

1. The broad, flat portion of a leaf is the (a)_____; the stalk that attaches the blade to the stem is the (b)_____. Some leaves also have (c)_____, which are leaflike outgrowths usually present in pairs at the base of the petiole.

2. Leaves may be (a)_____, having a single blade, or (b)_____, having a blade divided into two or more leaflets.

3. Leaves are arranged on a stem in one of three possible ways. These are
   (a)_____, with one leaf at each node;
   (b)_____, with two leaves at each node; and
   (c)_____, with three or more leaves at each node.

Leaf structure consists of an epidermis, photosynthetic ground tissue, and vascular tissue

4. The photosynthetic ground tissue of the leaf, called the (a)_____, is sandwiched between the upper epidermis and the lower epidermis. When this tissue is divided into two regions, the upper layer, nearest to the upper epidermis, is called the (b)_____, and the lower portion is called the (c)_____.

5. (a)_____ in veins of a leaf conduct water and essential minerals to the leaf, while (b)_____ in veins conducts sugar produced by photosynthesis to the rest of the plant. Veins may be surrounded by a (c)_____, consisting of parenchyma or sclerenchyma cells.

6. _____ venation is characteristic of monocot leaves.

Leaf structure is related to function

7. (a)_____, a raw material of photosynthesis, diffuses into the leaf through stomata, and the (b)_____ produced during photosynthesis diffuses rapidly out of the leaf through stomata.

STOMATAL OPENING AND CLOSING

8. When water moves into guard cells from surrounding cells, they become (a)_____ and bend, producing a pore. When water leaves the guard cells, they become (b)_____ and collapse against one another, closing the pore.

Blue light triggers stomatal opening

9. Light triggers an influx of potassium, malate, and chloride ions into the (a)_____ of (b)_____ cells, thereby increasing internal solute concentration. Water then moves into these cells, changing their shape, causing the pores to (c)_____(open or close?).

10. As evening approaches, the sucrose concentration in the guard cells declines, water leaves by osmosis, the guard cells lose their turgidity, and the pores _____(open or close?).

Additional factors affect stomatal opening and closing

11. Stomates open during the day and close at night. Factors that affect opening and closing of stomata include _____
_____.

TRANSPIRATION AND GUTTATION

12. Two benefits of transpiration are (a) _____ and (b)_____.

13. Transpiration is an important part of the _____ cycle, in which water cycles from the ocean and land to the atmosphere, and then back to the ocean and land.

Some plants exude liquid water

14. Loss of liquid water from leaves by force is known as _____.

LEAF ABSCISSION

15. Leaf abscission is a complex process that involves many physiological changes, all initiated and orchestrated by changing levels of plant hormones, particularly _____.

16. The brilliant colors found in autumn landscapes in temperate climates are due to the various combinations of (a)_____ and (b)_____.

In many leaves, abscission occurs at an abscission zone near the base of the petiole

17. The area where a petiole detaches from the stem is a structurally distinct area called the (a)_____; it is a weak area because it contains relatively few strengthening (b)_____.

18. The "cement" that holds the primary cell walls of adjacent cells together is called the _____.

MODIFIED LEAVES

19. Leaves are variously modified for a plethora of specialized functions. For example, the hard, pointed _____ on a cactus are leaves modified for protection.

20. _____ are specialized leaves that anchor long, climbing vines to the supporting structures on which they are growing.

Modified leaves of carnivorous plants capture insects

21. The traps of the pitcher plant are (a)_____, whereas those of the Venus flytrap are (b)_____.

BUILDING WORDS

Use combinations of prefixes and suffixes to build words for the definitions that follow.

Prefixes	The Meaning
ab-	from, away, apart
circ-	around
meso-	middle
trans-	across, beyond

Prefix	Suffix	Definition
_____	-scission	1. The separation and falling away from the plant stem of leaves, fruit and flowers.
_____	-adian	2. Pertains to something that cycles at approximately 24-hour intervals.
_____	-phyll	3. The photosynthetic tissue of the leaf sandwiched between (in the middle of) the upper and lower epidermis.
_____	-piration	4. The loss of water vapor from the plant body across leaf surfaces.

MATCHING

Terms:

a. Blade
b. Bundle sheath
c. Cuticle
d. Guard cell
e. Guttation

f. Petiole
g. Phloem
h. Spine
i. Stipule
j. Tendril

k. Trichome
l. Vascular bundle
m. Vein
n. Xylem

For each of these definitions, select the correct matching term from the list above.

_____ 1. A ring of cells surrounding the vascular bundle in monocot and dicot leaves.

_____ 2. A leaf or stem that is modified for holding or attaching to objects.

_____ 3. Vascular tissue that transports sugars produced by photosynthesis.

_____ 4. A waxy covering over the epidermis of the above-ground portion of plants.

_____ 5. A hard, pointed leaf that is modified for protection.

_____ 6. The part of a leaf that attaches to a stem.

_____ 7. One of two cells that collectively form a stoma.

_____ 8. In plants, vascular bundles in leaves.

_____ 9. The emergence of liquid water droplets on leaves, forced out through special structures.

MAKING COMPARISONS

Fill in the blanks.

Structure of Leaf	Function of Leaf Structure
Thin, flat shape	Maximizes light absorption, efficient gas diffusion
Ordered arrangement on stem	#1
Waxy cuticle	#2
#3	Allow gas exchange between plant and atmosphere
Relatively transparent epidermis	#4
#5	Allows for rapid diffusion of CO_2 to mesophyll cell surface
#6	Provide support to prevent leaf from collapsing
Xylem in veins	#7
#8	Transports sugar to other plant parts

MAKING CHOICES

Place your answer(s) in the space provided. Some questions may have more than one correct answer.

_____ 1. Loss of water by evaporation from aerial plant parts is called
    a. transpiration.   d. activation.
    b. guttation.   e. aspiration.
    c. abscission.

_____ 2. The "cement" that holds the primary cell walls of adjacent cells together is called the
    a. abscission zone.   d. middle lamella.
    b. leaf scar.   e. glial substance.
    c. adhesion layer.

_____ 3. The portion of mesophyll that is usually composed of loosely and irregularly arranged cells is
    a. the palisade layer.   d. toward the leaf's upperside.
    b. the spongy layer.   e. an area of photosynthesis.
    c. toward the leaf's underside.

_____ 4. The process by which plants secrete water as a liquid is

a. evaporation.

b. transpiration.

c. guttation.

d. the potassium ion mechanism.

e. found only in monocots.

_____ 5. The leaves of eudicots usually have

a. bulliform cells.

b. netted venation.

c. differentiated palisade and spongy tissues.

d. special subsidiary cells.

e. bean-shaped guard cells.

_____ 6. Trichomes are found on/in the

a. palisade layer.

b. spongy layer.

c. entire mesophyll.

d. epidermis.

e. guard cells.

_____ 7. Facilitated diffusion of potassium ions into guard cells

a. requires ATP.

b. closes stomates.

c. causes guard cells to shrink and collapse.

d. occurs more in daylight than at night.

e. indirectly causes pores to open.

_____ 8. In general, leaf epidermal cells

a. are living.

b. lack chloroplasts.

c. protect mesophyll cells from sunlight.

d. are absent on the lower leaf surface.

e. have a cuticle.

_____ 9. Factors that tend to affect the amount of transpiration include

a. the cuticle.

b. wind velocity.

c. ambient temperature.

d. relative humidity.

e. amount of light.

_____ 10. Guard cells generally

a. have chloroplasts.

b. form a pore.

c. are found in the epidermis.

d. are found only in monocots.

e. are found only in dicots.

_____ 11. Red water-soluble pigments in leaves are

a. carotenoids.

b. xanthophylls.

c. anthocyanins.

d. rhodophylls.

e. found only in monocots.

_____ 12. Proton pumps in the plasma membranes of guard cells are activated by

a. blue light.

b. ultraviolet light.

c. red light.

d. wavelengths in the 800–1,000 nm range.

e. wavelengths in the 400–500 nm range.

_____ 13. Water diffuses into guard cells by means of

a. voltage-activated ion channels.

b. proton pumps.

c. active transport.

d. facilitated diffusion.

e. osmosis.

VISUAL FOUNDATIONS

Color the parts of the illustration below as indicated. Also label the location of stomata. Using brackets, indicate the structures on the list below that would be components of the dermal tissue system, the ground tissue system, and the vascular tissue system.

RED ☐ stoma

YELLOW ☐ upper epidermis

ORANGE ☐ lower epidermis

VIOLET ☐ cuticle

BLUE ☐ palisade mesophyll

GREEN ☐ spongy mesophyll

PINK ☐ xylem

PURPLE ☐ phloem

BROWN ☐ bundle sheath

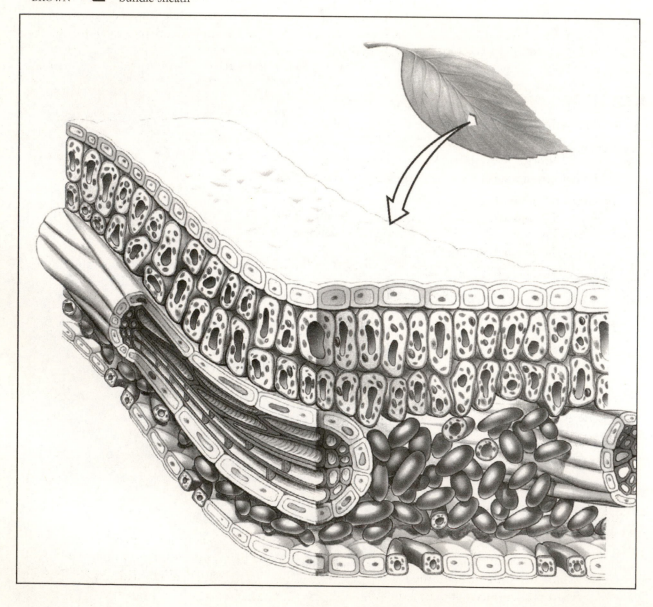

CHAPTER 34

❑

Stems and Transport in Vascular Plants

This chapter, the third of six addressing the integration of plant structure and function, discusses the structural and physiological adaptations of stems. Stems function to support leaves and reproductive structures, conduct materials absorbed by the roots and produced in the leaves to other parts of the plant, and produce new tissue throughout the life of the plant. Specialized stems have other functions as well. Although all herbaceous, or nonwoody, stems have the same basic tissues, the arrangement of the tissues in the monocot stem differs from that in the eudicot stem. The stems of herbaceous plants increase only in length. The stems of all gymnosperms and some eudicots increase in both length and girth. Increases in the length of plants result from mitotic activity in apical meristems, and increases in the girth of plants result from mitotic activity in lateral meristems. Simple physical forces are responsible for the movement of food, water, and minerals in multicellular plants.

REVIEWING CONCEPTS
Fill in the blanks.

INTRODUCTION

1. The three main parts of a vegetative vascular plant are the

    _____.

2. The primary functions of stems are to

    _____.

EXTERNAL STEM STRUCTURE IN WOODY TWIGS

3. All stems have (a)_____, which are embryonic shoots. A
    (b)_____ is the embryonic shoot located at the tip of a stem. The
    dormant apical meristem of a terminal bud is covered and protected by an outer
    layer of (c)_____, which are modified leaves.

4. A _____ shows where each leaf was attached on the stem.

STEM GROWTH AND STRUCTURE
Herbaceous eudicot and monocot stems differ in internal structure

5. Vascular tissues in herbaceous eudicots are located in _____.

6. Each vascular bundle contains two vascular tissues, the
    (a)_____, and a single layer of cells, the
    (b)_____, sandwiched between them.

Woody plants have stems with secondary growth

7. Secondary growth occurs in woody plants as a result of cell divisions in two lateral
    meristems, the (a)_____, which gives rise to secondary
    xylem and phloem, and the (b)_____, which, along with the
    tissues it produces, is called the (c)_____.

8. The vascular cambium produces secondary (a)_____ and secondary (b)_____ to replace the primary conducting and supporting tissues.

9. Lateral transport takes place in (a)_____, which are chains of (b)_____ cells that radiate out from the center of the woody stem or root.

10. Cork cambium cells divide to form new tissues toward the inside and the outside. The layer toward the outside consists of _____, heavily waterproofed cells that protect the plant.

11. To the inside, cork cambium forms the _____ that stores water and starch.

12. The older wood in the center of a tree is called (a)_____, and the younger, lighter-colored, more peripheral wood is (b)_____.

13. Annual rings are composed of two types of cells arranged in alternating concentric circles, each layer appropriately named for the season in which it developed. The (a)_____ has large-diameter conducting cells and few fibers, whereas the (b)_____ has narrower conducting cells and numerous fibers.

TRANSPORT IN THE PLANT BODY

14. In contrast to internal circulation in animals, in plants movement of materials throughout the organism is driven largely by _____.

Water and minerals are transported in xylem

15. One of the principal forces behind water movements through plants is a function of a cell's ability to absorb water by osmosis, also know as the "free energy of water," or the (a)_____. Water containing solutes has (b)_____(more or less?) free energy than pure water. Water moves from a region of (c)_____ water potential to a region of (d)_____ water potential.

16. Under normal conditions, the water potential of the root is more _____ (negative or positive?) than the water potential of the soil. Thus water moves by osmosis from the soil into the root.

17. According to the _____, water is pulled up the plant as a result of a tension produced at the top of the plant.

18. The tension results from the evaporative pull of _____.

Sugar in solution is translocated in phloem

19. Dissolved food, predominantly sucrose, is transported up or down in the phloem. Movement of materials in the phloem is explained by the (a)_____ hypothesis. Sugar is actively loaded into the sieve tubes at the (b)_____, causing water to move into sieve tubes by osmosis, then sugar is actively unloaded from the sieve tubes at the (c)_____, causing water to leave sieve tubes by osmosis.

BUILDING WORDS

Use combinations of prefixes and suffixes to build words for the definitions that follow.

Prefixes	The Meaning	Suffixes	The Meaning
inter-	between, among	-derm	skin
peri-	around		
trans-	across, beyond		

Prefix	Suffix	Definition
_____	-location	1. The movement of materials (across distances) in the vascular tissues of a plant.
_____	_____	2. Layers of cells covering the surface of woody stems and roots (i.e., the outer bark); the "skin" of woody dicots and cone-bearing gymnosperms.
_____	-node	3. The region of a stem between two successive nodes.

MATCHING

Terms:

a. Apical meristem
b. Cork cambium
c. Ground tissue
d. Lateral bud
e. Lateral meristem

f. Lenticels
g. Periderm
h. Pith
i. Ray
j. Root pressure

k. Terminal bud
l. Vascular cambium
m. Xylem

For each of these definitions, select the correct matching term from the list above.

_____ 1. An area of dividing tissue located at the tips of plant stems and roots.

_____ 2. Technical term for the outer bark of woody stems and roots.

_____ 3. Large, thin-walled parenchyma cells found in the innermost tissue in many plants.

_____ 4. A lateral meristem in plants that produces cork cells and cork parenchyma.

_____ 5. The cortex and the pith are parts of this tissue system.

_____ 6. Lateral meristem that gives rise to secondary vascular tissues.

_____ 7. The vascular tissue responsible for transporting water and dissolved minerals in plants.

_____ 8. The positive pressure in the root tissues of plants.

_____ 9. A chain of parenchyma cells that functions for lateral transport of food, water, and minerals in woody plants.

MAKING COMPARISONS

Fill in the blanks.

Tissue	Source	Location	Function
Secondary xylem	Produced by vascular cambium	Wood	Conducts water and dissolved minerals
#1	Produced by vascular cambium	Inner bark	#2
Cork parenchyma	#3	Periderm	#4
#5	Produced by cork cambium	#6	Replacement for epidermis
#7	A lateral meristem produced by procambium tissue	Between wood and inner bark in vascular bundles	#8
Cork cambium	#9	In epidermis or outer cortex	#10

MAKING CHOICES

Place your answer(s) in the space provided. Some questions may have more than one correct answer.

_____ 1. Secondary xylem and phloem are derived from
    a. cork parenchyma.   d. periderm.
    b. cork cambium.   e. epidermis.
    c. vascular cambium.

_____ 2. The outer portion of bark is formed primarily from
    a. cork parenchyma.   d. periderm.
    b. cork cambium.   e. epidermis.
    c. vascular cambium.

_____ 3. If soil water contains 0.1% dissolved materials and root water contains 0.2% dissolved materials, one would expect
    a. a negative water potential in soil water.   d. water to flow from soil into root.
    b. a negative water potential in root water.   e. water to flow from root into soil.
    c. less water potential in roots than in soil.

_____ 4. When water is plentiful, wood formed by the vascular cambium is
    a. springwood.   d. composed of large diameter conducting cells.
    b. summerwood.   e. composed of thick-walled vessels.
    c. late summerwood.

_____ 5. One daughter cell from a mother cell in the vascular cambium remains as part of the vascular cambium, the other divides to form
    a. secondary xylem or phloem.   d. secondary tissue.
    b. wood or inner bark.   e. primary xylem and phloem.
    c. outer bark.

_____ 6. If a plant is placed in a beaker of pure distilled water, then

a. water will move out of the plant.

b. water will move into the plant.

c. water pressure in the plant is positive relative to water in the beaker.

d. water potential in the beaker is zero.

e. water potential in the plant is less than zero.

_____ 7. The pull of water up through a plant is due in part to

a. cohesion.

b. adhesion.

c. the low water potential in the atmosphere.

d. transpiration.

e. a water potential gradient between the soil and the root.

_____ 8. Which of the following is/are true of monocot stems?

a. stem covered with epidermis

b. vascular tissues embedded in ground tissue

c. stem has distinct cortex and pith

d. vascular bundles scattered through the stem

e. vascular bundles arranged in circles

_____ 9. Which of the following is/are true of eudicot stems?

a. stem covered with epidermis

b. vascular tissues embedded in ground tissue

c. stem has distinct cortex and pith

d. vascular bundles scattered through the stem

e. vascular bundles arranged in circles

_____ 10. The area on a stem where each leaf is attached is called the

a. bud scale.

b. node.

c. lateral bud.

d. leaf scar.

e. lenticel.

_____ 11. Sites of loosely arranged cells along the bark of a woody twig allow gases to diffuse into the stem. These are called

a. bud scales.

b. nodes.

c. lateral buds.

d. leaf scars.

e. lenticels.

VISUAL FOUNDATIONS

Color the parts of the illustration below as indicated.

RED ☐ vascular cambium
GREEN ☐ secondary phloem
YELLOW ☐ pith
BLUE ☐ primary phloem
ORANGE ☐ secondary xylem
BROWN ☐ primary xylem
PINK ☐ epidermis
PURPLE ☐ cortex
TAN ☐ periderm

Color the parts of the illustration below as indicated. Also label xylem and phloem.

RED ☐ path of sucrose

YELLOW ☐ companion cell

BLUE ☐ path of water

ORANGE ☐ sink

BROWN ☐ sieve tube element

PINK ☐ source

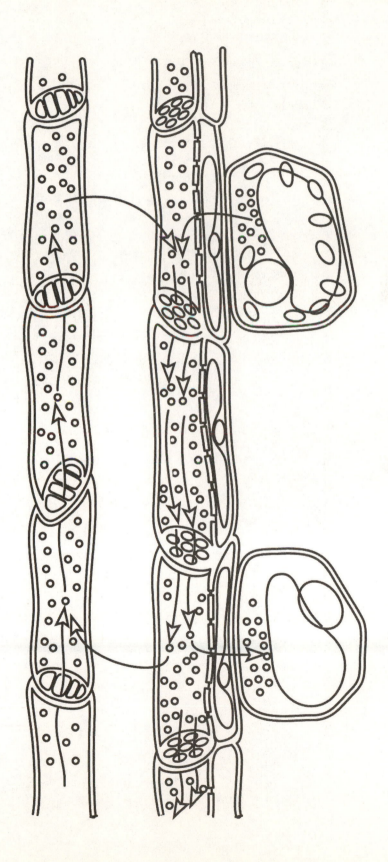

CHAPTER 35

Roots and Mineral Nutrition

The previous two chapters armed you with an understanding of the structural and physiological adaptations of the plant's shoot system, i.e., the leaves and stems. This chapter addresses the structural and physiological adaptations of the remaining portion of the plant, i.e., the root system. The roots of most plants function to anchor plants to the ground, to absorb water and dissolved minerals from the soil, and, in some cases, to store food. Additionally, the roots of some plants are modified for support, aeration, and/or photosynthesis. Although herbaceous roots have the same tissues and structures found in stems, they also have other tissues and structures as well. Various factors influence soil formation. Soil is composed of inorganic minerals, organic matter, soil organisms, soil atmosphere, and soil water. Plants require at least 19 essential nutrients for normal growth, development, and reproduction. Some human practices, such as harvesting food crops, depletes the soil of certain essential elements, making it necessary to return these elements to the soil in the form of organic or inorganic fertilizer.

REVIEWING CONCEPTS
Fill in the blanks.

INTRODUCTION

1. The primary functions of roots include: _____
_____.

ROOT STRUCTURE AND FUNCTION

2. Two types of root systems occur in plants; they are the
(a)_____ and the (b)_____.

3. Main roots of a fibrous root system that arise from the stem and not pre-existing
roots are called _____.

Roots have root caps and root hairs

4. Each root tip is covered by a _____, a protective, , thimble-like layer
of many cells that covers the apical meristem.

5. Short-lived tubular extensions of epidermal cells located just behind the growing
root tip are called _____.

The arrangement of vascular tissues distinguishes the roots of herbaceous eudicots and monocots

6. The inner layer of cortex in a eudicot root is the (a)_____, the cells
of which possess a bandlike region on their radial and transverse walls. This
"band," or (b)_____ as it's called, contains suberin.

7. Just inside the endodermis is a layer of parenchyma cells called the
(a)_____. The central portion of eudicot roots is occupied by
(b)_____ tissue.

8. Most of the water that enters roots first moves along the (a)_____ rather than entering the cells. This movement is made possible due to (b)_____ in the cell walls. Later, when water reaches the waterproof walls of the (c)_____, it does permeate cell membranes. Lateral movement of water through root tissues can be summarized as: root hair $\rightarrow$ (d)_____ $\rightarrow$ cortex $\rightarrow$ (e)_____ $\rightarrow$ pericycle $\rightarrow$ (f)_____, where it is then transported upward.

9. The primary function of the root cortex is (a)_____. Multicellular lateral roots originate in the (b)_____.

10. In most herbaceous eudicot roots, the central core of vascular tissue lacks _____,

Woody plants have roots with secondary growth

11. During growth, the root epidermis is replaced by (a)_____, composed of cork cells and cork parenchyma, both produced by the (b)_____.

Some roots are specialized for unusual functions

12. Roots that grow from unusual places on plants, such as aerial roots growing laterally from a stem, are known as _____ roots.

13. Two other types of such roots are (a)_____ roots that support the plant, like those in corn, and (b)_____ roots that pull bulbs deeper into the ground.

ROOT ASSOCIATIONS WITH FUNGI AND BACTERIA
Mycorrhizae facilitate the uptake of essential minerals by roots

14. Subterranean associations between roots and soil fungi are known as _____.

Rhizobial bacteria fix nitrogen in the roots of leguminous plants

15. Plants and nitrogen-fixing bacteria use _____ to establish contact and develop nodules.

THE SOIL ENVIRONMENT
Soil is composed of inorganic minerals, organic matter, air, and water

16. The four components of soil are _____ _____.

17. Good, loamy agricultural soil contains about 40% each of (a)_____ and about 20% of (b)_____. The partly decayed organic portion of soil, or (c)_____, is also an important component. The spaces between soil particles are filled with (d)_____.

The organisms living in the soil form a complex ecosystem

18. Bits of soil that have passed through the gut of an earthworm are called _____.

Soil pH affects soil characteristics and plant growth

    19. Air pollution in which sulfuric and nitric acids produced by human activities fall to the ground as acid rain, sleet, snow, or fog is known as

    _____ .

Soil provides most of the minerals found in plants

    20. The 19 elements that plants require for normal growth, development, and reproduction are considered (a)_____ for plant growth. Of these, ten are needed in relatively large amounts. These macronutrients include: (b)_____

    _____

    _____

    The remaining 9 elements are needed in trace amounts; these micronutrients include: (c)_____

    _____ .

    21. In order to identify elements that are essential for plant growth, investigators must reduce the number of variables to a minimum. To do this, since soil is very complex, plants are grown in aerated water containing known quantities of elements, a process known as _____ .

Soil can be damaged by human mismanagement

    22. The three elements that most often limit plant growth are

    _____ .

    23. The wearing away or removal of soil from the land is known as

    _____ .

    24. Irrigation sometimes causes salt to accumulate in the soil, a process called

    _____ .

BUILDING WORDS

Use combinations of prefixes and suffixes to build words for the definitions that follow.

Prefixes	The Meaning
endo-	within
hydro-	water
macro-	large, long, great, excessive
micro-	small

Prefix	Suffix	Definition
_____	-ponics	1. Growing plants in water (not soil) containing dissolved inorganic minerals.
_____	-nutrient	2. An essential element that is required in fairly large amounts for normal plant growth.
_____	-nutrient	3. An essential element that is required in trace (small) amounts for normal plant growth.
_____	-dermis	4. The innermost layer of the cortex in the plant root.

MATCHING

Terms:

a.   Adventitious root
b.   Apoplast
c.   Casparian strip
d.   Endodermis
e.   Fibrous root system

f.   Graft
g.   Humus
h.   Leaching
i.   Mycorrhizae
j.   Pneumatophore

k.   Prop root
l.   Root cap
m.   Root hair
n.   Stele
o.   Taproot system

For each of these definitions, select the correct matching term from the list above.

_____ 1. Mutualistic associations of fungi and plant roots that aid in the plant's absorption of essential minerals from the soil.

_____ 2. Organic matter in various stages of decomposition in the soil.

_____ 3. The innermost layer of the cortex in the plant root.

_____ 4. A root that arises in an unusual position on a plant

_____ 5. A band of waterproof material around the radial and transverse walls of endodermal root cells.

_____ 6. An adventitious root that arises from the stem and provides additional support for plants.

_____ 7. Aerial "breathing roots" of some plants in swampy and tidal environments that assist in getting oxygen to submerged roots.

_____ 8. The type of root system present in a plant that has several roots of the same size developing from the end of the stem, with lateral roots of various sizes branching off these roots..

_____ 9. An extension of an epidermal cell in roots, which increases the absorptive capacity of the roots.

_____10. A covering of cells over the root tip that protects delicate meristematic tissue beneath it.

MAKING COMPARISONS

Fill in the blanks.

Root Structure	Function of the Root Structure
Root cap	Protects the apical meristem, may orient the root downward
Root apical meristem	#1
#2	Absorption of water and minerals
Epidermis	#3
Cortex	#4
#5	Controls mineral uptake into root xylem
#6	Gives rise to lateral roots and lateral meristems
#7	Conducts water and dissolved minerals
Phloem	#8

MAKING CHOICES

Place your answer(s) in the space provided. Some questions may have more than one correct answer.

_____ 1. Structures found in primary roots that are also found in stems include the

    a. cortex.   d. apical meristem cap.

    b. cuticle.   e. epidermis.

    c. vascular tissues.

_____ 2. The three groups of organisms in soil that are the most important in decomposition are

    a. insects.   d. algae.

    b. fungi.   e. bacteria.

    c. soil protozoa.

_____ 3. The origin of multicellular branch roots is the

    a. cambium.   d. parenchyma.

    b. Casparian strip.   e. cortex.

    c. pericycle.

_____ 4. When water first enters a root, it usually

    a. enters parenchymal cells.   d. moves along cell walls.

    b. is absorbed by cellulose.   e. enters cells in the Casparian strip.

    c. moves from a water negative potential to a positive water potential.

_____ 5. The function(s) performed by all roots is/are

    a. absorption of water.   d. anchorage.

    b. absorption of minerals.   e. aeration.

    c. food storage.

_____ 6. The principal function(s) of the root cortex is/are

    a. conduction.   d. water absorption.

    b. storage.   e. mineral absorption.

    c. production of root hairs.

_____ 7. Essential macronutrients include

    a. phosphorus.   d. hydrogen.

    b. potassium.   e. magnesium.

    c. iron.

_____ 8. Essential micronutrients include

    a. phosphorus.   d. hydrogen.

    b. potassium.   e. magnesium.

    c. iron.

_____ 9. In general, the vascular tissues in monocot roots

    a. form a solid cylinder.   d. contain a vascular cambium.

    b. are absent.   e. continue into root hairs.

    c. are in bundles arranged around the central path.

_____10. The inorganic materials in soil come from

    a. fertilizers.   d. the atmosphere.

    b. water runoff.   e. percolating water.

    c. weathered rock.

_____11. Roots produced in unusual places on the plants, often as aerial roots, are called _____ roots.

    a. secondary   d. contractile

    b. enhancement   e. aerial water-absorbing

    c. adventitious

VISUAL FOUNDATIONS
Color the parts of the illustration below as indicated.

RED ☐ primary xylem
GREEN ☐ cortex
BLUE ☐ vascular cambium
ORANGE ☐ secondary xylem
YELLOW ☐ epidermis
BROWN ☐ periderm
TAN ☐ pericycle
PINK ☐ secondary phloem
VIOLET ☐ primary phloem

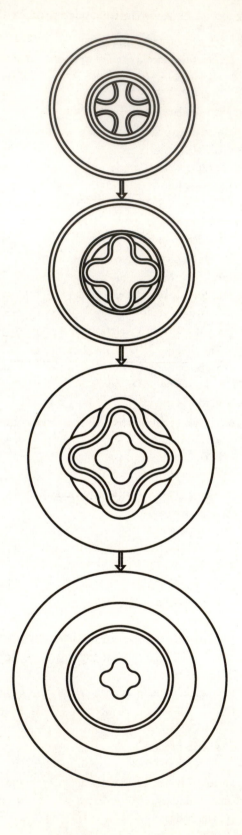

CHAPTER 36

Reproduction in Flowering Plants

This chapter, the fifth in a series of six addressing the integration of plant structure and function, discusses reproduction in flowering plants. As you recall from the first chapter in the series, flowering plants are the largest and most successful group of plants. All flowering plants reproduce sexually; some also reproduce asexually. Sexual reproduction involves flower formation, pollination, fertilization within the flower ovary, and seed and fruit formation. The offspring resulting from sexual reproduction exhibit a great deal of individual variation, due to gene recombination and the union of dissimilar gametes. Although sexual reproduction has some disadvantages, it offers the advantage of new combinations of genes that might make an individual plant better suited to its environment. Asexual reproduction involves only one parent; the offspring are genetically identical to the parent and each other. Asexual reproduction, therefore, is advantageous when the parent is well adapted to its environment. The stems, leaves, and roots of many flowering plants are modified for asexual reproduction. Also, in some plants, seeds and fruits are produced asexually without meiosis or the fusion of gametes.

REVIEWING CONCEPTS

Fill in the blanks.

INTRODUCTION

1. In flowering plants, the diploid (a)_____ is larger and nutritionally independent. The haploid (b)_____, which is located in the flower, is microscopic and nutritionally dependent on the sporophyte.

THE FLOWERING PLANT LIFE CYCLE

Flowers develop at apical meristems

2. The Flowering Locus C gene codes for a transcription factor that represses _____.

Each part of a flower has a specific function

3. Flowers are reproductive shoots, usually consisting of four kinds of organs, the _____, arranged in whorls on the end of a flower stalk.

4. The collective term for all the sepals of a flower is (a)_____. The collective term for all the petals of a flower is (b)_____.

Female gametophytes are produced in the ovary, male gametophytes in the anther

5. Pollen sacs within the anther contain numerous diploid cells called (a)_____, each of which undergoes meiosis to produce four haploid cells called (b)_____, each of which divides mitotically to produce an immature male gametophyte called a

(c)_____. The pollen grain becomes mature when its generative cell divides to form two nonmotile (d)_____.

POLLINATION

6. The transfer of pollen grains from (a)_____ to (b)_____ is known as pollination.

Many plants have mechanisms to prevent self-pollination

7. The mating of genetically similar individuals is known as _____.

8. A genetic condition in which the pollen is ineffective in fertilizing the same flower or other flowers on the same plant is known as _____.

Flowering plants and their animal pollinators have coevolved

9. Plants pollinated by insects often have (a)_____ (color?) petals, but usually not (b)_____ petals.

Some flowering plants depend on wind to disperse pollen

10. Wind-pollinated plants produce many small, inconspicuous _____.

FERTILIZATION AND SEED/FRUIT DEVELOPMENT

11. Once pollen grains have been transferred from anther to stigma, the tube cell grows a pollen tube down through the (a)_____ and into an (b)_____ in the ovary.

A unique double fertilization process occurs in flowering plants

12. The tissue with nutritive and hormonal functions that surrounds the developing embryonic plant in the seed is called _____.

Embryonic development in seeds is orderly and predictable

13. Embryonic tissue that anchors the developing embryo and aids in nutrient uptake from the endosperm in called the_____.

The mature seed contains an embryonic plant and storage materials

14. The mature embryo within the seed consists of a short embryonic root, or (a)_____; an embryonic shoot; and one or two seed leaves, or (b)_____.

15. The short portion of the embryonic shoot connecting the radicle to one or two cotyledons is the _____.

Fruits are mature, ripened ovaries

16. The four basic types of fruits are _____
_____.

17. One type of fruit, the (a)_____ fruit, develops from a single pistil. It may be fleshy or dry. Fruits that have soft tissues throughout, like those in tomatoes and grapes, are called (b)_____, while fleshy fruits that have a hard, stony pit surrounding a single seed are known as (c)_____. Of fruits that split open at maturity, some, like the milkweed (d)_____, split along one suture, while (e)_____ split along two sutures, and (f)_____ split along multiple sutures.

18. (a)_____ fruits result from the fusion of several developing ovaries in a single flower. The raspberry is an example. Similarly, (b)_____ fruits form from the fusion of several ovaries of many flowers that grow in proximity on a common floral stalk. The pineapple is an example.

19. (a)_____ fruits contain plant tissue in addition to ovary tissue. When one eats a strawberry, for example, he or she is eating the fleshy (b)_____, and when eating apples and pears, one is consuming the (c)_____ that surrounds the ovary.

Seed dispersal is highly varied

20. Flowering plant seeds and fruits are adapted for various means of dispersal, among which four common means are

_____

_____.

GERMINATION AND EARLY GROWTH

Some seeds do not germinate immediately

21. A process of scratching or scarring the seed coat before sowing it is called _____.

Eudicots and monocots exhibit characteristic patterns of early growth

22. Corn and grasses have a special sheath of cells called a _____ that surrounds and protects the young shoot.

ASEXUAL REPRODUCTION IN FLOWERING PLANTS

23. Various vegetative structures may be involved in asexual reproduction. Some of them, for example, are (a)_____, like those in bamboo and grasses that have horizontal underground stems; (b)_____, underground stems greatly enlarged for food storage, like those in potatoes; (c)_____, short underground stems with fleshy storage leaves, as in onions and tulips; (d)_____, thick underground stems covered with papery scales; and (e)_____ or horizontal, aboveground stems with long internodes, like those in strawberries.

24. Some plants reproduce asexually by producing _____, aboveground shoots that develop from adventitious buds on roots.

Apomixis is the production of seeds without the sexual process

25. The advantage of apomixis over other methods of asexual reproduction is that the seeds and fruits produced can be _____ by methods associated with asexual reproduction.

COMPARISON OF SEXUAL AND ASEXUAL REPRODUCTION

Sexual reproduction has some disadvantages

26. The many adaptations of flowers for different modes of pollination represent one cost of _____.

BUILDING WORDS

Use combinations of prefixes and suffixes to build words for the definitions that follow.

Prefixes	The Meaning
co-	with, together, in association
endo-	within
hypo-	under, below

Prefix	Suffix	Definition
_____	-sperm	1. Nutritive tissue within seeds.
_____	-cotyl	2. The part of a plant embryo or seedling below the point of attachment of the cotyledons.
_____	-evolution	3. Evolutionary changes that result from close interaction and reciprocal adaptations between two species.

MATCHING

Terms:

a. Aggregate fruit
b. Apomixis
c. Carpel
d. Corolla
e. Cotyledon

f. Drupe
g. Fruit
h. Ovary
i. Ovule
j. Pollen

k. Seed
l. Sepal
m. Stolon

For each of these definitions, select the correct matching term from the list above.

_____ 1. A collective term for the petals of a flower

_____ 2. A ripened ovary.

_____ 3. The outermost parts of a flower, usually leaf-like in appearance, that protect the flower as a bud.

_____ 4. A type of reproduction in which fruits and seeds are formed asexually.

_____ 5. A fruit that develops from a single flower with many separate carpels, such as a raspberry.

_____ 6. A plant reproductive body composed of a young embryo and nutritive tissue.

_____ 7. The male gametophyte in plants.

_____ 8. An above-ground, horizontal stem with long internodes.

_____ 9. The female reproductive unit of a flower.

_____10. A simple fleshy fruit that contains a hard and stony pit surrounding a single seed; such as a plum or peach.

MAKING COMPARISONS

Fill in the blanks.

Examples of Fruit	Main Type of Fruit	Subtype	Description
Bean	Simple	Dry: opens to release seeds; legume	Splits along the 2 sutures
Acorn	Simple	#1	#2
Corn	Simple	Dry: does not open; grain	#3

Examples of Fruit	Main Type of Fruit	Subtype	Description
Tomato	#4	#5	Soft and fleshy throughout, usually with few to many seeds
Strawberry	#6	None	Other plant tissues and ovary tissue comprise the fruit
Pineapple	#7	None	#8
Raspberry	#9	None	#10
Peach	#11	#12	#13

MAKING CHOICES

Place your answer(s) in the space provided. Some questions may have more than one correct answer.

_____ 1. A modified underground bud in which fleshy storage leaves are attached to a stem is a
a. corm.
b. stolon.
c. bulb.
d. fruit.
e. capsule.

_____ 2. A simple, dry fruit that develops from two or more fused carpels is a
a. corm.
b. stolon.
c. bulb.
d. fruit.
e. capsule.

_____ 3. A short, erect underground stem is a
a. corm.
b. stolon.
c. bulb.
d. fruit.
e. capsule.

_____ 4. A horizontal, above ground stem with long internodes is a
a. corm.
b. stolon.
c. bulb.
d. fruit.
e. capsule.

_____ 5. Which of the following is/are somehow related to simple dry fruits?
a. drupe
b. legume
c. grain
d. capsule
e. green bean

_____ 6. The process by which an embryo develops from a diploid cell in the ovary without fusion of haploid gametes is called
a. lateral bud generation.
b. dehiscence.
c. asexual reproduction by means of suckers.
d. apomixis.
e. spontaneous generation.

_____ 7. The tomato is actually a
a. drupe.
b. berry.
c. fleshy lateral bud.
d. fruit.
e. mature ovary.

_____ 8. The "eyes" of a potato are
   a. diploid gametes.   d. axillary buds.
   b. parts of a stem.   e. rhizomes.
   c. capable of producing complete plants.

_____ 9. Horizontal, asexually reproducing stems that run above ground are
   a. diploid gametes.   d. corms.
   b. rhizomes.   e. stolons.
   c. tubers.

_____10. Raspberries and blackberries are examples of
   a. follicles.   d. aggregate fruits.
   b. achenes.   e. multiple fruits.
   c. drupes.

_____11. A fruit that forms from many separate carpels in a single flower is a/an
   a. follicle.   d. aggregate fruit.
   b. achene.   e. accessory fruit.
   c. drupe.

_____12. A fruit composed of ovary tissue and other plant parts is a
   a. follicle.   d. aggregate fruit.
   b. achene.   e. accessory fruit.
   c. drupe.

_____13. In flowering plants, the fusion of two gametes
   a. produces the gametophyte generation.   d. produces a diploid generation.
   b. produces the sporophyte generation.   e. is called fertilization.
   c. produces a haploid generation.

_____14. In flowering plants, haploid spores are produced by
   a. the sporophyte generation.   d. gametes.
   b. the gametophyte generation.   e. self fertilization.
   c. meiosis.

VISUAL FOUNDATIONS

Color the parts of the illustration below as indicated.

RED ❑ ovule

GREEN ❑ sepal

YELLOW ❑ ovary (and derived from ovary)

BLUE ❑ stamen

ORANGE ❑ style

BROWN ❑ stigma

TAN ❑ seed

PINK ❑ floral tube (and derived from floral tube)

VIOLET ❑ petal

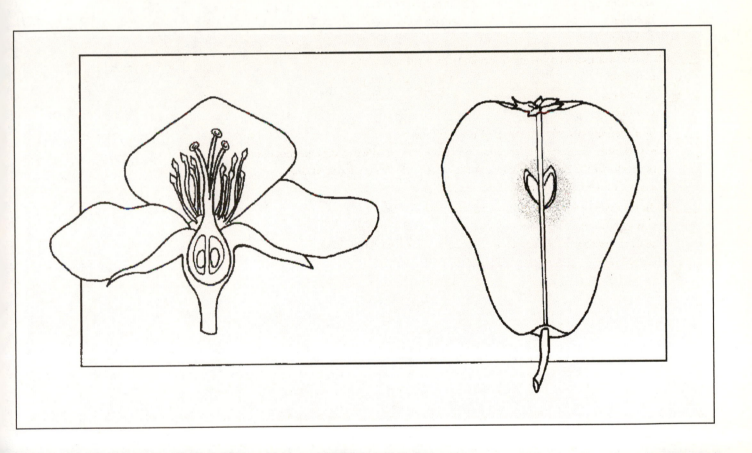

C H A P T E R   37

❑

Plant Growth and Development

This is the last chapter in a series of six addressing the integration of plant structure and function. You learned in Chapter 17 that an organism's genes control all of the changes that take place during its life. How a particular gene is expressed is determined by a variety of factors, including signals from other genes and from the environment. Environmental cues, such as temperature, light, growth, and touch, influence plant growth and development. All aspects of plant growth and development are affected by hormones, organic compounds present in very low concentrations that act as highly specific chemical signals between cells. Hormones interact in complex ways with one another to produce a variety of responses in plants. The auxins are involved in cell elongation, phototropism, gravitropism, apical dominance, and fruit development; the gibberellins are involved in stem elongation, flowering, and seed germination; the cytokinins promote cell division and differentiation, delay senescence, and interact with auxins in apical dominance; ethylene has a role in the ripening of fruits, apical dominance, leaf abscission, and wound response; and abscisic acid is involved in stomatal closure due to water stress, and bud and seed dormancy. Other chemicals have been implicated in plant growth and development, and are the subject of ongoing research.

REVIEWING CONCEPTS
Fill in the blanks.

INTRODUCTION

1. Growth and development of a plant are controlled by its _____ organic compounds present in low concentration in the plant's tissues that act as chemical signals between cells.

TROPISMS

2. For an organism to have a biological response to light, it must contain a light-sensitive substance, called a _____, to absorb the light.

3. Tropisms are categorized according to the stimulus that causes them to occur. For example, growth or movement initiated by light is (a)_____, response to gravity is (b)_____, and movement caused by a mechanical stimulus is (c)_____.

PLANT HORMONES AND DEVELOPMENT

4. The five major classes of hormones that regulate responses in plants are

_____.

PLANT HORMONES ACT BY SIGNAL TRANSDUCTION

5. Many plant hormones bind to _____, located in the plasma membrane where they trigger enzymatic reactions.

AUXINS PROMOTE CELL ELONGATION

6. _____ is the most common and physiologically important auxin.

7. Auxins always move in one direction, specifically from the (a)_____ toward the roots. Such unidirectional movement is called (b)_____. Auxins are also involved in phototropism, gravitropism, apical dominance, and fruit development.

GIBBERELLINS PROMOTE STEM ELONGATION

8. Seedling disease is caused by a _____ that produces gibberellin.

9. In addition to influencing stem elongation, gibberellins are also involved in _____.

CYTOKININS PROMOTE CELL DIVISION

10. Cytokinins mainly promote _____.

11. They also delay _____(aging), and they interact with auxins in apical dominance.

ETHYLENE PROMOTES ABSCISSION AND FRUIT RIPENING

12. A developmental response to a mechanical stimulus is known as _____.

ABSCISIC ACID PROMOTES SEED DORMANCY

13. _____ is the temporary state of reduced physiological activity in flowering plants.

14. The seeds of many desert plants contain high concentrations of _____,.which inhibits germination under unfavorable conditions.

15. In seeds, the level of abscisic acid decreases during the winter, and the level of _____ increases.

ADDITIONAL SIGNALING MOLECULES AFFECT GROWTH AND DEVELOPMENT, INCLUDING PLANT DEFENSES

16. The five groups of signaling molecules involved in defense responses to plant diseases are _____ _____.

PROGRESS IS BEING MADE IN IDENTIFYING THE ELUSIVE FLOWER-PROMOTING SIGNAL

17. _____ is a flower-promoting substance.

LIGHT SIGNALS AND PLANT DEVELOPMENT

18. _____.is any response of a plant to the relative lengths of daylight and darkness.

19. (a)_____ plants flower when the night length is equal to or greater than some critical period; (b)_____ plants

flower when the night length is equal to or less than some critical period;
(c)_____ plants do not flower when night length
is either too long or too short; (d)_____ plants do not
initiate flowering in response to seasonal changes in the period of daylight and
darkness but instead respond to some other type of stimulus, external or internal.

Phytochrome detects day length

20. _____ is the main photoreceptor for photoperiodism and
many other light-initiated plant responses.

Competition for sunlight among shade-avoiding plants involves phytochrome

21. Plants tend to grow taller when closely surrounded by other plants, a
phenomenon known as _____.

Phytochrome is involved in other responses to light, including germination

22. Seeds with a light requirement must be exposed to light containing (a)_____
wavelengths to convert (b)_____ to (c)_____.

Phytochrome acts by signal transduction

23. A mechanical stimulus may initiate an electrical "impulse" that causes changes in
cell membrane permeability. When this occurs, (a)_____ ions flow
out of the affected cells, resulting in the net movement of water out of these cells.
This sudden change in water volume causes the leaf to move. Such movements,
known as (b)_____, are temporary.

Light influences circadian rhythms

24. Internal cycles, known as circadian rhythms, help organisms detect
_____.

BUILDING WORDS

Use combinations of prefixes and suffixes to build words for the definitions that follow.

Prefixes	The Meaning		Suffixes	The Meaning
photo-	light		-chrom(e)	color
phyto-	plant		-tropism	turn, turning

Prefix	Suffix	Definition
_____	_____	1. The growth response of an organism to light; usually the turning toward or away from the light source.
gravi-	_____	2. The growth response of an organism to gravity; usually the turning toward or away from the direction of gravity.
_____	_____	3. A blue-green, proteinaceous pigment involved in photoperiodism and a number of other light-initiated physiological responses of plants.
_____	periodism	4. The physiological response of organisms to variations of light and darkness.

MATCHING

Terms:

a. Abscisic acid
b. Apical dominance
c. Auxin
d. Coleoptile
e. Cytokinin

f. Ethylene
g. Imbibition
h. Nastic movements
i. Pulvinus
j. Senescence

k. Statolith
l. Thigmotropism
m. Vernalization

For each of these definitions, select the correct matching term from the list above.

_____ 1. The inhibition of axillary bud growth by the apical meristem.

_____ 2. A plant hormone involved in dormancy and responses to stress.

_____ 3. The aging process in plants.

_____ 4. A protective sheath that encloses the stem in certain monocots.

_____ 5. A plant hormone involved in apical dominance and cell elongation.

_____ 6. A plant hormone that promotes fruit ripening.

_____ 7. Plant growth in response to contact with mechanical stimuli, such as a solid object.

_____ 8. A plant hormone that promotes rapid cell division and is involved in other aspects of plant growth and development.

MAKING COMPARISONS

Fill in the blanks.

Principal Action	Hormone(s)
Regulates growth by promoting cell elongation	Auxin and gibberellin
Promotes apical dominance, stem elongation, root initiation, fruit development	#1
Delays leaf senescence, inhibition of apical dominance, embryo development	#2
Promotes seed germination, stem elongation, flowering, fruit development	#3
Fruit ripening, seed germination, root initiation, abscission	#4
Promotes cell division	#5
Promotes seed dormancy	#6

MAKING CHOICES

Place your answer(s) in the space provided. Some questions may have more than one correct answer.

_____ 1. Growth in response to gravity is known as

    a. a tropism.

    b. phototropism.

    c. gravitropism.

    d. thigmotropism.

    e. turgor.

_____ 2. The hormone(s) that is/are involved in rapid stem elongation just prior to flowering is/are

    a. auxin.

    b. gibberellin.

    c. cytokinin.

    d. ethylene.

    e. abscisic acid.

_____ 3. Which of the following is/are correct about hormones?

    a. They are effective in very small amounts.

    b. They are organic compounds.

    c. For the most part, their effects occur near the area where they are produced.

    d. The effects of different hormones overlap.

    e. Each plant hormone has multiple effects.

_____ 4. The hormone(s) principally responsible for cell division and differentiation is/are

    a. auxin.

    b. gibberellin.

    c. cytokinin.

    d. ethylene.

    e. abscisic acid.

_____ 5. The hormone(s) principally responsible for the growth of a coleoptile toward light is/are

    a. auxin.

    b. gibberellin.

    c. cytokinin.

    d. ethylene.

    e. abscisic acid.

_____ 6. If a plant that touches your house continues to grow toward and attach itself to the house, the plant is exhibiting a

    a. tropism.

    b. phototropism.

    c. gravitropism.

    d. thigmotropism.

    e. turgor movement.

_____ 7. The plant hormone(s) about which Charles Darwin gathered information is/are

    a. auxins.

    b. gibberellin.

    c. cytokinin.

    d. ethylene.

    e. abscisic acid.

_____ 8. The rapid elongation of a floral stalk during the initiation of flowering is know as

    a. acceleration.

    b. bolting.

    c. blooming.

    d. floral enhancement.

    e. internodal elongation.

_____ 9. The aging process is called

    a. dormancy.

    b. bolting.

    c. abscissionation.

    d. systemination.

    e. senescence.

VISUAL FOUNDATIONS

Color the parts of the illustration below as indicated. Also label transcription and protein targeted for destruction.

RED	☐	auxin
PINK	☐	ubiquitin
TAN	☐	nucleus
YELLOW	☐	cell wall
ORANGE	☐	protein

GREEN	☐	receptor
GREY	☐	cytoplasm
BROWN	☐	nuclear envelope
BLUE	☐	plasma membrane
VIOLET	☐	DNA

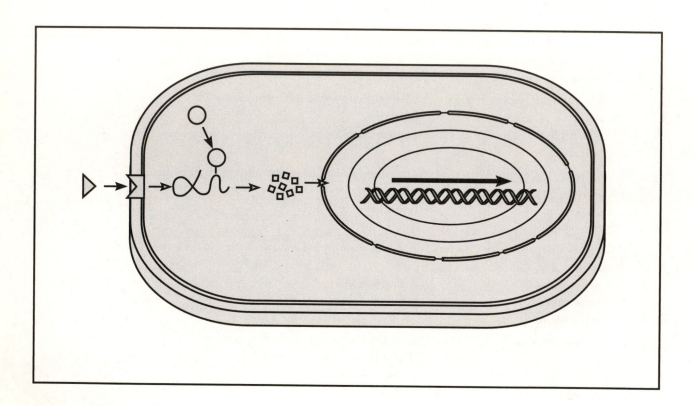

Structure and Life Processes in Animals

CHAPTER 38

Animal Structure and Function: An Introduction

This is the first of 14 chapters that examine the structural, functional, and behavioral adaptations that help animals meet environmental challenges. This chapter examines the architecture of the animal body. In most animals, cells are organized into tissues, tissues into organs, and organs into organ systems. The principal animal tissues are epithelial, connective, muscular, and nervous. Epithelial tissues are characterized by tight-fitting cells and the presence of a basement membrane. Covering body surfaces and lining cavities, they function in protection, absorption, secretion, and sensation. Connective tissue joins together other tissues, supports the body and its organs, and protects underlying organs. There are many different types of connective tissues, consisting of a variety of cell types. Muscle tissue is composed of cells that are specialized to contract. There are three major types of muscle tissue — cardiac, smooth, and skeletal. Nervous tissue is composed of cells that are specialized for conducting impulses and those that support and nourish the conducting cells. Organs are comprised of two or more kinds of tissues. Complex animals have many organs and organ systems. The organ systems work together to maintain the body's homeostasis.

REVIEWING CONCEPTS

Fill in the blanks.

INTRODUCTION

1. New cells formed by cell division remain associated in multicellular animals. The size of an animal is determined by the _____ of cells that make up its body, not the size of the individual cells.

2. A group of cells that carry out a specific function is called a (a)_____, and these groups associate to form (b)_____, which in turn are grouped into the (c)_____ of the body.

TISSUES

Epithelial tissues cover the body and line its cavities

3. Epithelial tissues cover body surfaces and cavities, forming (a)_____ of tightly joined cells that are attached to underlying tissues by a fibrous noncellular (b)_____.

4. Four of the major functions of epithelial cells include

   _____

   _____.

5. Epithelial cells can be distinguished on the basis of their shape. (a)_____ cells are flat, (b)_____ cells resemble dice, and (c)_____ cells are tall, slender cells shaped like cylinders.

6. Epithelial tissues vary in the number of cell layers and the shapes of their cells. For example, (a)_____ epithelium is made up of only one layer of cells and (b)_____ epithelium has two or more cell layers.

Connective tissues support other body structures

7. Characteristically, connective tissues contain very few cells, an (a)_____ in which cells and (b)_____ are embedded, and a (c)_____ that is secreted by the cells.

8. The _____ associated with connective tissue is largely responsible for the nature and function of each kind of connective tissue.

9. There are three types of fibers in connective tissues. Collagen fibers are numerous, strong fibers composed of the protein (a)_____. Elastic fibers are composed of (b)_____ and can stretch. Reticular fibers, composed of (c)_____, form delicate networks of connective tissue.

10. Fibroblast cells produce _____ in connective tissues.

11. _____ function as the body's scavenger cells.

12. Connective tissues join, support, and protect other tissues. There are many kinds of connective tissues, the main types of which are _____ _____ _____.

13. The most widely distributed connective tissue in the vertebrate body is the _____.

14. Loose connective tissue is found in spaces between body parts, around muscles, nerves, and blood vessels, and under the skin where it forms the _____.

15. Dense connective tissues are predominantly composed of _____ fibers.

16. (a)_____ are the cords that connect muscles to bones. (b)_____ are the cables that connect bones to one another.

17. All vertebrates have an internal supporting structure called the _____ that is composed of cartilage and/or bone.

18. The (a)_____ that provides support in the vertebrate embryo is largely replaced by (b)_____ in most vertebrate adults.

19. Cartilage cells called (a)_____ are surrounded by a rubbery (b)_____ in which collagen fibers are embedded. "Imprisoned" by their own secretions, the cells are isolated in holes called (c)_____. Nutrients diffuse through the matrix to the cells.

20. Bone is similar to cartilage in that it too has a (a)_____ containing lacunae in which cells called (b)_____ are "imprisoned." Unlike cartilage however, bone is highly (c)_____, a term that refers to their abundant supply of blood vessels.

21. Osteocytes communicate with one another through small channels called _____.

22. Compact bone consists of spindle-shaped units called (a)_____. Osteocytes are arranged in concentric layers called (b)_____. The lamellae surround central microscopic channels known as (c)_____, through which blood vessels and nerves pass.

23. The noncellular component of blood is the _____.

24. Red blood cells function to _____.

25. White blood cells function to _____.

26. Platelets are small cell fragments that originate in _____.

Muscle tissue is specialized to contract

27. Each muscle cell is called a _____.

28. Contractile proteins called (a)_____ are contained within the elongated (b)_____ of muscle cells.

29. Vertebrates have three kinds of muscles: (a)_____ muscle that attaches to bones and causes body movements; (b)_____ muscle occurring in the walls of the digesive tract, uterus, blood vessels, and many other internal organs; and heart or (c)_____ muscle.

Nervous tissue controls muscles and glands

30. Cells in nervous tissue that conduct impulses are (a)_____, and support cells in this tissue are called (b)_____ cells.

31. Neurons communicate with one another at cellular junctions called _____.

32. A _____ is a collection of neurons bound together by connective tissue.

33. Neurons generally contain three functionally and anatomically distinct regions: the (a)_____, which contains the nucleus; the (b)_____, which receive incoming impulses; and the (c)_____, which carry impulses away from the cell body.

ORGANS AND ORGAN SYSTEMS

The body maintains homeostasis

34. The control processes that maintain a balanced internal environment are called _____.

35. The normal homeostatic state of an organism is also referred to as its _____.

REGULATING BODY TEMPERATURE

36. _____ is the ability to maintain body temperature within certain limits, even when the temperature of the environment changes a lot.

Ectotherms absorb heat from their surroundings

37. An ectotherm's metabolic rate tends to change with the weather, therefore, they have a much _____ daily energy expenditure than endotherms.

Endotherms derive heat from metabolic processes

38. Endotherms can have a metabolic rate as much as _____(#?) times as high as that of ectotherms.

Many animals adjust to challenging temperature changes

39. When stressed by cold, many animals sink into (a)_____, a decrease in body temperature below normal levels. (b)_____ is long-term torpor in response to winter cold and scarcity of food. (c)_____ is a state of torpor caused by lack of food or water during periods of high temperature.

BUILDING WORDS

Use combinations of prefixes and suffixes to build words for the definitions that follow.

Prefixes	The Meaning		Suffixes	The Meaning
chondro-	cartilage		-blast	embryo
fibro-	fiber		-cyte	cell
homeo-	similar, "constant"		-phage	eat, devour
inter-	between, among		-stasis	equilibrium
macro-	large, long, great, excessive			
multi-	many			
myo-	muscle			
osteo-	bone			
pseudo-	false			

Prefix	Suffix	Definition
_____	-cellular	1. Composed of many cells.
_____	-stratified	2. An arrangement of epithelial cells in which the cells falsely appear to be stratified.
_____	_____	3. A cell, especially active in developing ("embryonic") tissue and healing wounds, that produces connective tissue fibers.
_____	-cellular	4. Situated between or among cells.
_____	_____	5. A large cell, common in connective tissues, that phagocytizes ("eats") foreign matter including bacteria.
_____	_____	6. A cartilage cell.
_____	_____	7. A bone cell.
_____	-fibril	8. A thin longitudinal contractile fiber inside a muscle cell.
_____	_____	9. Maintaining a constant internal environment.

MATCHING

Terms:

a. Adipose tissue
b. Cardiac muscle
c. Cartilage
d. Collagen
e. Endocrine gland

f. Gland
g. Glial cell
h. Matrix
i. Organ
j. Organ system

k. Osteon
l. Skeletal muscle
m. Stressor
n. Ectotherm
o. Endotherm

For each of these definitions, select the correct matching term from the list above.

_____ 1. Striated (voluntary) muscle.

_____ 2. Tissue in which fat is stored, or the fat itself.

_____ 3. Spindle-shaped unit of bone composed of concentric layers of osteocytes.

_____ 4. Supporting skeleton in the embryonic stages of all vertebrates.

_____ 5. A specialized structure made up of tissues and adapted to perform a specific function or group of functions.

_____ 6. Glands that secrete products directly into the blood or tissue fluid instead of into ducts.

_____ 7. Tissue characterized by the presence of intercalated discs.

_____ 8. A cell that supports and nourishes neurons.

_____ 9. General term for a body cell or organ specialized for secretion.

_____10. A protein in connective tissue fibers.

MAKING COMPARISONS

Fill in the blanks.

Principle Tissue	Tissue	Location	Function
Epithelial tissue	Stratified squamous epithelium	Skin, mouth lining, vaginal lining	Protection; outer layer continuously sloughed off and replaced from below
Epithelial tissue	#1	#2	Allows for transport of materials, especially by diffusion
#3	Pseudostratified epithelium	#4	#5
#6	Adipose tissue	Subcutaneous layer, pads certain internal organs	#7
Connective tissue	#8	Forms skeletal structure in most vertebrates	#9
#10	Blood	#11	#12
Muscle tissue	Cardiac muscle	#13	Contraction of the heart
Muscle tissue	#14	Attached to bones	Movement of the body
#15	Nervous tissue	Brain, spinal cord, nerves	Respond to stimuli, conduct impulses

MAKING CHOICES

Place your answer(s) in the space provided. Some questions may have more than one correct answer.

_____ 1. The nucleus of a neuron is typically found in the
a. cell body.　　　　　　　　　d. axon.
b. synapse.　　　　　　　　　　e. glial body.
c. dendrite.

_____ 2. The major classes of animal tissues include
a. epithelial.　　　　　　　　　d. skeletal.
b. nervous.　　　　　　　　　　e. connective.
c. muscular.

_____ 3. Myosin and actin are the main components of
a. cartilage.　　　　　　　　　　d. muscle.
b. blood.　　　　　　　　　　　e. adipose tissue.
c. collagen.

_____ 4. The basement membrane
a. lies beneath epithelium.　　　　　　d. is synonymous with plasma membrane.
b. contains polysaccharides.　　　　　e. is noncellular.
c. consists of cells that lie beneath the epithelium.

_____ 5. Epithelium cells that are flattened and thin are called
a. columnar.　　　　　　　　　d. stratified.
b. cuboidal.　　　　　　　　　　e. endothelium.
c. squamous.

_____ 6. Chondrocytes are
a. part of cartilage.　　　　　　　d. eventually found in the lacunae of a matrix.
b. found in bone.　　　　　　　　e. embryonic osteocytes.
c. immature osteocytes.

_____ 7. Large muscles attached to bones are
a. skeletal muscles.　　　　　　　d. nonstriated.
b. smooth muscles.　　　　　　　e. multinucleated.
c. composed largely of actin and myosin.

_____ 8. Platelets are
a. bone marrow cells.　　　　　　d. a type of RBCs.
b. fragments of large cells.　　　　　e. derived from bone marrow cells.
c. a type of white blood cells.

_____ 9. Blood is one type of
a. endothelium.　　　　　　　　d. connective tissue.
b. collagen.　　　　　　　　　　e. cardiac tissue.
c. mesenchyme.

_____ 10. Epithelium located in the ducts of glands would most likely be
a. simple columnar.　　　　　　　d. stratified squamous.
b. simple cuboidal.　　　　　　　e. stratified cuboidal.
c. simple squamous.

_____11. The portion of the neuron that is specialized to receive a nerve impulse is the
    a. cell body.
    b. synapse.
    c. dendrite.
    d. axon.
    e. glial body.

_____12. Collagen is
    a. part of blood.
    b. part of bone.
    c. part of connective tissue.
    d. composed of fibroblasts.
    e. fibrous.

_____13. The connective tissue matrix is
    a. noncellular.
    b. mostly lipids.
    c. a gel.
    d. a polysaccharide.
    e. fibrous.

_____14. Haversian canals
    a. contain nerves.
    b. run through cartilage.
    c. run through bone.
    d. are matrix lacunae.
    e. are synonymous with canaliculi.

_____15. Adipose tissue is a type of
    a. cartilage.
    b. epithelium.
    c. connective tissue.
    d. modified blood tissue.
    e. marrow.

_____16. Cells that nourish and support nerve cells are known as
    a. neurons.
    b. glial cells.
    c. neuroblasts.
    d. synaptic cells.
    e. neurocytes.

_____17. A group of closely associated cells that carries out a specific function is a/an
    a. organ.
    b. system.
    c. organ system.
    d. tissue.
    e. clone.

_____18. Cells lining internal cavities that secrete a lubricating mucus are
    a. goblet-cells.
    b. pseudostratified.
    c. exocrine glands.
    d. epithelial.
    e. part of connective tissue.

_____19. Cells that make up the outer layer of skin are
    a. the columnar layer.
    b. the cuboidal layer.
    c. stratified squamous epithelium.
    d. basement membrane cells.
    e. endothelium.

VISUAL FOUNDATIONS

Color the parts of the illustration below as indicated. Also label compact bone and spongy bone.

RED ☐ blood vessel

GREEN ☐ lacuna

YELLOW ☐ osteon

BLUE ☐ osteocyte

ORANGE ☐ Haversian canal

VIOLET ☐ cytoplasmic extensions

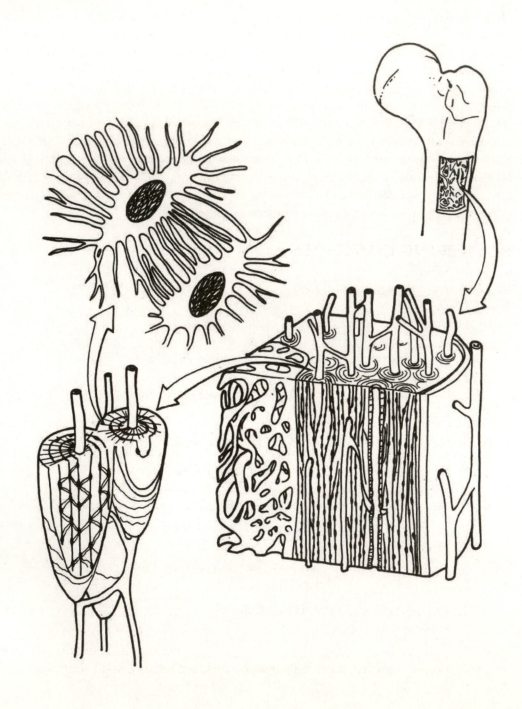

CHAPTER 39

Protection, Support, and Movement

The next several chapters discuss how animals carry out life processes and how the organ systems of a complex animal work together to maintain homeostasis of the organism as a whole. This chapter focuses on epithelial coverings, skeleton, and muscle – systems that are closely interrelated in function and significance. Epithelium covers all external and internal body surfaces. In invertebrates, the external epithelium may contain secretory cells that produce a protective cuticle, secrete lubricants or adhesives, produce odorous or poisonous substances, or produce threads for nests or webs. In vertebrates, specifically humans, the external epithelium (skin) includes nails, hair, sweat glands, oil glands, and sensory receptors. In other vertebrates, it may include feathers, scales, mucous, and pigmentation. The skeleton supports and protects the body and transmits mechanical forces generated by contractile cells. Among the invertebrates are found hydrostatic skeletons and exoskeletons. Exoskeletons, composed of nonliving material above the epidermis, prevent growth, necessitating periodic molting. Endoskeletons on the other hand, extensive in echinoderms and chordates, are composed of living tissue that can grow. All animals have the ability to move. Muscle tissue is found in most invertebrates and all vertebrates.

REVIEWING CONCEPTS

Fill in the blanks.

INTRODUCTION

1. Epithelial coverings of animals function to _____
_____.

2. Muscle can contract. When it is anchored to the _____, muscle contraction causes movement (locomotion).

EPITHELIAL COVERINGS

Invertebrate epithelium may function in secretion or gas exchange

3. In insects, the outer epithelium secretes a protective and supportive outer covering of nonliving material called the _____.

Vertebrate skin functions in protection and temperature regulation

4. Primate _____ includes nails, hair, sweat glands, oil glands, and sensory receptors.

5. Feathers, scales, mucus coverings, and poison glands are all structures associated with the _____ of vertebrates.

6. The outer layer of skin is called the (a)_____. It consists of several sublayers, or (b)_____.

7. Cells in the _____ continuously divide. As they are pushed upward, these cells mature, produce keratin, and eventually die.

8. Epithelial cells produce an insoluble, elaborately coiled protein called (a)_____, which functions in the skin for (b)_____.

9. The dermis consists of dense, fibrous (a)_____ resting on a layer of (b)_____ tissue composed largely of fat.

SKELETAL SYSTEMS

In hydrostatic skeletons, body fluids transmit force

10. The hydrostatic skeleton transmits forces generated by _____.

11. Many invertebrates (e.g., hydra) have a (a)_____ in which fluid is used to transmit forces generated by contractile cells or muscle. In these animals, contractile cells are arranged in two layers, an outer (b)_____ oriented layer and an inner (c)_____ arranged layer. When the (d)_____ (outer or inner?) layer contracts, the animal becomes shorter and thicker, and when the (e)_____ (outer or inner?) contracts, the animal becomes longer and thinner.

12. Annelid worms have sophisticated hydrostatic skeletons. The body cavity is divided by transverse partitions called _____, creating isolated, fluid-filled segments that can operate independently.

Mollusks and arthropods have nonliving exoskeletons

13. The support system in some organisms is an _____ consisting of a non-living substance overlying the epidermis.

14. The main function of the mollusk exoskeleton is for _____ _____.

15. Arthropod exoskeletons, composed mainly of chiton, are jointed for flexibility. This nonliving skeleton prevents growth, necessitating periodic _____.

Internal skeletons are capable of growth

16. Internal skeletons, called (a)_____, are extensively developed only in the echinoderms and the (b)_____.

17. The endoskeletons of _____ are internal "shells" formed from spicules and plates composed of calcium salts.

The vertebrate skeleton has two main divisions

18. The two main divisions of the vertebrate endoskeleton are the (a)_____ skeleton along the long axis of the body and the (b)_____ skeleton comprised of the bones of the limbs and girdles.

19. Components of the axial skeleton include the (a)_____, which consists of cranial and facial bones; the (b)_____, made up of a series of vertebrae; and the (c)_____, consisting of the sternum and ribs.

20. Components of the appendicular skeleton include the (a)_____, consisting of clavicles and scapulas; the (b)_____, which consists of large, fused hipbones; and the (c)_____, each of which terminates in digits.

21. The long bones of the limbs, such as the radius, have characteristic structures. They are covered by a connective tissue layer called the (a)_____. The ends of these bones are called the (b)_____, and the shaft is the (c)_____. A cartilaginous "growth center" in children called the (d)_____ becomes an (e)_____ in adults. (f)_____, made up of osteons, is a thin, dense, and hard outer shell.

22. Long bones develop from cartilage templates in a process called (a)_____ bone development. Other bones develop from noncartilage connective tissue templates in a process called (b)_____ bone development.

23. Junctions between bones are called (a)_____. They are classified according to the degree of their movement: (b)_____, such as sutures, are tightly bound by fibers; (c)_____, like those between vertebrae; and the most common type of joint, the (d)_____.

MUSCLE CONTRACTION

24. (a)_____ is a ubiquitous contractile protein in eukaryotic cells. In many cells, it functions in association with the contractile protein (b)_____.

Invertebrate muscle varies among groups

25. Most animals move. Most invertebrates have specialized muscles for movements. For example, bivalve mollusks have both smooth and striated muscle. The (a)_____ muscle acts to hold their shells closed for long periods of time, whereas the (b)_____ muscle enables them to shut their shells rapidly.

Insect flight muscles are adapted for rapid contraction

26. _____ contractions are muscle contractions that are not synchronized with signals from motor neurons.

Vertebrate skeletal muscles act antagonistically to one another

27. Skeletal muscles pull on cords of connective tissue called _____, which then pull on bones.

28. Usually, muscles cross a joint and are attached to opposite sides of the joint. At most joints, muscles act (a)_____, which means that the action of one muscle (the agonist) is opposed by the action of the other muscle, called the (b)_____.

A vertebrate muscle may consist of thousands of muscle fibers

29. Each skeletal muscle of a vertebrate consists of bundles of muscle fibers called _____.

30. A unit of actin and myosin filaments makes up a _____.

Contraction occurs when actin and myosin filaments slide past one another

31. A motor neuron releases (a)_____ into the synaptic cleft between the motor neuron and each muscle fiber, where it binds with receptors on

the surface of the muscle fiber, depolarizing the sarcolemma and initiating an
(b)_____.

32. Depolarization of the T tubules stimulates calcium release.
(a)_____ bind to a protein in the troponin complex on the
(b)_____ filaments.

33. During the power stroke of muscle contraction the actin filament is pulled
toward the _____ of the sarcomere.

ATP powers muscle contraction

34. _____ is the immediate energy source for muscle contraction.

35. _____ is the backup energy storage compound in
muscle cells.

36. _____ is the chemical energy stored in muscle fibers.

The strength of muscle contraction varies

37. The more motor units that are recruited for a task, the _____ the
contraction will be.

Muscle fibers may be specialized for slow or quick responses

38. Skeletal muscle contains two sets of specialized fibers: the (a)_____, which
are specialized for slow responses and the (b)_____, specialized for
fast responses.

Smooth muscle and cardiac muscle are involuntary

39. (a)_____ muscle is capable of slow, sustained contractions and
(b)_____ muscle contracts rhythmically.

BUILDING WORDS

Use combinations of prefixes and suffixes to build words for the definitions that follow.

Prefixes	The Meaning	Suffixes	The Meaning
endo-	within	-blast	embryo, "formative cell"
epi-	upon, over, on	-chondr(al)	cartilage
myo-	muscle	-dermis	skin
osteo-	bone	-oste(um)	bone
peri-	about, around, beyond		

Prefix	Suffix	Definition
_____	_____	1. The outermost layer of skin, resting on the dermis.
_____	_____	2. Connective tissue membrane on the surface of bone that is capable of forming bone.
_____	_____	3. Pertains to the occurrence or formation of "something" within cartilage.
_____	_____	4. A bone-forming cell.
_____	-clast	5. Cells that break down bone.
_____	-filament	6. Subunit of myofibril consisting of either actin or myosin.

MATCHING
Terms:

a. Actin
b. Axial skeleton
c. Endoskeleton
d. Exoskeleton
e. Joint

f. Keratin
g. Ligament
h. Motor unit
i. Oxygen debt
j. Sarcomere

k. Skeletal muscle
l. Stratum basale
m. Tendon
n. Vertebral column

For each of these definitions, select the correct matching term from the list above.

_____ 1. A motor neuron and the muscle fibers it stimulates.

_____ 2. A water-insoluble, elaborately coiled protein manufactured by epidermal cells that gives skin mechanical strength and flexibility.

_____ 3. A segment of a striated muscle cell that serves as the basic unit of muscle contraction.

_____ 4. The deepest layer of the epidermis, consisting of cells that continuously divide.

_____ 5. A band of connective tissue that connects bones and limits movement at the joints.

_____ 6. The skull, vertebral column, sternum, and ribs.

_____ 7. The hard exterior covering of certain invertebrates.

_____ 8. Tough cords of connective tissue that anchor muscles to bone.

_____ 9. The oxygen necessary to metabolize the lactic acid produced during strenuous exercise.

_____10. The protein composing one type of myofilament.

MAKING COMPARISONS
Fill in the blanks.

Organism	Skeletal Material	Kind of Skeleton	Characteristics
Many invertebrates	Fluid	Hydrostatic	Contractile tissue generates force that moves the body
Arthropods	#1	#2	#3
#4	#5	Endoskeleton	Internal shell, some have spines that project to the outer surface
Chordates	#6	#7	#8

MAKING CHOICES
Place your answer(s) in the space provided. Some questions may have more than one correct answer.

_____ 1. A major component of the arthropod skeleton is the polysaccharide

    a. keratin.
    b. melanin.
    c. chitin.

    d. actin.
    e. myosin.

_____ 2. Actin filaments contain

    a. actin.
    b. tropomyosin.
    c. troponin complex.

    d. myosin.
    e. keratin.

_____ 3. The H zone in a sarcomere consists of
   a. actin.
   b. tropomyosin.
   c. troponin complex.
   d. myosin.
   e. keratin.

_____ 4. The human skull is part of the
   a. cervical complex.
   b. girdle.
   c. axial skeleton.
   d. appendicular skeleton.
   e. atlas.

_____ 5. Most of the mechanical strength of bone is due to
   a. spongy bone.
   b. osteocytes.
   c. intramembranous bone.
   d. endochondral bone.
   e. marrow.

_____ 6. You perceive touch, pain, and temperature through sense organs in your
   a. epidermis.
   b. stratum basale.
   c. stratum corneum.
   d. epithelium.
   e. dermis.

_____ 7. Myofilaments are composed of
   a. myofibrils.
   b. fibers.
   c. actin.
   d. myosin.
   e. sarcoplasmic reticulum.

_____ 8. The outer layer of a vertebrate's skin is the
   a. epidermis.
   b. stratum basale.
   c. stratum corneum.
   d. epithelium.
   e. dermis.

_____ 9. The tissue around bones that lays down new layers of bone is the
   a. metaphysis.
   b. epiphysis.
   c. marrow.
   d. periosteum.
   e. endosteum.

_____ 10. The human axial skeleton includes the
   a. ulna.
   b. shoulder blades.
   c. skull.
   d. femur.
   e. breastbone.

_____ 11. Cartilaginous growth centers in children are called
   a. metaphysis.
   b. epiphysis.
   c. marrow.
   d. periosteum.
   e. endosteum.

_____ 12. The human appendicular skeleton includes the
   a. ulna.
   b. shoulder blades.
   c. centrum.
   d. femur.
   e. breastbone.

_____ 13. Vertebrate appendages are connected to
   a. the cervical complex.
   b. girdles.
   c. the axial skeleton.
   d. the appendicular skeleton.
   e. the atlas.

_____14. Skeletons that operate entirely by hydrostatic pressure are found in
    a. annelids.   d. lobsters and crayfish.
    b. echinoderms.   e. vertebrates.
    c. hydra.

_____15. An opposable digit is found in
    a. lobsters.   d. some insects.
    b. great apes.   e. humans.
    c. crayfish.

_____16. Internal skeletons are found in
    a. annelids.   d. lobsters and crayfish.
    b. echinoderms.   e. chordates.
    c. hydra.

_____17. External skeletons are found in
    a. annelids.   d. insects.
    b. echinoderms.   e. chordates.
    c. hydra.

_____ 18. The "space" between the terminal axon of one neuron and a dendrite of the next neuron is called the
    a. synaptic cleft.   d. Z line.
    b. chemotrasmitter zone.   e. binding site.
    c. cross bridge.

VISUAL FOUNDATIONS

Color the parts of the illustration below as indicated. Also label H zone, A band, I band, sarcomere, and muscle fiber.

RED   ❑   Z line
GREEN   ❑   mitochondria
YELLOW   ❑   sarcoplasmic reticulum
BLUE   ❑   T tubule
ORANGE   ❑   myofibrils
TAN   ❑   sarcomere
PINK   ❑   sarcolemma
VIOLET   ❑   nucleus

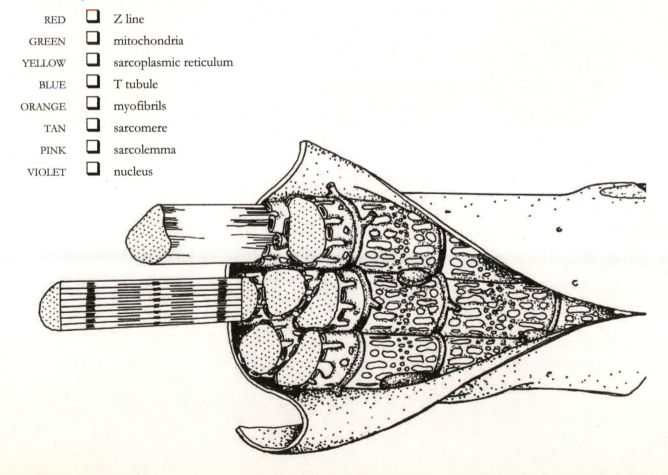

CHAPTER 40

Neural Signaling

Behavior and physiological processes in animals are regulated by the endocrine and nervous systems. Endocrine regulation is generally slow and long lasting, whereas nervous regulation is typically rapid and brief. Changes within an animal or in the outside world are detected by receptors and transmitted as electrical signals to the central nervous system where the information is sorted and interpreted. An appropriate response is then sent to muscle or glands where the response occurs. The structural and functional unit that carries the impulses is the neuron. This chapter focuses on the role of the neuron in nervous regulation. The typical neuron consists of a cell body, dendrites, and an axon. The cell body houses the nucleus and has branched cytoplasmic projections called dendrites that receive stimuli. The axon is a single, long structure that transmits impulses to an adjacent neuron, muscle, or gland. Transmission involves changes in ion distribution between the inside and outside of the neuron. Transmission from one neuron to the next generally involves the release of chemicals into the space between neurons. These chemicals diffuse across the space and bring about a response in the adjacent neuron, muscle, or gland. Hundreds of messages may arrive at a single neuron at the same time, messages that must be integrated before the neuron can respond. After the neuron has completed its integration, it may or may not initiate an impulse along its axon.

REVIEWING CONCEPTS

Fill in the blanks

INTRODUCTION

1. The ability of an organism to survive is largely dependent on its ability to detect and respond appropriately to _____, which are defined as changes in either the internal or external environments.

2. In complex animals, two systems are responsible for the regulation of physiological processes and behavior. Regulation by the (a)_____ system is slow and long-lasting, whereas regulation by the (b)_____ system is rapid.

INFORMATION FLOW THROUGH THE NERVOUS SYSTEM

3. Information flow through the nervous system begins with the detection of a stimulus by a "sensory device," a process called (a)_____. The "sensory device" is associated with sensory or (b)_____ neurons that (c)_____ or relay the information to the central nervous system (CNS).

4. Association neurons, also called (a)_____, in the CNS "manage" the information, sorting and interpreting the information in a process called (b)_____. Once the appropriate response is selected, the information is transmitted from the CNS by motor or (c)_____

neurons to muscles and/or glands, the (d)_____, which carry out the response.

NEURONS AND GLIAL CELLS

A typical neuron consists of a cell body, dendrites, and an axon

5. Neurons are highly specialized cells with a distinctive structure consisting of a cell body and two types of cytoplasmic extensions. Numerous, short (a)_____ receive impulses and conduct them to the (b)_____, which integrates impulses. Once integrated, the impulse is conducted away by the (c)_____, a long process that terminates at another neuron or effector.

6. Branches at the ends of axons, called (a)_____, form tiny (b)_____ that release a transmitter chemical called a (c)_____.

7. Many axons outside of the vertebrate CNS are surrounded by a series of (a)_____ that form an insulating covering called the (b)_____. Gaps in this insulating covering, called (c)_____, occur between successive Schwann cells.

8. A (a)_____ is a bundle of axons outside the CNS, whereas a (b)_____ is a bundle of axons within the CNS.

9. Masses of cell bodies outside the CNS are called (a)_____, whereas masses of cell bodies within the CNS are called (b)_____.

Glial cells provide metabolic and structural support

10. (a)_____ are star-shaped glial cells that provide physical support for neurons, while (b)_____ are glial cells that secrete growth factors such as nerve growth factor.

TRANSMITING INFORMATION ALONG THE NEURON

The neuron membrane has a resting potential

11. The membrane potential in a resting neuron or muscle cell is its (a)_____. Its value is expressed as (b)_____.

12. When a neuron is not conducting an electrical impulse, the _____(inner or outer?) surface of the plasma membrane is negatively charged compared to the extracellular fluid.

13. In most cells, potassium ion concentration is highest (a)_____ (inside or outside?) the cell and sodium ion concentration is highest) (b)_____ (inside or outside?) the cell.

14. The ionic balance across the plasma membrane of neurons is a result of several factors. The _____ is an active transport system that moves sodium out of the cell and potassium into the cell.

Graded local signals vary in magnitude

15. Depolarization is described as an (a)_____ action as compared to hyperpolarization, which is described as an (b)_____ action.

An action potential is generated by an influx of Na⁺ and an efflux of K⁺

16. If a stimulus is strong enough, it will provoke a response from the neuron called an (a)_____, which is an alteration of the resting potential caused by an increase in membrane permeability to (b)_____.

17. (a)_____ in the plasma membrane of the neuron open when they detect a critical level of voltage change in the membrane potential. This value, known as the (b)_____, is generally given as about (c)_____.

18. A _____ is a sharp rise and fall of an action potential.

19. When depolarization reaches threshold level, an action potential may be generated. An action potential is a _____ that spreads along the axon.

20. As the action potential moves down the axon, _____ occurs behind it.

21. In saltatory conduction in myelinated neurons, depolarization skips along the axon from one _____ to the next.

22. Once a stimulus depolarizes a neuron to the threshold, the neuron will respond fully with an action potential. All action potentials have the same strength. A response that always occurs at a given value, or it won't occur at all, is referred to as an _____ response.

NEURAL SIGNALING ACROSS SYNAPSES

23. A junction between two neurons, or between a neuron and an effector, is called a _____.

24. The neuron that terminates at a synapse is known as the (a)_____ neuron, and the neuron that begins at that synapse is called the (b)_____ neuron.

Signals across synapses can be electrical or chemical

25. In (a)_____, the pre- and postsynaptic neurons occur very close together and form gap junctions. Most synapses are (b)_____ that involve (c)_____ that cross the (d)_____ between neurons.

Neurons use neurotransmitters to signal other cells

26. Cells that release (a)_____ are called cholinergic neurons. This neurotransmitter functions at (b)_____ junctions and some synapses in the brain and autonomic nervous system.

27. (a)_____ release norepinephrine, which, like dopamine, is a (b)_____.

Neurotransmitters bind with receptors on postsynaptic cells

28. Neurotransmitters, or (a)_____, are stored in the synaptic terminals within small membrane-bounded sacs called (b)_____.

Activated receptors can send excitatory or inhibitory signals

29. A depolarization of the postsynaptic membrane that brings the neuron closer to firing is called an (a)_____. A hyperpolarization of the postsynaptic membrane that reduces the probability that the neuron will fire is called an (b)_____.

NEURAL INTEGRATION

30. Local responses in the postsynaptic membrane that vary in magnitude, fade over distance, and can be summated are called _____. EPSPs and IPSPs are examples.

31. Graded potentials can be added together in a process called (a)_____. When a second EPSP occurs before the depolarization caused by the first EPSP has decayed, (b)_____ occurs. In (c)_____, several synapses in the same area generate EPSPs simultaneously. Adding EPSPs together can bring the neuron to threshold.

NEURAL CIRCUITS

32. A _____ circuit is a neural pathway arranged so that an axon collateral synapses with an interneuron.

BUILDING WORDS

Use combinations of prefixes and suffixes to build words for the definitions that follow.

Prefixes	The Meaning	Suffixes	The Meaning
inter-	between, among	-neuro(n)	nerve
multi-	many, much, multiple		
neuro-	nerve		
post-	behind, after		
pre-	before, prior to, in advance of		

Prefix	Suffix	Definition
_____	_____	1. A nerve cell that carries impulses from one nerve cell to another, and is between a sense receptor and an effector.
_____	-glia	2. Cells providing support and protection for neurons.
_____	-transmitter	3. Substance used by neurons to transmit impulses across a synapse.
_____	-polar	4. Pertains to a neuron with more than two (often many) processes or projections.
_____	-synaptic	5. Pertains to a neuron that begins after a specific synapse.
_____	-synaptic	6. Pertains to a neuron that ends before a specific synapse.

MATCHING

Terms:

a. Action potential
b. Axon
c. Convergence
d. Dendrite
e. Facilitation

f. Ganglion
g. Integration
h. Myelin sheath
i. Nerve
j. Neuron

k. Refractory period
l. Schwann cell
m. Summation
n. Synapse
o. Threshold level

For each of these definitions, select the correct matching term from the list above.

_____ 1. One of many projections from a nerve cell that conducts a nerve impulse toward the cell body.

_____ 2. A nerve cell; a conducting cell of the nervous system that typically consists of a cell body, dendrites, and an axon.

_____ 3. A mass of neuron cell bodies located outside the central nervous system.

_____ 4. The long extension of the neuron that transmits nerve impulses away from the cell body.

_____ 5. A large bundle of axons wrapped together in connective tissue that are located outside the CNS.

_____ 6. The brief period of time that must elapse after the response of a neuron or muscle cell, during which it cannot respond to another stimulus.

_____ 7. The electrical activity developed in a muscle or nerve cell during activity.

_____ 8. An insulating covering around axons of certain neurons.

_____ 9. The junction between two neurons or between a neuron and an effector.

MAKING COMPARISONS

Fill in the blanks.

Kind of Potential	mV	Membrane Response	Ion Channel Activity
Membrane potential	-70mV	Stable	Sodium-potassium pump active, passive sodium and potassium channels open
Resting potential	#1	#2	#3
#4	From -50 to -55mV	Depolarization	#5
Action potential	#6	#7	Voltage-activated sodium ion channels and potassium ion channels open
EPSP	Increasingly positive	#8	#9
IPSP	Increasingly negative	#10	Neurotransmitter-receptor combination may open potassium ion channels or chloride ion channels

MAKING CHOICES

Place your answer(s) in the space provided. Some questions may have more than one correct answer.

_____ 1. The inner surface of a resting neuron is _____ compared with the outside.

    a. positively charged   d. 70 mV

    b. negatively charged   e. −70 mV

    c. polarized

_____ 2. The part of the neuron that transmits an impulse from the cell body to an effector cell is the

    a. dendrite.   d. collateral.

    b. axon.   e. hillock.

    c. Schwann body.

_____ 3. Saltatory conduction

    a. occurs between nodes of Ranvier.   d. requires less energy than continuous conduction.

    b. occurs only in the CNS.   e. involves depolarization at nodes of Ranvier.

    c. is more rapid than the continuous type.

_____ 4. The nodes of Ranvier are

    a. on the cell body.   d. gaps between adjacent Schwann cells.

    b. on the dendrites.   e. insulated with myelin.

    c. on the axon.

_____ 5. A neuron that begins at a synapse is called a

    a. presynaptic neuron.   d. neurotransmitter.

    b. postsynaptic neuron.   e. acetylcholine releaser.

    c. synapsing neuron.

_____ 6. Which of the following correctly expresses the movement of ions by the sodium-potassium pump?

    a. sodium out, potassium in   d. more sodium out than potassium in

    b. sodium in, potassium out   e. less sodium out than potassium in

    c. sodium and potassium both in and out, but in different amounts

_____ 7. A nerve pathway or tract in the CNS is

    a. one neuron.   d. a ganglion.

    b. a group of nerves.   e. a bundle of cell bodies.

    c. a bundle of axons.

_____ 8. An axon cannot transmit an action potential no matter how great a stimulus is applied when it is

    a. hyperpolarized.   d. in the relative refractory period.

    b. depolarized.   e. in the resting state.

    c. in the absolute refractory period.

_____ 9. If a neurotransmitter hyperpolarizes a postsynaptic membrane, the change in potential is referred to as

    a. spatial summation.   d. IPSP.

    b. temporal summation.   e. EPSP.

    c. a threshold level impulse.

_____ 10. Aside from the sodium-potassium pump, the resting potential is due mainly to

    a. inward diffusion of chloride ions.   d. open sodium channels in the membrane.

    b. inward diffusion of sodium ions.   e. large protein anions inside the cell.

    c. outward diffusion of potassium ions.

_____ 11. Branchlike extensions of the cell body involved in receiving stimuli are

    a. dendrites.   d. collaterals.

    b. axons.   e. hillocks.

    c. Schwann bodies.

VISUAL FOUNDATIONS

Color the parts of the illustration below as indicated. Also label nodes of Ranvier and Schwann cell.

RED	❑	synaptic terminals
GREEN	❑	terminal branches
YELLOW	❑	myelin sheath
BLUE	❑	axon
ORANGE	❑	cellular sheath
PINK	❑	cell body and dendrite
VIOLET	❑	nucleus of neuron

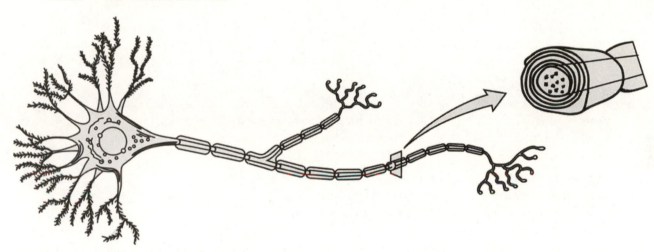

Color the parts of the illustration below as indicated. Also label extracellular fluid and cytoplasm.

RED	❑	sodium ion	BROWN	❑	potassium channel
GREEN	❑	potassium ion	TAN	❑	large anions
YELLOW	❑	plasma membrane	PINK	❑	sodium-potassium pump
BLUE	❑	arrows indicating diffusion into the cell	VIOLET	❑	arrows for diffusion out of the cell
ORANGE	❑	sodium channel			

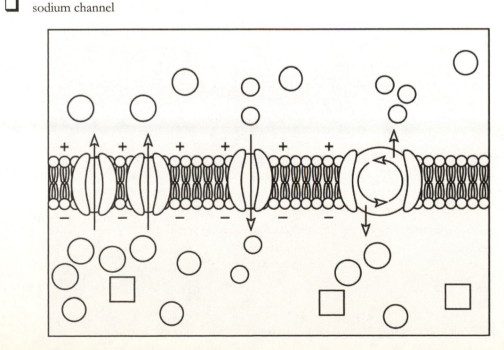

□

CHAPTER 41

Neural Regulation

This chapter compares several animal nervous systems, and then examines the vertebrate nervous system, with emphasis on the human brain. Also examined are the cellular mechanisms of memory and learning. The simplest organized nervous system is the nerve net found in cnidarians. A nerve net consists of nerve cells scattered throughout the body; there is no central control organ and no definite nervous pathways. The nervous system of echinoderms is more complex, with a nerve ring and nerves that extend into various parts of the body. Bilaterally symmetric animals have more complex nervous systems. The vertebrate nervous system has two main divisions: a central nervous system (CNS) consisting of a complex tubular brain that is continuous with a tubular nerve cord, and a peripheral nervous system (PNS). The CNS provides centralized control, integrating incoming information and determining appropriate responses. The PNS consists of sensory receptors and cranial and spinal nerves. It functions to regulate the organism's internal environment and to help it adjust to its external environment. Learning is a function of the brain, involving the storage of information and its retrieval. Many drugs affect the nervous system, some by changing the levels of neurotransmitters within the brain.

REVIEWING CONCEPTS

Fill in the blanks.

INTRODUCTION

INVERTEBRATE NERVOUS SYSTEMS

1. In general, the lifestyle of an animal is closely related to the design and complexity of its nervous system. For example, the simplest nervous system, called a (a)_____, is found in (b)_____, which includes sessile animals such as Hydra.

2. Echinoderms have a modified nerve net. In this system, a _____ surrounds the mouth, and large, radial nerves extend into each arm. Branches from these nerves coordinate the animal's movements.

3. Planarian worms have a (a)_____ nervous system with concentrations of nerve cells in the head called (b)_____; these serve as a primitive brain.

4. Annelids have a ventral nerve cord with the cell bodies of many of the neurons massed into _____.

ORGANIZATION OF THE VERTEBRATE NERVOUS SYSTEM

5. The vertebrate nervous system is divided into a (a)_____ system consisting of the brain and spinal cord, and a (b)_____ system consisting of sensory receptors and nerves.

6. The PNS system is divided into the somatic division, which is concerned with changes in the external environment, and the autonomic division, which regulates the internal environment. The autonomic division has two efferent pathways made up of _____ nerves.

EVOLUTION OF THE VERTEBRATE BRAIN

7. The _____ of the vertebrate embryo differentiates anteriorly into the brain and posteriorly into the spinal cord.

The hindbrain develops into the medulla, pons, and cerebellum

8. The (a)_____, which coordinates muscle activity, and the (b)_____, which connects the spinal cord and medulla with upper parts of the brain, are formed from the (c)_____. The (d)_____, made up largely of nerve tracks, is formed from the (e)_____.

9. The _____ is made up of the medulla, pons, and midbrain.

The midbrain is prominent in fishes and amphibians

10. The midbrain is the most prominent part of the brain in fish and amphibians. It is their main _____ area, linking sensory input and motor output.

11. In mammals, the midbrain consists of the (a)_____ which regulate visual reflexes, and the (b)_____, which regulate some auditory reflexes. It also contains the (c)_____ that integrates information about muscle tone and posture.

The forebrain gives rise to the thalamus, hypothalamus, and cerebrum

12. The forebrain, or proencephalon, differentiates into the diencephalon and the telencephalon, which in turn develop into the (a)_____ and (b)_____ respectively.

13. The right and left cerebral (a)_____ are predominantly comprised of axons, which are collectively referred to as (b)_____. Mammals and most reptiles have a layer of grey matter called the (c)_____ that makes up the outer portion of the cerebrum. In complex mammals the surface area of the cortex is increased by numerous folds called (d)_____.

THE HUMAN CENTRAL NERVOUS SYSTEM

14. The human central nervous system consists of the brain and spinal cord. Both are protected by bone and three meninges, the (a)_____ _____, and both are bathed by cerebrospinal fluid produced in a specialized capillary bed called the (b)_____.

The spinal cord transmits impulses to and from the brain

15. The spinal cord extends from the base of the brain to the _____ _____ vertebra.

16. Grey matter in a cross section of the spinal cord is shaped like the letter "H." It is surrounded by (a)_____ that contains nerve pathways, or (b)_____.

The most prominent part of the human brain is the cerebrum

17. The human cerebral cortex consists of three functional areas. (a)_____ areas receive incoming sensory information, (b)_____ areas control voluntary movement, and (c)_____ areas link the other two areas.

18. Each hemisphere of the human cerebrum is divided into lobes. (a)_____ contain primary motor areas and are separated from the primary sensory areas in the (b)_____ lobes by a groove called the (c)_____.

Brain activity cycles in a sleep-wake pattern

19. The reticular activating system (RAS) is responsible for maintaining consciousness. It is located within the _____.

20. Electrical activity in the form of brain waves can be detected by a device that produces brain wave tracings called an (a)_____. The slow (b)_____ waves are associated with certain stages of sleep, while (c)_____ waves are associated with relaxation, and (d)_____ waves are generated by an active, thinking brain.

21. There are two main stages of sleep: (a)_____ sleep, which is associated with dreaming, and (b)_____ sleep, which is characterized by theta and delta waves.

The limbic system affects emotional aspects of behavior

22. The limbic system affects the emotional aspects of behavior, evaluates rewards, and is important in motivation. It consists of parts of the _____ as well as parts of the thalamus, and hypothalamus, several nuclei in the midbrain, and the neural pathways that connect these structures.

INFORMATION PROCESSING

Learning involves the storage of information and its retrieval

23. (a)_____ is the process by which information is encoded, stored, and retrieved; (b)_____ is unconscious memory for perceptual and motor skills; and (c)_____ involves factual knowledge of people, places, or objects, and requires conscious recall of the information.

Language involves comprehension and expression

24. Wernicke's area is located in the _____ and is an important center for language comprehension.

THE PERIPHERAL NERVOUS SYSTEM

The somatic division helps the body adjust to the external environment

25. The somatic nervous system consists of (a)_____ that detect changes in the external environment, (b)_____ that transmit information to the CNS, and (c)_____ that adjust the positions of the skeletal muscles that help maintain the body's posture and balance.

26. Neurons are organized into nerves. The 12 pairs of (a)_____ nerves connect to the brain and are largely involved with sense receptors. Thirty-one pairs of (b)_____ nerves connect to the spinal cord.

The autonomic division regulates the internal environment

27. The efferent component of the autonomic division is divided into two systems. The (a)_____ system frequently operates to stimulate organs and to mobilize energy. The (b)_____ system influences organs to conserve and restore energy.

28. The autonomic system uses two neurons between the CNS and the effector. The first neuron is called the (a)_____ and the second is called the (b)_____.

EFFECTS OF DRUGS ON THE NERVOUS SYSTEM

29. Habitual use of almost any mood-altering drug can result in _____, in which the user becomes emotionally dependent on the drug.

BUILDING WORDS

Use combinations of prefixes and suffixes to build words for the definitions that follow.

Prefixes	**The Meaning**
hypo-	under, below
para-	beside, near
post-	behind, after
pre-	before, prior to, in advance of

Prefix	**Suffix**	**Definition**
_____	-thalamus	1. Part of the brain located below the thalamus; principal integration center for the regulation of the viscera.
_____	-ganglionic	2. Pertains to a neuron located distal to (after) a ganglion.
_____	-ganglionic	3. Pertains to a neuron located proximal to (before) a ganglion.
_____	-vertebral	4. Pertains to structures located beside the vertebral column.

MATCHING

Terms:

a. Autonomic nervous system
b. Central nervous system
c. Cerebellum
d. Cerebral cortex
e. Corpus callosum
f. Gyrus
g. Limbic system
h. Meninges
i. Parasympathetic nervous system
j. Peripheral nervous system
k. Pons
l. Sensory areas
m. Somatic nervous system
n. Thalamus
o. Brain ventricle

For each of these definitions, select the correct matching term from the list above.

_____ 1. An action system of the brain that plays a role in emotional responses.

_____ 2. The outer layer of the cerebrum, composed of gray matter and consisting of densely-packed nerve cells.

_____ 3. The receptors and nerves that lie outside of the central nervous system.

_____ 4. The subdivision of the brain that coordinates muscle activity and is responsible for muscle tone, posture, and equilibrium.

_____ 5. The connective tissue that envelops the brain and spinal cord.

_____ 6. The portion of the peripheral nervous system that controls the visceral functions of the body.

_____ 7. A large bundle of nerve fibers interconnecting the two cerebral hemispheres.

_____ 8. The bulge on the anterior surface of the brainstem between the medulla and midbrain; connects various parts of the brain.

_____ 9. The nervous system consisting of the brain and spinal column.

_____10. A division of the autonomic nervous system concerned primarily with storing and restoring energy.

MAKING COMPARISONS
Fill in the blanks.

Human Brain Structure	Subdivision or Location of Structure	Function of Structure
Medulla	Myelencephalon	Contains vital centers and other reflex centers
#1	Metencephalon	Connects various parts of the brain
#2	Mesencephalon	#3
Thalamus	#4	#5
#6	#7	Controls autonomic functions; links nervous and endocrine systems; controls temperature, appetite, and fluid balance; involved in some emotional and sexual responses
Cerebellum	#8	Responsible for muscle tone, posture, and equilibrium
#9	Telencephalon	Complex association functions
#10	Brain stem and thalamus	Arousal system
#11	Certain structures of the cerebrum, diencephalon	Affect emotional aspects of behavior, motivation, sexual behavior, autonomic responses, and biological rhythms

MAKING CHOICES

Place your answer(s) in the space provided. Some questions may have more than one correct answer.

_____ 1. The largest part of the human brain is the
   a. cerebrum.
   b. cerebellum.
   c. myelencephalon.
   d. medulla oblongata.
   e. midbrain.

_____ 2. The part of the human brain that is the center for intellect, memory, and language is the
   a. cerebrum.
   b. cerebellum.
   c. myelencephalon.
   d. medulla oblongata.
   e. midbrain.

_____ 3. Skeletal muscles are controlled by the
   a. parietal lobes.
   b. frontal lobes.
   c. central sulcus.
   d. temporal lobes.
   e. mesolimbic ganglia.

_____ 4. Cerebrospinal fluid (CSF)
   a. is produced by choroid plexi.
   b. circulates through ventricles.
   c. is between the arachnoid and pia mater.
   d. is within layers of the meninges.
   e. is in the subarachnoid space.

_____ 5. In vertebrates, the cerebrum is typically
   a. divided into two hemispheres.
   b. mostly gray matter.
   c. mainly cell bodies and some sensory neurons.
   d. mostly white matter.
   e. mainly axons connecting parts of the brain.

_____ 6. Regulation of body temperature under ordinary circumstances is under the control of the
   a. sympathetic n.s.
   b. autonomic n.s.
   c. forebrain.
   d. cranial nerve VI.
   e. dorsal root ganglion.

_____ 7. When you are eating one of your favorite foods, the cranial nerve(s) directly involved in your perception of this pleasurable experience is/are
   a. facial.
   b. VII.
   c. X.
   d. XIX.
   e. XI.

_____ 8. The regulation of body temperature, appetite, and fluid balance is mainly controlled by the
   a. cerebellum.
   b. cerebrum.
   c. thalamus.
   d. hypothalamus.
   e. red nucleus.

_____ 9. In non-REM sleep, as compared to REM sleep, there is/are
   a. higher-amplitude delta waves.
   b. faster breathing.
   c. lower blood pressure.
   d. more dream consciousness.
   e. more norepinephrine released.

_____10. The brain and spinal cord are wrapped in connective tissue called
   a. gray matter.
   b. meninges.
   c. gyri.
   d. sulcus.
   e. neopallium.

_____11. The part of the mammalian brain that integrates information about posture and muscle tone is the
    a. cerebrum.
    d. medulla oblongata.
    b. cerebellum.
    e. midbrain.
    c. myelencephalon.

_____12. The part(s) of the brain that coordinate(s) muscular activity is/are the
    a. cerebrum.
    d. medulla oblongata.
    b. cerebellum.
    e. midbrain.
    c. myelencephalon.

_____13. If you are having strong sexual feelings one minute and enraged over little or nothing the next minute, chances are very good that you have just had a very active
    a. frontal lobe.
    d. reticular activating system (RAS).
    b. limbic system.
    e. cerebellum.
    c. thalamus.

_____14. If only one temporal lobe is damaged, one might expect
    a. blindness in one eye.
    d. partial loss of hearing in both ears.
    b. total blindness.
    e. inability to detect taste.
    c. decrease in hearing acuity in both ears.

_____15. The limbic system
    a. is important in motivation.
    d. affects sexual behavior.
    b. is involved in biological rhythms.
    e. plays a role in autonomic responses.
    c. is present in all mammals.

VISUAL FOUNDATIONS

Color the parts of the illustration below as indicated. Also label sensory neuron, interneuron, and motor neuron.

    RED   ☐ muscle
    YELLOW ☐ receptor
    ORANGE ☐ location of cell bodies of sensory neuron
    TAN ☐ central nervous system

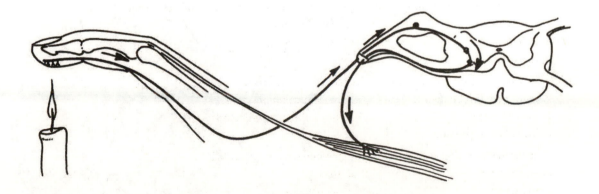

Color the parts of the illustration below as indicated. Also label sulcus. Indicate the vertebrate class represented by each brain illustration.

RED	☐	cerebrum
GREEN	☐	olfactory bulb, tract, and lobe
YELLOW	☐	corpus striatum
BLUE	☐	cerebellum
ORANGE	☐	optic lobe
BROWN	☐	epiphysis
TAN	☐	medulla
VIOLET	☐	diencephalon

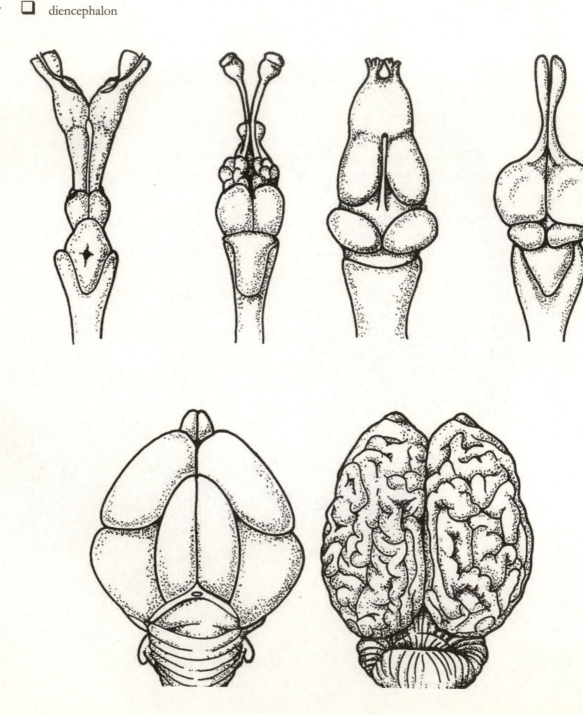

Color the parts of the illustration below as indicated. Also label diencephalon and midbrain.

RED ❑ pituitary
GREEN ❑ thalamus
YELLOW ❑ spinal cord
BLUE ❑ hypothalamus
ORANGE ❑ medulla and pons
BROWN ❑ cerebrum
TAN ❑ cerebellum
PINK ❑ corpus callosum
VIOLET ❑ pineal body

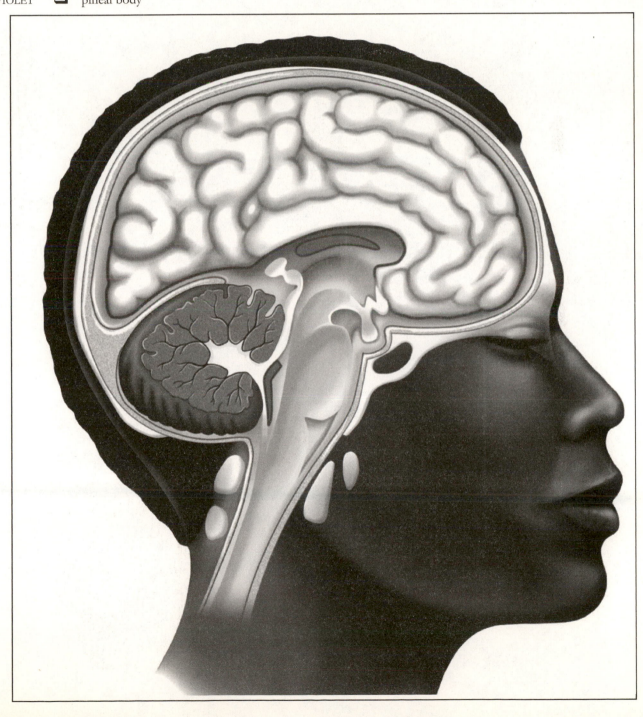

CHAPTER 42

Sensory Systems

Sense organs are specialized structures with receptor cells that detect changes in the internal or external environment and transmit the information to the nervous system. Receptor cells may be neuron endings or specialized cells in contact with neurons. Sensory receptors can be classified according to the location of the stimuli to which they respond or to the type of energy they transduce. They work by absorbing energy and converting it into electrical energy that depolarizes the receptor cell. When the state of depolarization reaches a threshold level, an impulse is generated in the axon, thereby transmitting the information to the CNS. Sense organs that respond to heat and cold provide important cues about body temperature, and help some animals locate a warm-blooded host. Those that detect electrical energy are found in some fish; they can detect electric currents in water, and in some species can generate a shock for defense or to stun prey. Sense organs that detect magnetic fields are used for orientation and migration, and those that detect physical force are responsible for pain reception. Sense organs that respond to mechanical energy include the various tactile receptors in the skin, the lateral line organs in fish, the receptors that respond continuously to tension and movement in muscles and joints, the receptors that respond to body position, and the receptors responsible for hearing. Sense organs that respond to chemical energy are responsible for the sense of taste and smell, and those that respond to light energy are responsible for vision.

REVIEWING CONCEPTS

Fill in the blanks.

INTRODUCTION

1. Dolphins, bats, and a few other vertebrates detect distant objects by _____, sometimes called bisonar.

HOW SENSORY SYSTEMS WORK

Sensory receptors receive information

2. In the process of _____, sensory receptors absorb a small amount of energy from some stimulus.

Sensory receptors transduce energy

3. Various forms of energy are transduced into (a)_____ energy by receptor cells that then produce a (b)_____ potential.

4. A receptor potential is a kind of (a)_____ response. If it (b)_____ the sensory neuron to threshold, it initiates an (c)_____ that travels along the (d)_____ neuron to the central nervous system.

Sensory input is integrated at many levels

5. Sensation is determined by the part of the _____ that receives the message from the sense organ.

6. Impulses from sense organs are encoded into messages based upon the (a)_____ of neurons (or fibers) carrying the message and the (b)_____ of action potentials carried by the neuron (fiber).

7. A decrease in the frequency of action potentials in a sensory neuron, even though the stimulus is maintained, is called _____.

TYPES OF SENSORY RECEPTORS

8. Sense receptors can also be classified according to the source of the stimuli to which they respond. Those that detect stimuli in the external environment are called (a)_____; and (b)_____ detect changes in pH, osmotic pressure, body temperature, and the chemical composition of the blood.

THERMORECEPTORS

9. Two types of snakes, the _____ and _____, use thermoreceptors to locate prey.

10. Thermoreceptors that detect internal temperature changes in mammals are found in the _____ of the brain.

ELECTRORECEPTORS AND ELECTROMAGNETIC RECEPTORS

11. Some electroreceptors, called _____, are sensitive enough to detect Earth's magnetic field.

NOCICEPTORS

12. Nociceptors transmit signals through sensory neurons to interneurons in the (a)_____. The sensory neurons release the neurotransmitter (b)_____ and several neuropeptides, including (c)_____.

MECHANORECEPTORS

Touch receptors are located in the skin

13. The tactile receptors in the skin are _____ that respond to mechanical displacement of hairs or of the receptor cells themselves.

14. _____ are sensitive to deep pressure that causes rapid movement of tissues.

15. In human skin, the mechanoreceptors that all sense touch in one form or another are (b)_____; free nerve endings detect touch, pressure, and (c)_____.

Proprioceptors help coordinate muscle movement

16. Vertebrates have three categories of proprioceptors:
(a)_____, which detect muscle movement;
(b)_____, which respond to tension in contracting muscles and in the tendons that attach muscle to bone; and
(c)_____, which detect movement in ligaments.

Many invertebrates have gravity receptors called statocysts

17. (a)_____ are tiny granules that stimulate (b)_____ when pulled down by gravity.

18. Gravity receptors in some invertebrates, e.g., jelly fish and crayfish, are called _____.

Hair cells are characterized by stereocilia

19. A _____ is a true cilium with a 9X2 arrangement of microtubules.

Lateral line organs supplement vision in fishes

20. Lateral line organs detect _____ in the water.

21. Lateral line organs consist of a (a)_____ lined with (b)_____ that runs the length of the animal's lateral surface. Water disturbances move the (c)_____ on the end of hairs in the receptor cells, generating an electrical response.

The vestibular apparatus maintains equilibrium

22. Gravity detectors in the form of ear stones called (a)_____ are housed in the (b)_____ of the vestibular apparatus.

23. Turning movements, referred to as (a)_____, cause movement of a fluid called (b)_____ in the semicircular canals, which in turn stimulate the hair cells of the (c)_____.

Auditory receptors are located in the cochlea

24. Auditory receptors in an inner ear structure of birds and mammals, called the (a)_____, contain (b)_____ that detect pressure waves.

25. In terrestrial vertebrates, sound waves first initiate vibrations in the eardrum, or (a)_____. Three tiny ear bones, the (b)_____, _____, and _____ transmit the vibration to fluids in the inner ear through an opening called the (c)_____.

26. Fluids in the cochlea, in response to vibrations from the oval window, initiate vibrations in the (a)_____, which in turn cause stimulation of hair cells in the (b)_____. The hair cells initiate impulses in the (c)_____.

CHEMORECEPTORS

Taste receptors detect dissolved food molecules

27. The four basic tastes that are generally recognized are _____ _____.

28. Although the idea of a fifth taste is somewhat controversial, a fifth taste known as _____ has been reported.

The olfactory epithelium is responsible for the sense of smell

29. Most invertebrates depend on _____, or the detection of odors, as their main sensory modality.

Many animals communicate with pheromones

30. Animals within many species communicate with one another by releasing _____, small volatile molecules that are secreted into the environment.

PHOTORECEPTORS

31. Cephalopod mollusks, arthropods, and vertebrates all have photosensitive pigments called _____ in their eyes.

Invertebrate photoreceptors include eyespots, simple eyes, and compound eyes

32. Some flatworms have _____, which are photoreceptive organs capable of differentiating the intensity of light.

33. The compound eye in insects and crustaceans consists of _____, which collectively form a mosaic image.

Vertebrate eyes form sharp images

34. The tough outer coat of the mammalian eye, called the (a)_____, helps maintain the (b)_____ of the eyeball.

35. The anterior, transparent part of this coat, which allows the entry of light, is called the _____.

The retina contains light-sensitive rods and cones

36. Light is focused by the (a)_____ on the (b)_____, which contains the photoreceptive cells. (c)_____, containing the pigment (d)_____, are concentrated in the periphery of the retina. They function best in dim light and perceive black and white.

37. (a)_____ are responsible for color vision and are densest in the (b)_____ in the center of the retina.

BUILDING WORDS

Use combinations of prefixes and suffixes to build words for the definitions that follow.

Prefixes	The Meaning		Suffixes	The Meaning
chemo-	chemical		-lith	stone
endo-	within			
inter(o)-	between, among			
oto-	ear			
proprio-	one's own			
thermo-	heat, warm			

Prefix	Suffix	Definition
_____	-ceptor	1. Sense organs in muscles, tendons, and joints that enable the animal to perceive the position of its own body parts.
_____	_____	2. Calcium carbonate "stone" in the inner ear of vertebrates.
_____	-lymph	3. Fluid within the semicircular canals of the vertebrate ear.
_____	-receptor	4. A sense organ or sensory cell that responds to chemical stimuli.

_____ -receptor   5. A sensory receptor that provides information about body temperature.

_____ -ceptor   6. A sensory receptor within (among) body organs that helps maintain homeostasis.

MATCHING

Terms:

a. Chemoreceptor
b. Cochlea
c. Cone
d. Cornea
e. Electroreceptor

f. Exteroceptor
g. Fovea
h. Interoceptor
i. Iris
j. Mechanoreceptor

k. Pacinian corpuscle
l. Rhodopsin
m. Statocyst
n. Thermoreceptor
o. Tympanic membrane

For each of these definitions, select the correct matching term from the list above.

_____ 1. A light-sensitive pigment found in rod cells of the vertebrate eye.

_____ 2. The structure of the inner ear of mammals that contains the auditory receptors.

_____ 3. A sensory receptor that responds to mechanical energy.

_____ 4. The structure of the vertebrate eye that regulates the size of the pupil.

_____ 5. Sensory receptor that provides information about body temperature.

_____ 6. A mechanoreceptor that is sensitive to deep pressure touches.

_____ 7. The "ear drum."

_____ 8. The transparent anterior covering of the eye.

_____ 9. A sense organ or sensory cell that responds to chemical stimuli.

_____10. A conical photoreceptive cell of the retina that is particularly sensitive to bright light, and, by light of various wave lengths, mediates color vision.

MAKING COMPARISONS

Fill in the blanks.

Stimuli	Receptor Classification: Type of Stimuli	Receptor Classification: Location	Example
Light touch	Mechanoreceptor	Exteroceptor	Meissner's corpuscle in skin
Electrical currents in water	#1	#2	Organ in skin of some fish
#3	#4	Exteroceptor	Organ of Corti in birds
Change in internal body temperature	Thermoreceptor	#5	#6
#7	Mechanoreceptor	#8	Muscle spindle in human skeletal muscle
Heat from prey	#9	Exteroceptor	#10
#11	#12	Exteroceptor	Ocelli of flatworms
Food dissolved in saliva	#13	Exteroceptor	#14

Stimuli	Receptor Classification: Type of Stimuli	Receptor Classification: Location	Example
Pheromones	#15	#16	Vomeronasal organ in the epithelia of terrestrial vertebrates

MAKING CHOICES

Place your answer(s) in the space provided. Some questions may have more than one correct answer.

_____ 1. The process by which sensory receptors change the form of energy of a stimulus to a form of energy that can be transmitted along neurons is called
   a. sensory adaptation.   d. receptor potential.
   b. conversion.   e. integration.
   c. energy transduction.

_____ 2. A change in ion distribution that causes a change in the voltage across the membrane of a sensory receptor is a
   a. sensory adaptation.   d. receptor potential.
   b. conversion.   e. integration.
   c. transduction.

_____ 3. If the release of a neurotransmitter from a presynaptic terminal decreases during a constant input of action potentials, _____ is said to have occurred.
   a. sensory adaptation.   d. receptor potential.
   b. conversion.   e. integration.
   c. transduction.

_____ 4. In humans, visual stimuli are interpreted in the
   a. brain.   d. retina.
   b. photoreceptors.   e. optic nerves.
   c. postsynaptic photoganglia.

_____ 5. The state of depolarization or hyperpolarization in a receptor neuron that is caused by a stimulus is the
   a. depolarization potential.   d. action potential.
   b. resting potential.   e. threshold potential.
   c. receptor potential.

_____ 6. Variations in the quality of sound are recognized by the
   a. number of hair cells stimulated.   d. intensity of stimulation.
   b. pattern of hair cells stimulated.   e. amplitude of response.
   c. frequency of nerve impulses.

_____ 7. The membrane at the opening of the inner ear that is in contact with the stapes is in the
   a. tectorial membrane.   d. eardrum.
   b. basilar membrane.   e. round window.
   c. oval window.

_____ 8. Receptors within muscles, tendons, and joints that perceive position and body orientation are
   a. exteroceptors.   d. mechanoreceptors.
   b. proprioceptors.   e. electroreceptors.
   c. interoceptors.

_____ 9. Movement in ligaments is detected by

a. muscle spindles.   d. proprioceptors.

b. joint receptors.   e. tonic sense organs.

c. Golgi tendon organs.

_____10. Sense organs that detect changes in pH, osmotic pressure, and temperature within body organs are

a. exteroceptors.   d. mechanoreceptors.

b. proprioceptors.   e. electroreceptors.

c. interoceptors.

_____11. Eyespots that detect light but do not form clear images are called

a. simple eyes.   d. ocelli.

b. ommatidia.   e. ciliary bodies.

c. facets.

_____12. In the human eye, the "posterior cavity," between the lens and the retina, is filled with

a. aqueous fluid.   d. the vitreous humor.

b. aqueous humor.   e. the vitreous body.

c. rods and cones.

_____13. Accommodation involves

a. changing the shape of the lens.   d. ciliary muscle contractions.

b. changing the shape of the fovia.   e. redistribution of photoreceptors in the retina.

c. flattening of the retina.

VISUAL FOUNDATIONS

Color the parts of the illustrations below as indicated. Also label the locations for endolymph, perilymph, tympanic canal, and vestibular canal.

RED	☐ hair cells		ORANGE	☐ basilar membrane
GREEN	☐ cilia		BROWN	☐ tympanic canal and vestibular canal
YELLOW	☐ cochlear nerve		TAN	☐ bone
BLUE	☐ tectorial membrane		PINK	☐ cochlear duct

Color the parts of the illustrations below as indicated. Also label the light rays and circle the optic nerve fibers.

RED ☐ rod cell
GREEN ☐ cone cell
ORANGE ☐ bipolar cell
WHITE ☐ ganglion cell
VIOLET ☐ optic nerve fibers
TAN ☐ choroid layer and sclera
PINK ☐ horizontal cell
BLUE ☐ amacrine cell
YELLOW ☐ vitreous body
BROWN ☐ pigmented epithelium

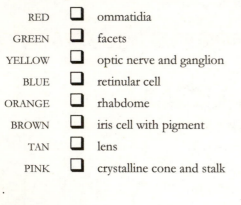

Color the parts of the illustrations below as indicated.

RED ☐ ommatidia
GREEN ☐ facets
YELLOW ☐ optic nerve and ganglion
BLUE ☐ retinular cell
ORANGE ☐ rhabdome
BROWN ☐ iris cell with pigment
TAN ☐ lens
PINK ☐ crystalline cone and stalk

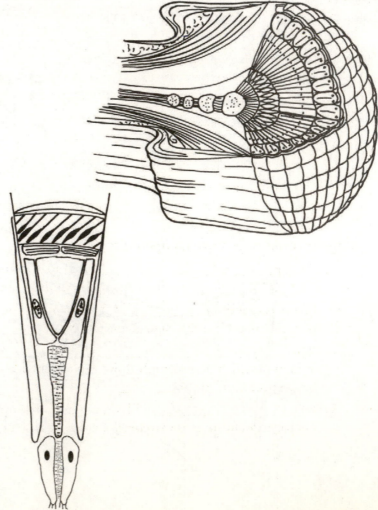

□

CHAPTER 43

Internal Transport

This chapter discusses how materials get to and from the body cells of animals. Some animals are so small that diffusion alone is effective at transporting materials. Larger animals, however, require a circulatory system. Some invertebrates have an open circulatory system in which blood is pumped from a heart into vessels that have open ends. Blood spills out of the vessels into the body cavity and baths the tissues directly. The blood then passes back into the heart, either directly through openings in the heart (arthropods) or indirectly, passing first through open vessels that lead to the heart (some mollusks). Other animals have a closed circulatory system in which blood flows through a continuous circuit of blood vessels. The walls of the smallest vessels are thin enough to permit exchange of materials between the vessels and the extracellular fluid that baths the tissue cells. The vertebrate circulatory system consists of a muscular heart that pumps blood into a system of blood vessels. The heart consists of one or two chambers that receive blood, and one or two that pump blood into the arteries. The lymphatic system, a subsystem of the circulatory system, collects extracellular fluid and returns it to the blood.

REVIEWING CONCEPTS

Fill in the blanks.

INTRODUCTION

1. High-density lipoproteins play a protective role, removing excess
   _____ from the blood and tissues.

TYPES OF CIRCULATORY SYSTEMS

2. _____ are
   animal groups that have no specialized circulatory structures.

3. Components of a circulatory system include a fluid, the (a)_____, that is
   usually pumped by a (b)_____ through a system of spaces or blood
   (c)_____.

Many invertebrates have an open circulatory system

4. An (a)_____ system consists of a blood cavity, or
   (b)_____, a heart, and open-ended vessels.

5. Blood is called (a)_____ in animals with open circulatory systems
   because blood is indistinguishable from (b)_____ fluid.

6. (a)_____ have an open circulatory
   system in which hemolymph flows out the open ends of the blood vessels, filling
   large spaces, called (b)_____.

7. (a)_____, a blood pigment that imparts a bluish color to
   hemolymph in some invertebrates, contains the metal (b)_____.

Some invertebrates have a closed circulatory system

8. Annelids have a (a)_____ blood vessel that conducts blood anteriorly, a (b)_____ blood vessel that conducts blood posteriorly, and, in the anterior part of the worm, (c)_____(#?) pairs of contractile blood vessels connecting the two.

9. Hemoglobin is found in the _____ of earthworm blood.

Vertebrates have a closed circulatory system

10. The circulatory system of all vertebrates is a/an (a)_____(open or closed?) system with a muscular (b)_____ that pumps (c)_____ through blood vessels.

11. The vertebrate circulatory system transports (a)_____
_____. Other functions include
(b)_____.

VERTEBRATE BLOOD

12. Human blood consists of (a)_____ suspended in
(b)_____.

Plasma is the fluid component of blood

13. Plasma is in dynamic equilibrium with (a)_____ fluid, which bathes cells, and with (b)_____ fluid.

14. (a)_____, a protein involved in blood clotting,
(b)_____, which are involved in immunity, and albumin are all (c)_____.

Red blood cells transport oxygen

15. Red blood cells (erythrocytes) are specialized to transport _____.

16. RBCs are produced in (a)_____, contain the respiratory pigment (b)_____, and live about (c)_____ days.

White blood cells defend the body against disease organisms

17. White blood cells (leukocytes) called (a)_____ are the principle phagocytic cells in blood. These cells, together with
(b)_____, are called
(c)_____ because of the distinctive granules in their cytoplasm.

18. Two types of agranular leukocytes are (a)_____ which secrete antibodies, and (b)_____, which develop into phagocytic
(c)_____.

Platelets function in blood clotting

19. The blood cells responsible for blood clotting in vertebrates other than mammals are (a)_____; in mammals, they are the
(b)_____.

VERTEBRATE BLOOD VESSELS

20. (a)_____ carry blood away from the heart; (b)_____ return blood to the heart.

21. The exchange of nutrients and waste products takes place across the thin wall of
_____.

EVOLUTION OF THE VERTEBRATE CARDIOVASCULAR SYSTEM

22. Chambers in the heart that pump blood into arteries are called
(a)_____, and chambers that receive blood from veins are called
(b)_____.

23. Fish hearts have (a)_____(#?) atrium(ia) and (b)_____ ventricle(s).

24. In the (a)_____(#?)-chambered amphibian heart, the
(b)_____ pumps venous blood into the right atrium, and the
(c)_____ helps to separate oxygen-rich blood from oxygen-
poor blood.

25. Because oxygen-rich blood can be isolated from oxygen-poor blood in a four
chambered heart, tissues receive (a)_____(more or less?) oxygen, and can
maintain a (b)_____(higher or lower?) metabolic rate.

THE HUMAN HEART

26. The human heart is enclosed by a (a)_____, creating a
(b)_____ filled with fluid that serves to reduce friction.

27. The (a)_____ septum separates the two ventricles, and the
(b)_____ separates the two atria in a four-chambered heart.

28. (a)_____(#?) valves control the flow of blood in the human heart. The
(b)_____ prevents backflow into the atria
during ventricular contractions. The valve on the right is also known as the
(c)_____, and the valve on the left is known as the
(d)_____. The (e)_____
"guard" the exits from the heart.

Each heartbeat is initiated by a pacemaker

29. The conduction system of the heart contains a pacemaker called the
(a)_____, an (b)_____ which
links the atria to the (c)_____, which
divides, sending a branch into each ventricle.

30. The portion of the (a)_____ cycle in which contraction occurs is called
(b)_____, and the portion in which relaxation occurs is called
(c)_____.

31. Of the two main heart sounds, the (a)_____ occurs first and is associated
with closure of the (b)_____.

32. The (a)_____ sound is produced by the closure of the (b)_____, which
marks the beginning of ventricular diastole.

The nervous system regulates heart rate

33. _____ are drugs that block the action of
norepinephrine on the heart; they are used to treat hypertension and other heart
diseases.

Stroke volume depends on venous return

34. According to _____, if veins deliver more blood to the heart, the heart pumps more blood.

Cardiac output varies with the body's need

35. (a)_____ is the volume of blood pumped by one ventricle in one minute. It is determined by multiplying the number of ventricular beats per minute times the amount of blood pumped by one ventricle per beat, a value referred to as the (b)_____.

36. The cardiac output of a resting adult is about _____.

BLOOD PRESSURE

37. High blood pressure is called (a)_____. It can be caused by an (b)_____(increase or decrease?) in blood volume such as frequently occurs with a high dietary intake of (c)_____.

38. The most important factor in determining peripheral resistance to blood flow is the _____.

Blood pressure varies in different blood vessels

39. Flow rate can be maintained in veins at low pressure because they are _____ vessels.

Blood pressure is carefully regulated

40. _____ are sense organs in the walls of some arteries.

41. Sense organs in arteries detect (a)_____ and send the perceived information to (b)_____ in the medulla of the brain.

THE PATTERN OF CIRCULATION

42. A double circulatory system consists of a (a)_____ between the heart and lungs, and (b)_____ between the heart and body.

The pulmonary circulation oxygenates the blood

43. Pulmonary veins carry oxygen-(a)_____(rich or poor?) blood to the (b)_____ atrium of the heart.

The systemic circulation delivers blood to the tissues

44. Arteries in the systemic circuit branch from the (a)_____, the largest artery in the body, and serve major body areas. For example, the carotid arteries feed the (b)_____, subclavian arteries supply the (c)_____, and iliac arteries feed the (d)_____.

45. Veins returning blood from the head and neck empty into the large (a)_____ _____ vein, while venous return from the lower body empties into the (b)_____ vein.

THE LYMPHATIC SYSTEM

The lymphatic system consists of lymphatic vessels and lymph tissue

46. Considered an accessory circulatory system in vertebrates, the lymphatic system returns _____ to the blood.

47. (a)_____, the fluid contained within lymphatic vessels, is formed from (b)_____, filtered by (c)_____, and emptied into (d)_____ veins by the (e)_____ duct on the left side and the (f)_____ duct on the right.

The lymphatic system plays an important role in fluid homeostasis

48. The obstruction of the lymphatic vessels causes _____, swelling from excessive accumulation of interstitial fluid.

BUILDING WORDS

Use combinations of prefixes and suffixes to build words for the definitions that follow.

Prefixes	The Meaning	Suffixes	The Meaning
baro-	pressure	-cardium	heart
erythro-	red	-coel	cavity
hemo-	blood	-cyte	cell
leuk(o)-	white (without color)	-lunar	moon
neutro-	neutral	-phil	loving, friendly, lover
peri-	about, around, beyond		
semi-	half		
vaso-	vessel		

Prefix	Suffix	Definition
_____	_____	1. The blood cavity that comprises the open circulatory system of arthropods and some mollusks.
_____	-cyanin	2. The copper-containing blood pigment in some mollusks and arthropods.
_____	_____	3. Red blood cell.
_____	_____	4. General term for all of the body's white blood cells.
_____	_____	5. The principal phagocytic cell in the blood that has an affinity for neutral dyes.
eosino-	_____	6. WBC with granules that have an affinity for eosin.
baso-	_____	7. WBC with granules that have an affinity for basic dyes.
_____	-emia	8. A form of cancer in which WBCs multiply rapidly within the bone marrow.
_____	-constriction	9. Constriction of a blood vessel.
_____	-dilation	10. Relaxation of a blood vessel.
_____	_____	11. The tough connective tissue sac around the heart.
_____	_____	12. Shaped like a half-moon.
_____	-receptor	13. Pressure receptor.

MATCHING

Terms:

a. Aorta
b. Arterioles
c. Artery
d. Atrium
e. Blood pressure

f. Interstitial fluid
g. Lymph
h. Lymphatic system
i. Mitral valve
j. Plasma

k. Pulse
l. Systole
m. Vein

For each of these definitions, select the correct matching term from the list above.

_____ 1. Network of fluid-carrying vessels and associated organs that participate in immunity and in the return of tissue fluid to the main circulation.

_____ 2. The heart valve located between the left atrium and the left ventricle.

_____ 3. The fluid in lymphatic vessels.

_____ 4. The fluid portion of the blood consisting of a pale, yellowish fluid.

_____ 5. A contracting chamber of the heart that forces blood into the ventricle.

_____ 6. The rhythmic expansion and recoil of an artery that may be felt with the finger.

_____ 7. The smallest arteries, which carry blood to the capillary beds.

_____ 8. Contraction of the heart muscle, especially that of the ventricle, during which the heart pumps blood into arteries.

_____ 9. Largest blood vessel in the body through which blood leaves the heart and enters the systemic circulation.

_____ 10. A blood vessel that carries blood from the tissues toward the heart.

MAKING COMPARISONS

Fill in the blanks.

Specific Protein or Cell Type	Blood Component	Function
Fibrinogen	Plasma	Involved in blood clotting
#1	#2	Transport fats, cholesterol
Erythrocytes	#3	#4
Thrombocytes	Cell Component	#5
Monocytes	#6	#7
#8	#9	Help regulate the distribution of fluid between plasma and interstitial fluid
Neutrophils	Cell Component	#10

MAKING CHOICES

Place your answer(s) in the space provided. Some questions may have more than one correct answer.

_____ 1. The pericardium surrounds
    a. atria only.   d. the heart.
    b. ventricles only.   e. clusters of some blood cells.
    c. blood vessels.

_____ 2. Very small vessels that deliver oxygenated blood to capillaries are called
    a. arterioles.   d. veins.
    b. venules.   e. setum vessels.
    c. arteries.

_____ 3. One would expect an ostrich to have a heart most like a
    a. lizard.   d. human.
    b. snail.   e. guppy.
    c. earthworm.

_____ 4. Fibrinogen is
    a. a plasma lipid.   d. a gamma globulin.
    b. in serum.   e. involved in clotting.
    c. a protein.

_____ 5. Three chambered hearts generally consist of the following number of atria/ventricles:
    a. 2/1.   d. 0/3.
    b. 1/2.   e. 3/0.
    c. 1/1 and an accessory chamber.

_____ 6. Blood enters the right atrium from the
    a. right ventricle.   d. pulmonary artery.
    b. inferior vena cava.   e. jugular vein.
    c. superior vena cava.

_____ 7. In general, blood circulates through vessels in the following order:
    a. veins, capillaries, arteries, arterioles.   d. arteries, arterioles, capillaries, veins.
    b. veins, capillaries, arterioles, arteries.   e. capillaries, arteries, arterioles, veins.
    c. veins, arteries, arterioles, capillaries.

_____ 8. Prothrombin
    a. is produced in liver.   d. requires vitamin K for its production.
    b. is a precursor to thrombin.   e. is a toxic by-product of metabolism.
    c. catalyzes conversion of fibrinogen to fibrin.

_____ 9. Erythrocytes are
    a. also called RBCs.   d. one kind of leucocyte.
    b. spherical.   e. carriers of oxygen.
    c. produced in bone marrow.

_____10. Monocytes
    a. are RBCs.   d. are large cells.
    b. are WBCs.   e. can become macrophages.
    c. are produced in the spleen.

_____11. The blood-filled cavity of an open circulatory system is called the
    a. lymphocoel.   d. sinus.
    b. hemocoel.   e. ostium.
    c. atracoel.

_____12. Pulmonary arteries carry blood that is
    a. low in oxygen.   d. high in CO_2.
    b. low in CO_2.   e. on its way to the lungs.
    c. high in oxygen.

_____13. The semilunar valve(s) is/are found between a ventricle and
    a. an atrium.   d. the pulmonary vein.
    b. another ventricle.   e. the pulmonary artery.
    c. the aorta.

_____14. Phenomena that tend to be associated with hypertension include
    a. obesity.   d. decrease in ventricular size.
    b. increased vascular resistance.   e. deteriorating heart function.
    c. increased workload on the heart.

_____15. A heart murmur may result when
    a. the heart is punctured.   d. diastole is too low.
    b. systole is too high.   e. diastole is too low or too high.
    c. semilunar valves are injured.

_____16. Ventricles receive stimuli for contraction directly from the
    a. SA node.   d. atria.
    b. AV node.   e. intercalated disks.
    c. Purkinje fibers.

_____17. If a person's heart rate is 65 and one ventricle pumps 75 ml with each contraction, what is that person's cardiac output in liters?
    a. 3.575   d. 4.875
    b. 4.000   e. 5.350
    c. 4.525

VISUAL FOUNDATIONS

Color the parts of the illustration below as indicated. Also circle the capillary bed.

RED	☐	artery and arrows indicating flow of blood away from the heart
PINK	☐	arteriole
VIOLET	☐	capillaries
PURPLE	☐	venuole
BLUE	☐	vein and arrows indicating flow of blood to the heart
GREEN	☐	lymphatic, lymph capillaries
YELLOW	☐	lymph node

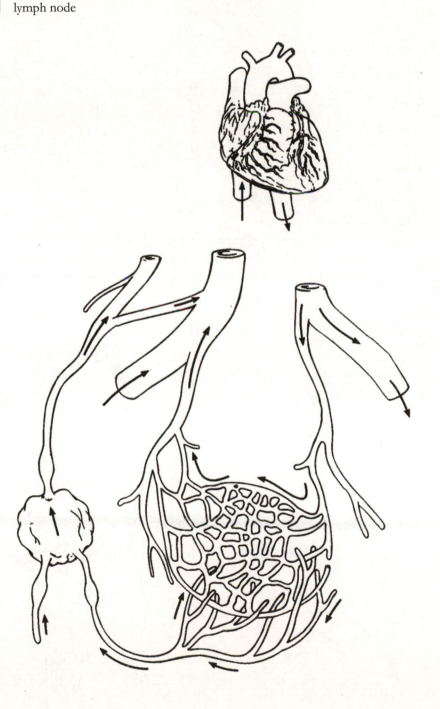

Color the parts of the illustration below as indicated. Label artery, vein, and capillary

RED ☐ smooth muscle

GREEN ☐ outer coat (connective tissue)

YELLOW ☐ endothelium

Color the parts of the illustration below as indicated.

RED ☐ aorta and pulmonary artery	ORANGE ☐ valve	
GREEN ☐ ventricle	BROWN ☐ partition	
YELLOW ☐ atrium	TAN ☐ conus	
BLUE ☐ vein from body and pulmonary vein	PINK ☐ sinus venosus	

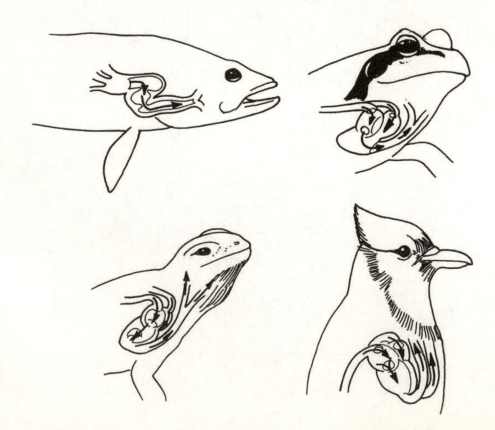

Color the parts of the illustration below as indicated. Also label interventricular septum, superior vena cava, inferior vena cava, pulmonary veins, aorta, and pulmonary artery.

RED ☐ chordae tendineae
GREEN ☐ tricuspid valve
YELLOW ☐ pulmonary semilunar valve
BLUE ☐ mitral valve
ORANGE ☐ aortic semilunar valve
BROWN ☐ papillary muscle
TAN ☐ atrium
PINK ☐ ventricle

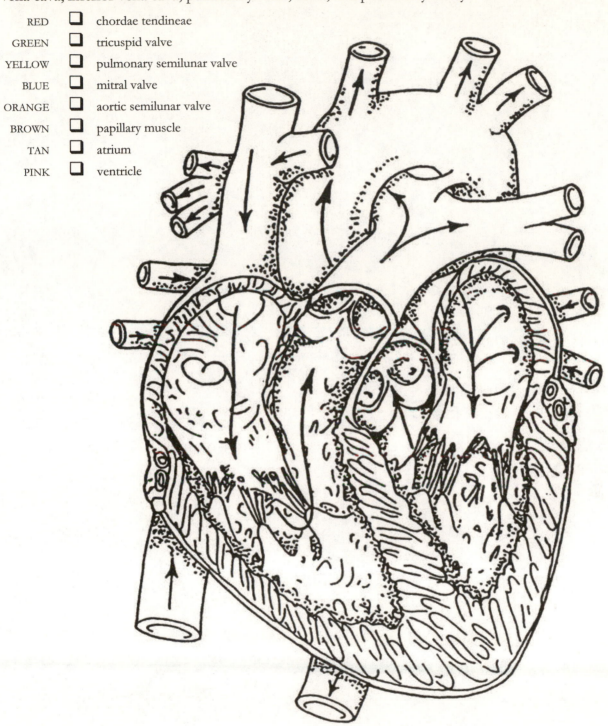

CHAPTER 44

The Immune System: Internal Defense

All animals have the ability to prevent disease-causing microorganisms from entering their body. Because these external defense mechanisms sometimes fail, most animals have developed internal defense mechanisms as well. Before an animal can attack an invader, it must be able to recognize its presence and to distinguish between its own cells and the invader. Whereas most animals are capable of nonspecific responses, such as phagocytosis, only vertebrates are capable of specific responses, such as the production of antibodies that target specific pathogens. This type of internal defense is characterized by a more rapid and intense response the second time the organism is exposed to the pathogen. In this manner, the organism develops resistance to the pathogen. Resistance to many pathogens can be induced artificially by either the injection of the pathogen in a weakened, killed, or otherwise altered state, or the injection of antibodies produced by another person or animal. The immune system is constantly surveying the body for abnormal cells and destroys them whenever they arise; failure results in cancer. This same surveillance system is responsible for graft rejection by reading the foreign cells as nonself cells and destroying them. Sometimes the system falsely reads self cells as nonself and destroys them, resulting in an autoimmune disease. Allergies are another example of abnormal immune responses.

REVIEWING CONCEPTS

Fill in the blanks.

INTRODUCTION

1. _____ is the study of internal defense systems.

NONSPECIFIC AND SPECIFIC IMMUNITY: AN OVERVIEW

2. The two main types of immune responses are _____
_____, and
_____.

3. A substance that the immune system specifically recognizes as foreign is called an
_____.

4. _____ are highly specific proteins that recognize and bind to specific antigens.

The immune system responds to danger signals

5. The danger model hypothesizes that the immune system responds to _____
_____ from injured tissues, such as proteins released when cell membranes are damaged.

Invertebrates launch nonspecific immune responses

6. Most invertebrate coelomates have amoeba-like cells that engulf and destroy bacteria and other foreign matter in a process known as _____.

7. Another important nonspecific immune response in invertebrates involves the production of _____ in response to pathogens that enter the body.

Vertebrates launch nonspecific and specific immune responses

8. The specialized lymphatic system that has evolved in vertebrates includes cells such as (a)_____, white blood cells specialized to carry out immune responses, and organs such as (b)_____.

NONSPECIFIC IMMUNE RESPONSES

9. The first line of defense in animals is the (a)_____. Other nonspecific defense mechanisms that prevent entrance of pathogens include the, (b)_____ in the stomach, and the (c)_____ of the respiratory passageways.

Phagocytes and natural killer cells destroy pathogens

10. _____ are the main phagocytes in the body.

Cytokines and complement mediate immune responses

11. Interferons inhibit _____.

Inflammation is a protective response

12. The inflammatory response includes three main processes. These are _____ _____.

SPECIFIC IMMUNE RESPONSES

Many types of cells are involved in specific immune responses

13. All lymphocytes develop from (a)_____ cells in the (b)_____.

14. Two main types of T cells are: the (a)_____ T cells, or killer T cells, that recognize and destroy cells with foreign antigens on their surfaces, and the (b)_____ T cells that secrete cytokines that activate B cells and macrophages.

15. The thymus makes T cells _____, that is, capable of immunological response.

16. Macrophages are sometimes referred to as APCs, which stands for _____.

The major histocompatibility complex is responsible for recognition of self

17. The MHC genes encode _____, or self-antigens, that differ in chemical structure, function, and tissue distribution.

CELL-MEDIATED IMMUNITY

18. After a cytotoxic T cell combines with an antigen on the surface of a target cell, it secretes _____ that perforate the plasma membrane of the target cell and induce it to kill itself by apoptosis.

ANTIBODY-MEDIATED IMMUNITY

A typical antibody consists of four polypeptide chains

19. Antibodies are also called (a)_____, abbreviated as (b)_____. These molecules "recognize" specific amino acid sequences on antigens called (c)_____. The antibody molecule is Y-shaped, the two arms functioning as (d)_____.

Antibodies are grouped in five classes

20. Antibodies are grouped into five classes, which are abbreviated as _____.

21. In humans, about 75% of the circulating antibodies are (a)_____. (b)_____ predominate in secretions, (c)_____ is an important immunoglobulin on the B cell surface, and (d)_____ mediates the release of histamine from mast cells.

Antigen-antibody binding activates other defenses

22. The antigen-antibody complex may stimulate _____ to ingest the pathogen.

The immune system responds to millions of different antigens

23. _____ is the process whereby a lymphocyte responds to a specific antigen by repeatedly dividing, giving rise to a clone of cells withy identical receptors.

Monoclonal antibodies are highly specific

24. Identical antibodies produced by cells cloned from a single cell are called _____.

IMMUNOLOGICAL MEMORY

A secondary immune response is more effective than a primary response

25. The principal antibody synthesized in the primary response is _____.

26. A second exposure to an antigen evokes a secondary immune response, which is more rapid and more intense than the primary response. The principal antibody synthesized in the secondary response is _____.

Immunization induces active immunity

27. Researchers are developing _____ to induce active immunity made from a part of the pathogen's genetic material.

Passive immunity is borrowed immunity

28. Passive immunity is characterized by the fact that its effects are only _____.

THE IMMUNE SYSTEM AND DISEASE

Cancer cells evade the immune system

29. _____ are used to identify cancer subtypes by detecting patterns of gene expression in cancer cells.

Immunodeficiency disease can be inherited or acquired

30. The leading cause of acquired immunodeficiency in children is _____
_____.

HIV is the major cause of acquired immunodeficiency in adults

31. _____ block the viral enzyme protease, resulting in
viral copies that cannot infect new cells.

HARMFUL IMMUNE RESPONSES

Graft rejection is an immune response against transplanted tissue

32. In graft rejection, the first stage, or the (a)_____,
occurs when T cells recognize the antigens, and in the second stage, the
(b)_____, the cells attack the transplanted tissue.

Rh incompatibility can result in hypersensitivity

33. Rh incompatibility can cause _____,
which destroys red blood cells in a fetus and may cause death.

Allergic reactions are directed against ordinary environmental antigens

34. In an allergic response, an (a)_____ stimulates production of
(b)_____ type antibody.

35. Mast cells then release _____ and other substances, causing
inflammation and other symptoms of allergy.

In an autoimmune disease, the body attacks its own tissues

36. A lymphocyte that has the potential to be _____ may
launch an immune response against self tissues.

BUILDING WORDS

Use combinations of prefixes and suffixes to build words for the definitions that follow.

Prefixes	The Meaning		Suffixes	The Meaning
anti-	against, opposite of		-cyte	cell
auto-	self, same			
lyso-	loosening, decomposition			
mono-	alone, single, one			

Prefix	Suffix	Definition
_____	-body	1. A specific protein that acts against pathogens and helps destroy them.
_____	-histamine	2. A drug that acts against (blocks) the effects of histamine.
lympho-	_____	3. A white blood cell strategically positioned in the lymphoid tissue; the main "warrior" in specific immune responses.
_____	-clonal	4. An adjective pertaining to a single clone of cells.
_____	-immune	5. An adjective pertaining to the situation wherein the body reacts immunologically against its own tissues (against "self").
_____	-zyme	6. An enzyme that attacks and degrades the cell walls of gram-positive bacteria.

MATCHING

Terms:

a. Allergen
b. Antigen
c. Complement
d. Histamine
e. Immune response

f. Immunoglobulin
g. Interferon
h. Interleukin
i. Mast cell
j. Memory cell

k. Passive immunity
l. Pathogen
m. Plasma cell
n. T cell
o. Thymus gland

For each of these definitions, select the correct matching term from the list above.

_____ 1. Any substance capable of stimulating an immune response; usually a protein or large carbohydrate that is foreign to the body.

_____ 2. A gland that functions as part of the lymphatic system.

_____ 3. Substance released from mast cells that is involved in allergic and inflammatory reactions.

_____ 4. A type of white blood cell responsible for cell-mediated immunity.

_____ 5. The process of recognizing foreign and dangerous macromolecules and responding to eliminate them.

_____ 6. A protein produced by animal cells when challenged by a virus.

_____ 7. Type of lymphocyte that secretes antibodies.

_____ 8. An organism capable of producing disease.

_____ 9. A type of cell found in connective tissue; contains histamine and is important in allergic reactions.

_____10. Temporary immunity derived from the immunoglobulins of another organism.

MAKING COMPARISONS

Fill in the blanks.

Cells	Nonspecific Immune Response	Specific Immune Response: Antibody-Mediated Immunity	Specific Immune Response: Cell-Mediated Immunity
B-lymphocytes	None	Activated by helper T cells and macrophages; multiply into a clone	None
#1	None	Secrete specific antibodies	None
#2	None	Long-term immunity	None
#3	None	None	Activated by helper T cells and macrophages; multiply into a clone
#4	None	Involved in B cell activation	Involved in T cell activation
#5	None	None	Chemically destroy cancer cells, foreign tissue grafts, and cells infected with viruses
#6	None	None	Long-term immunity
#7	Kill virus-infected cells & tumor cells	Kill virus-infected cells and tumor cells	None

Cells	Nonspecific Immune Response	Specific Immune Response: Antibody-Mediated Immunity	Specific Immune Response: Cell-Mediated Immunity
#8	Phagocytic, destroy bacteria	Antigen-presenting cells, activate helper T cells	Antigen-presenting cells; stimulates cloning
#9	Phagocytic, destroy bacteria	None	None

MAKING CHOICES

Place your answer(s) in the space provided. Some questions may have more than one correct answer.

_____ 1. T-cell receptors

    a. bind antigens.

    b. have no known function.

    c. are identical on all T cells.

    d. are found on killer T cells.

    e. stimulate antibody production.

_____ 2. Complement

    a. is a system of several proteins.

    b. is highly antigen-specific.

    c. is stimulated into action by an antibody-antigen complex.

    d. helps destroy pathogens.

    e. is an antibody.

_____ 3. The secondary response is due to

    a. killer T cells.

    b. memory cells.

    c. plasma cells.

    d. macrophages.

    e. helper T cells.

_____ 4. Active immunity can be artificially induced by

    a. transfusions.

    b. injecting vaccines.

    c. passing maternal antibodies to a fetus.

    d. injecting gamma globulin.

    e. stimulating macrophage growth.

_____ 5. B cells

    a. are granular.

    b. are lymphocytes.

    c. clone after contacting a targeted antigen.

    d. are derived from plasma cells.

    e. include many antigen-binding forms.

_____ 6. Which of the following is true of AIDS?

    a. HIV infects helper T cells.

    b. AIDS is not spread by casual contact.

    c. Both heterosexuals and homosexuals are at risk.

    d. HIV can be transmitted by sharing needles.

    e. There is currently no cure for AIDS.

_____ 7. T cells

    a. are lymphocytes.

    b. are called LGLs.

    c. are also called T lymphocytes.

    d. are platelets

    e. are involved in specific cell-mediated immunity.

_____ 8. An allergic reaction involves

    a. killer T cells.

    b. mast cells.

    c. interaction between an allergen and mast cells.

    d. production of IgE.

    e. histocompatibility antigens.

_____ 9. Which of the following is/are true of Ig?

    a. They are antibodies.

    b. They contain a C region.

    c. They are produced in response to specific antigens.

    d. They contain an antigenic determinant.

    e. They are also called immunoglobulins.

_____ 10. Nonspecific immune responses in vertebrates include

    a. skin.

    b. acid secretions.

    c. inflammation.

    d. phagocytes.

    e. antibody-mediated immunity.

_____ 11. The histocompatibility complex is

    a. found in cell nuclei.

    b. called HLA in humans.

    c. different in each individual.

    d. the same in individuals comprising a species.

    e. a group of closely linked genes.

VISUAL FOUNDATIONS

Color the parts of the illustration below as indicated. Label variable region and constant region, and circle the antigen-antibody complex.

RED ☐ antigenic determinants

GREEN ☐ antigen

YELLOW ☐ antibody-heavy chain

BLUE ☐ binding sites

ORANGE ☐ antibody-light chain

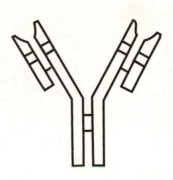

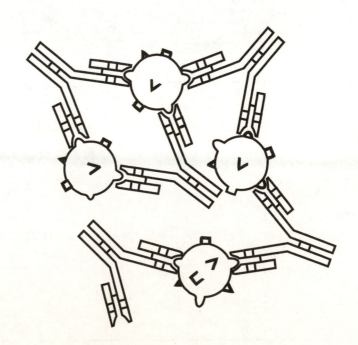

Color the parts of the illustration below as indicated. Label cell-mediated immunity and antibody-mediated immunity. Also label antigen presentation, cooperation, and migration to lymph nodes.

RED ❑ bone marrow

BROWN ❑ thymus

YELLOW ❑ T cell

ORANGE ❑ helper T cell

GREY ❑ cytotoxic T cell

BLUE ❑ memory cell

VIOLET ❑ B cell

PINK ❑ plasma cell

BROWN ❑ natural killer cell

TAN ❑ monocyte, macrophage, and dendritic cell

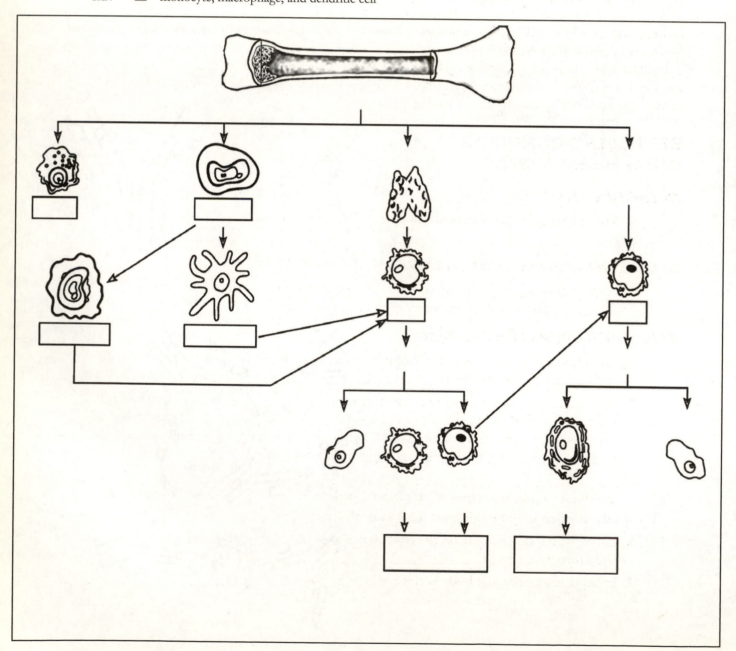

❑

CHAPTER 45

Gas Exchange

In chapter 8, you studied cellular respiration, the intracellular process that uses oxygen as the final electron acceptor in the mitochondrial electron transport chain and produces carbon dioxide as a waste product. This chapter discusses organismic respiration, that is, how animal cells take up oxygen from the environment in the first place and get it to the mitochondria, and how the carbon dioxide generated in the mitochondria gets out of the animal body. Oxygen exchange with air is more efficient than with water. This is because air contains more oxygen than water, and oxygen diffuses more rapidly in air than in water. In small, aquatic organisms, gases diffuse directly between the environment and all body cells. In larger, more complex organisms, specialized respiratory structures, such as tracheal tubes, gills, and lungs, are required. Respiratory pigments, such as hemoglobin and hemocyanin, greatly increase respiratory efficiency by increasing the capacity of blood to transport oxygen.

REVIEWING CONCEPTS

Fill in the blanks.

INTRODUCTION

1. The exchange of gases between an organism and the medium in which it lives is called _____.

ADAPTATIONS FOR GAS EXCHANGE IN AIR OR WATER

2. Air contains (a)_____(more or less?) oxygen than water, and oxygen diffuses (b)_____(faster or slower?) in air than in water.

TYPES OF RESPIRATORY SURFACES

3. Most respiratory systems actively move air or water over their respiratory surfaces, a process known as _____.

4. Specialized respiratory structures must have (a)_____ to facilitate the transfer of adequate volumes of gases, (b)_____ surfaces that can dissolve oxygen and carbon dioxide, and a rich supply of (c)_____ to transport respiratory gases.

5. _____ are the four principal types of respiratory structures.

The body surface may be adapted for gas exchange

6. Animals that rely entirely on gas exchange across the body surface are small and therefore have a large _____ ratio.

7. In aquatic animals, the body surface is kept moist by the _____.

Tracheal tube systems deliver air directly to the cells

8. Arthropods have a respiratory system consisting of a network of (a)_____; air enters the system through small openings called (b)_____.

9. The respiratory network of an arthropod terminates in small fluid-filled _____ where gas exchange takes place.

Gills are respiratory surfaces in many aquatic animals

10. The gills of bony fish have many thin _____ that extend out into the water, within which are numerous capillaries.

11. In bony fish, blood flows through respiratory structures in a direction opposite to water movement, an arrangement called the _____ _____, a system that maximizes the difference in oxygen concentrations between the animal's blood and the water.

Terrestrial vertebrates exchange gases through lungs

12. Lungs are respiratory structures that develop as ingrowths of the (a)_____ or from the wall of a (b)_____ such as the pharynx.

13. The _____ of spiders are within an inpocketing of the abdominal wall.

THE MAMMALIAN RESPIRATORY SYSTEM

The airway conducts air into the lungs

14. Both trachea and bronchi are lined by a mucous membrane containing _____.

Gas exchange occurs in the alveoli of the lungs

15. The lungs are located in the thoracic cavity. The right lung has (a)___(#?) lobes, and the left lung has (b)_____(#?) lobes.

16. The pathway of air as it passes through the human's "breathing apparatus" can be summarized as: nose and mouth → (a)_____ (throat region) → (b)_____ (voice box) → (c)_____ (wind pipe) → (d)_____ (tubes leading to lungs) → (e)_____ (small branching tubes) → (f)_____ (air sacs).

Ventilation is accomplished by breathing

17. During inspiration, the volume of the thoracic cavity (a)_____; during exhalation, it (b)_____.

The quantity of respired air can be measured

18. (a)_____ is the volume of air that is inhaled and exhaled with each normal resting breath. It averages about (b)_____(volume?).

19. _____ is the maximum volume of air that can be expelled after filling the lungs to the maximum extent.

Gas exchange takes place in the alveoli

20. The factor that determines the direction and rate of diffusion of a gas across a respiratory surface is the _____ of that gas.

21. _____ states that the total pressure in a mixture of gases is the sum of the pressures of the individual gases.

Gas exchange takes place in the tissues

22. The partial pressure of oxygen in arterial blood is (a)_____, while that in venous blood is (b)_____.

Respiratory pigments increase capacity for oxygen transport

23. The respiratory pigment of most vertebrates is hemoglobin, a compound containing a (a)_____ group bound to the protein (b)_____.

24. Oxygen combines with the element (a)_____ in the (b)_____ group of hemoglobin. This "association" can be illustrated as Hb + O2 → (c)_____. The result is a compound that carries oxygen and releases it where it is in lower concentrations.

25. HbO2 is prone to dissociation in a (a)_____(lower or higher?) pH. A change in the normal HbO2 dissociation curve caused by a change in pH is known as the (b)_____.

Carbon dioxide is transported mainly as bicarbonate ions

26. Most carbon dioxide moves through plasma as (a)_____ ions, the formation of which is catalyzed by an enzyme in RBCs called (b)_____.

Breathing is regulated by respiratory centers in the brain

27. Respiratory centers are groups of neurons in the _____ that regulate the rhythm of ventilation.

28. _____ in the walls of the aorta and carotid arteries are sensitive to changes in hydrogen ion concentration.

Hyperventilation reduces carbon dioxide concentration

29. A certain concentration of carbon dioxide is needed in the blood to maintain normal _____.

High flying or deep diving can disrupt homeostasis

30. _____ is a deficiency of oxygen that may cause drowsiness, mental fatigue, and headaches.

31. A sudden decrease in environmental pressure may cause the release of dissolved gases in the blood in the form of bubbles that block capillaries, causing a very painful syndrome which, in the vernacular, is called "the bends," and in the scientific community is called

_____.

Some mammals are adapted for diving

32. The diving reflex includes three physiological mechanisms; these are:

_____

_____.

BREATHING POLLUTED AIR

33. When dirty air is breathed into the lungs, _____ occurs, manifested by the narrowing of the bronchi.

BUILDING WORDS

Use combinations of prefixes and suffixes to build words for the definitions that follow.

Prefixes	The Meaning		Suffixes	The Meaning
hyper-	over		-ox(ia)	containing oxygen
hyp(o)-	under		-tion	the process of
ox(y)-	containing oxygen			
ventila-	to fan			

Prefix	Suffix	Definition
_____	-ventilation	1. Excessive rapid and deep breathing; a series of deep inhalations and exhalations.
_____	_____	2. Oxygen deficiency.
_____	_____	3. The process whereby animals actively move air or water over their respiratory surfaces.
_____	-hemoglobin	4. A complex of oxygen and hemoglobin.

MATCHING

Terms:

a.	Alveolus	f.	Gill	k.	Pleural cavity		
b.	Bronchioles	g.	Larynx	l.	Pleural membrane		
c.	Bronchus	h.	Lung	m.	Respiratory center		
d.	Diaphragm	i.	Operculum	n.	Trachea		
e.	Epiglottis	j.	Pharynx				

For each of these definitions, select the correct matching term from the list above.

_____ 1. "Windpipe."

_____ 2. An air sac of the lung through which gas exchange with the blood takes place.

_____ 3. A type of respiratory organ of aquatic animals.

_____ 4. The membrane that lines the thoracic cavity and envelopes the lungs.

_____ 5. One of the branches of the trachea and its immediate branches within the lung.

_____ 6. The throat region in humans.

_____ 7. Tiny air ducts of the lung that branch to form the alveoli.

_____ 8. The dome-shaped muscle that forms the floor of the thoracic cavity.

_____ 9. The bony plate covering the gills in fish.

_____10. The organ at the upper end of the trachea that contains the vocal cords.

MAKING COMPARISONS

Fill in the blanks.

Organism	Type of Gas Exchange Surface
Amphibians	Body surface and lungs
Fish	#1
Spiders	#2
Reptiles	#3
Birds	#4
Nudibranch mollusks	#5
Insects	#6
Sea stars	#7
Clams	#8
Mammals	#9

MAKING CHOICES

Place your answer(s) in the space provided. Some questions may have more than one correct answer.

_____ 1. A cockroach obtains oxygen for tissues located deep in its body by means of

   a. book gills.

   b. pseudolungs.

   c. circulation of hemolymph.

   d. a countercurrent exchange system.

   e. branching tracheal tubes.

_____ 2. Air passes through structures of a mammal's respiratory system in the following order:

   a. trachea, pharynx, bronchi, bronchioles.

   b. larynx, trachea, bronchioles, bronchi.

   c. pharynx, larynx, bronchi, bronchioles.

   d. larynx, trachea, bronchi, bronchioles.

   e. trachea, larynx, bronchi, bronchioles.

_____ 3. The ability of oxygen to be released from oxyhemoglobin is affected by

   a. temperature.

   b. pH.

   c. concentration of carbon dioxide.

   d. oxygen concentration in tissues.

   e. ambient oxygen concentration.

_____ 4. If a patient is found to have inelastic air sacs, trouble expiring air, an enlarged right ventricle, obstructed air flow, and large alveoli, he/she probably

   a. has lung cancer.

   b. has emphysema.

   c. has chronic obstructive pulmonary disease.

   d. lives at a high altitude.

   e. experiences chronic bronchitis.

_____ 5. If you are SCUBA diving for a lengthy time at 200 feet, you might get the bends if
    a. you surface quickly.   d. you come to the surface very slowly.
    b. you are using a helium mixture.   e. you stay at 200 feet even longer.
    c. nitrogen in your blood is rapidly absorbed by your tissues.

_____ 6. The diaphragm and rib or intercostal muscles alternately contract and relax, these actions causing respectively
    a. inspiration and expiration.   d. exhalation and inhalation.
    b. inhalation and exhalation.   e. oxygen intake and carbon dioxide output.
    c. expiration and inspiration.

_____ 7. Characteristics that all respiratory surfaces share in common include
    a. moist surfaces.   d. large surface to volume ratio.
    b. thin walls.   e. alveoli or spiracles.
    c. specialized structures such as tracheal tubes, gills, or lungs.

_____ 8. If the PO_2 in the tissue of your pet dog is 10, atmospheric PO_2 is 150, and arterial PO_2 is 110, you might expect the dog to (units are mm Hg)
    a. function normally.   d. die.
    b. accumulate carbon dioxide.   e. become dizzy from too much oxygen.
    c. have a serious, but not lethal, oxygen deficit.

_____ 9. Reasonable rates of expiration and inspiration fall in the range of
    a. 25/min.   d. 0.2/sec.
    b. 14/min.   e. 840/hr.
    c. 72/hr.

_____ 10. Most of the mucus produced by epithelial cells in the nasal cavities is disposed of by means of
    a. a large handkerchief.   d. absorption in surrounding lymph ducts.
    b. nose picking.   e. swallowing.
    c. evaporation.

VISUAL FOUNDATIONS

Color the parts of the illustration below as indicated.

RED ☐ lungs

GREEN ☐ gills

YELLOW ☐ body surface used for gas exchange

BLUE ☐ tracheal tube

ORANGE ☐ book lung

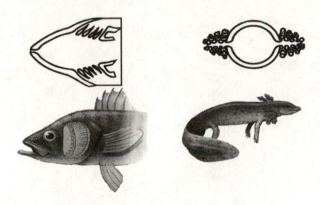

Color the parts of the illustration below as indicated.

RED ▢ red blood cell
BROWN ▢ bronchiole
YELLOW ▢ epithelial cell of alveolus
BLUE ▢ macrophage
PINK ▢ capillary

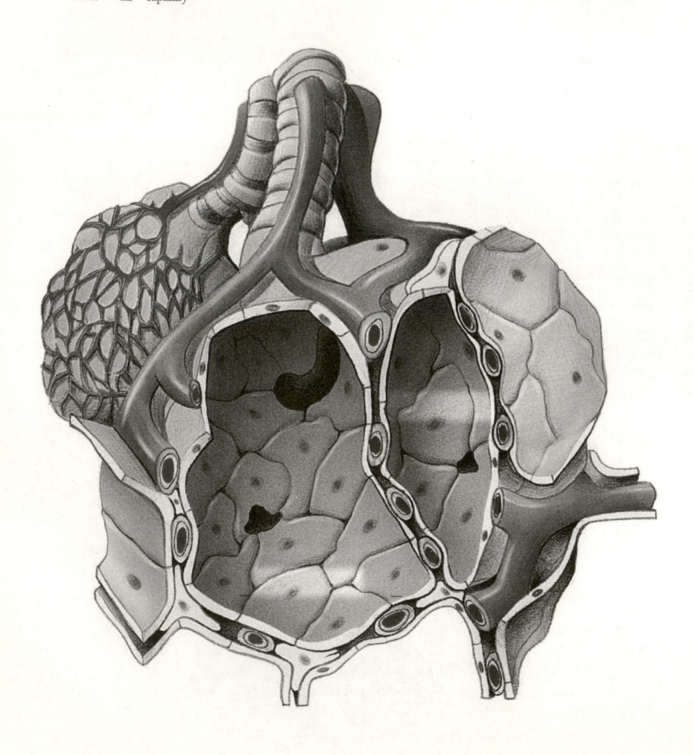

CHAPTER 46

Processing Food and Nutrition

The processing of food involves several steps — taking food into the body, breaking it down into its constituent nutrients, absorbing the nutrients, and eliminating the material that is not broken down and absorbed. Animals eat either plants, animals, or both. Some animals have no digestive systems, with digestion occurring intracellularly within food vacuoles. Others have incomplete digestive systems with only a single opening for both food to enter and wastes to exit. Still other animals have complete digestive systems, in which the digestive tract is a complete tube with two openings, a mouth where food enters and an anus where waste is expelled. The human digestive system has highly specialized structures for processing food. All animals require the same basic nutrients — carbohydrates, lipids, proteins, vitamins, and minerals. Carbohydrates are used by the body as fuel. Lipids are also used as fuel, and additionally, as components of cell membranes, and as substrates for the synthesis of steroid hormones and other lipid substances. Proteins serve as enzymes and as structural components of cells. Vitamins and minerals are needed for many biochemical processes. Serious nutritional problems result from eating too much food, eating too little food, or not eating a balanced diet.

REVIEWING CONCEPTS
Fill in the blanks.

INTRODUCTION

1. Organisms that obtain their main source of energy from organic molecules synthesized by other organisms are known as _____.

NUTRITIONAL STYLES AND ADAPTATIONS

2. Undigested material is removed from the digestive tracts of simpler animals by a process known as _____.

Animals are adapted to their mode of nutrition

3. Primary consumers, or _____, have special digestive processes to break down the plant material they consume.

4. (a)_____ have well-developed canine teeth, digestive enzymes that break down proteins, and an overall (b)_____(shorter or longer?) gastrointestinal tract.

5. Organisms that consume both animal and plant material are called _____.

Some invertebrates have a digestive cavity with a single opening

6. Gastrovascular cavities have a (a)_____ opening in which undigested wastes are egested through the (b)_____.

Most animal digestive systems have two openings

7. The vertebrate digestive system is a complete tube extending from the (a)_____ to the (b)_____.

THE VERTEBRATE DIGESTIVE SYSTEM

8. The specialized portions of the vertebrate digestive tube, in order, are the: mouth → (a)_____ → (b)_____ → stomach → (c)_____ → large intestine → anus.

Food processing begins in the mouth

9. Ingestion and the beginning of mechanical and enzymatic breakdown of food take place in the mouth. The teeth of mammals perform varied functions: (a)_____ are designed for biting, (b)_____ for tearing, and (c)_____ crush and grind food. Each tooth is covered by a coating of very hard (d)_____, under which is the main body of the tooth, the (e)_____, which resembles bone. The (f)_____ contains blood vessels and nerves.

10. The salivary glands of terrestrial vertebrates moisten food and release _____, an enzyme that initiates carbohydrate (starch) digestion.

The pharynx and esophagus conduct food to the stomach

11. Waves of muscular contractions called _____ move a lump of food through the esophagus into the stomach.

12. A lump of food passing through the esophagus is called a _____.

Food is mechanically and enzymatically digested in the stomach

13. In the stomach, food is mechanically broken down, the _____ in gastric juice initiates protein digestion, and food is reduced to chyme.

14. The stomach is lined with (a)_____ cells, which contain many gastric glands and specialized cells. For example, (b)_____ cells secrete hydrochloric acid and (c)_____ cells secrete pepsinogen, an enzyme precursor that converts to (d)_____.

Most enzymatic digestion takes place in the small intestine

15. The three regions of the small intestine are (a)_____. Most chemical digestion of food in vertebrates takes place in the (b)_____ portion.

16. The surface area of the small intestine is increased by small fingerlike projections called (a)_____, and by (b)_____, which are folds of cytoplasm on the exposed surface of the simple columnar epithelial cells of the villi.

The liver secretes bile

17. Among the many important functions performed by the liver are the secretion of bile, maintenance of homeostasis, the conversion of excess glucose to the carbohydrate storage molecule (a)_____, the conversion of excess amino acids to (b)_____, and the detoxifying of drugs and other poisons.

18. Bile produced by the liver is stored in the _____.

The pancreas secretes digestive enzymes

19. Pancreatic juice contains _____ that digests polypeptides to dipeptides.

20. The enzyme (a)_____ breaks down fats, and (b)_____ breaks down most carbohydrates.

Nutrients are digested as they move through the digestive tract

21. Enzymes reduce macromolecular polymers to the small subunits that comprise them. For example, carbohydrates are digested to _____.

22. Proteins are reduced to _____.

23. Fats are reduced to _____.

Nerves and hormones regulate digestion

24. Hormones such as secretin, gastrin, CCK, and GIP are polypeptides secreted by (a)_____ that help regulate the (b)_____ _____.

Absorption takes place mainly through the villi of the small intestine

25. Most digested nutrients are absorbed through the _____ of the small intestine.

26. Triacylglycerols, along with absorbed cholesterol and phospholipids, are packaged into protein-covered fat droplets, called _____.

The large intestine eliminates waste

27. The large intestine absorbs sodium and water, cultures bacteria, and eliminates wastes. It is made up of seven regions, which are, in order: the (a)_____, a blind pouch near the junction of the small and large intestines; the (b)_____; the (c)_____; the (d)_____; the (e)_____; the (f)_____, the last portion of the tube; and the (g)_____, the opening at the end of the tube.

28. (a)_____ is the process of getting rid of metabolic wastes, while (b)_____ is the process of getting rid of digestive wastes that never participated in metabolism.

REQUIRED NUTRIENTS

29. The amount of energy in food is expressed in terms of (a)_____, which are equivalent to (b)_____.

Carbohydrates provide energy

30. _____ is a mixture of cellulose and other indigestible carbohydrates derived from plants.

Lipids provide energy and are used to make biological molecules

31. Macromolecular complexes of cholesterol or triacylglycerols bound to proteins are called _____.

32. Two types of these important complexes are the (a)_____, which apparently decrease the risk of heart disease by transporting excess

cholesterol to the liver; and (b)_____,
which have been associated with coronary artery disease.

Proteins serve as enzymes and as structural components of cells

33. There are approximately (a)_____(#?) amino acids that cannot be synthesized
in adult humans. These are called (b)_____
_____ and are provided in a person's diet.

Vitamins are organic compounds essential for normal metabolism

34. Vitamins are divided into two broad groups: the (a)_____
vitamins such as A, D, E, and K, and the (b)_____ vitamins
that include (c)_____.

35. Surpluses from overdoses of the _____-type vitamins can
accumulate to harmful levels.

Minerals are inorganic nutrients

36. The essential minerals required by the body are _____
_____.

37. Minerals required in amounts of less than 100 mg per day are known as
_____.

Antioxidants protect against oxidants

38. Antioxidants destroy _____, or molecules
with one or more unpaired electrons.

Phytochemicals play important roles in maintaining health

39. Diets rich in (a)_____ may be more
important for health than diets with a low intake of (b)_____.

ENERGY METABOLISM

40. (a)_____ is a measure of the rate of
energy used during resting conditions. (b)_____
is the sum of an individuals' BMR and the energy needed to carry out daily
activities.

41. When energy input equals energy output, body weight remains constant. When
energy input exceeds energy output, body weight (a)_____(increases or
decreases?); and when energy input is less than output, the body draws on fuel
reserves (fat) and body weight (b)_____(increases or decreases?).

Undernutrition can cause serious health problems

42. Without _____, digestive
enzymes cannot be manufactured, so protein that is ingested cannot be digested.

Obesity is a serious nutritional problem

43. Leptin signals centers in the brain about the status of energy stores in the
_____.

BUILDING WORDS

Use combinations of prefixes and suffixes to build words for the definitions that follow.

Prefixes	The Meaning		Suffixes	The Meaning
epi-	upon, over, on		-itis	inflammation
micro-	small		-micro(n)	small, "tiny"
omni-	all		-vore	eating
sub-	under, below			

Prefix	Suffix	Definition
herbi-	_____	1. An animal that eats plants.
carni-	_____	2. An organism that eats flesh.
_____	_____	3. An organism that eats both plants and animals.
_____	-mucosa	4. A layer of connective tissue below the mucosa that binds it to the muscle layer beneath.
periton-	_____	5. Inflammation of the peritoneum.
_____	-glottis	6. A flap of tissue over the airway that prevents food and drink from entering the airway when swallowing.
_____	-villi	7. Small projections of the cell membrane that increase the surface area of the cell.
chylo-	_____	8. Tiny droplets of lipid that are absorbed from the intestine into the lymph circulation.

MATCHING

Terms:

a.	Absorption	f.	Gastrin	k.	Peristalsis
b.	Adventitia	g.	Mineral	l.	Rugae
c.	Bile	h.	Mucosa	m.	Stomach
d.	Digestion	i.	Pancreas	n.	Villus
e.	Elimination	j.	Pepsin	o.	Vitamin

For each of these definitions, select the correct matching term from the list above.

_____ 1. A cell layer that lines the digestive tract and secretes a lubricating layer of mucous.

_____ 2. The ejection of waste products, especially undigested food remnants, from the digestive tract.

_____ 3. The chief enzyme of gastric juice; hydrolyses proteins.

_____ 4. A large digestive gland located in the vertebrate abdominal cavity, having both exocrine and endocrine functions.

_____ 5. An organic compound necessary in small amounts for the normal metabolic functioning of a given organism; usually acts as a coenzyme.

_____ 6. Folds in the stomach wall, giving the inner lining a wrinkled appearance.

_____ 7. The taking up of a substance, as by the lining of the digestive tract.

_____ 8. Digestive juice produced by the liver and stored in the gallbladder.

_____ 9. Powerful, rhythmic waves of muscular contraction and relaxation in the walls of hollow tubular organs that aid in the movement of substances through the tube.

_____10. A minute "finger-like" projection from the surface of a membrane.

MAKING COMPARISONS

Fill in the blanks.

Enzyme	Function	Site of Production
Salivary amylase	Enzymatic digestion of starch	Salivary glands
Maltase	#1	#2
#3	Proteins to polypeptides	#4
#5	Polypeptides to dipeptides	#6
Ribonuclease	#7	#8
#9	Degrades fats	Pancreas
#10	Splits small peptides to amino acids	#11
Lactase	#12	Small intestine

MAKING CHOICES

Place your answer(s) in the space provided. Some questions may have more than one correct answer.

_____ 1. Some important functions of minerals include their role in/as

    a. cofactors.   d. maintaining fluid balance.

    b. nerve impulses.   e. digestive hormones.

    c. neurotransmitters.

_____ 2. The relative amounts of nutrient types consumed by impoverished societies, as compared to affluent societies, would likely be

    a. more protein, less carbohydrate.   d. more carbohydrate, less lipid.

    b. more lipid, less carbohydrate.   e. more vegetable matter, less animal matter.

    c. more carbohydrate, less protein.

_____ 3. Most absorption of macromolecular subunits occurs in the

    a. stomach.   d. small intestine.

    b. rectum.   e. colon.

    c. large intestine.

_____ 4. Products of complete fat digestion include

    a. fatty acids.   d. glycerol.

    b. amino acids.   e. triacylglycerol.

    c. chylomicrons.

_____ 5. Nerves and blood vessels in teeth are located in the

    a. pulp.   d. cementum.

    b. enamel.   e. canines only, not other teeth.

    c. dentin.

_____ 6. A herbivore is a/an

    a. plant consumer.   d. omnivore.

    b. primary consumer.   e. mutualistic symbiote.

    c. carnivore.

_____ 7. Which of the following is true concerning human nutrition?
    a. Water is essential.
    b. Excess nutrients are converted to fat.
    c. We cannot synthesize some required fatty acids.
    d. Proteins cannot be used for energy.
    e. The liver synthesizes amino acids.

_____ 8. The principal function(s) of villi is/are to
    a. absorb nutrients.
    b. secrete enzymes.
    c. increase surface area.
    d. stimulate digestion.
    e. secrete hormones.

_____ 9. All animals are
    a. omnivores.
    b. carnivores.
    c. primary consumers.
    d. consumers.
    e. heterotrophs.

_____ 10. Lipids are used to
    a. maintain body temperature.
    b. build membranes.
    c. provide energy.
    d. synthesize steroid hormones.
    e. create vitamins.

_____ 11. Bile is principally involved in the digestion of
    a. carbohydrates.
    b. lipids.
    c. starch.
    d. nucleic acids.
    e. proteins.

_____ 12. The single most crucial intermediate molecule in the metabolism of lipids and most other nutrients is
    a. keto acid.
    b. ATP.
    c. acetyl CoA.
    d. NAPH.
    e. glucose.

_____ 13. A lacteal is a
    a. lymph vessel.
    b. capillary bed.
    c. digestive gland.
    d. milk protein enzyme.
    e. salivary enzyme.

_____ 14. Peristalsis results from activity of tissues in the
    a. submucosa.
    b. mucosa.
    c. adventitia.
    d. muscle layer.
    e. peritoneum.

_____ 15. Which of the following is true regarding cellulose in the human diet?
    a. It's harmful.
    b. It's a major source of protein.
    c. We can't digest it.
    d. It's an important carbohydrate nutrient.
    e. It's an important source of fiber.

_____ 16. Pepsin is principally involved in the digestion of
    a. carbohydrates.
    b. lipids.
    c. starch.
    d. nucleic acids.
    e. proteins.

_____ 17. Important sources of energy in the human diet are
    a. proteins.
    b. lipids.
    c. carbohydrates.
    d. starches and sugars.
    e. meat.

_____ 18. The total metabolic rate
    a. encompasses the basal metabolic rate.
    b. is less than the basal metabolic rate.
    c. refers to metabolic rate after exercise.
    d. is the rate when at rest.
    e. is a nonexistent phrase.

_____19. Which of the following has/have a complete digestive system?

    a. birds.   d. jelly fish.

    b. earthworms.   e. fish.

    c. sponges.

_____20. A herbivore would likely have

    a. well-developed claws.   d. a short gut.

    b. flattened molars.   e. symbiotic microorganisms.

    c. large, sharp canines.

VISUAL FOUNDATIONS

Color the parts of the illustration below as indicated. Label the mucosa, submucosa, muscle layer, visceral peritoneum, intestinal glands, and villi.

RED   ❑ artery

GREEN   ❑ lymph

YELLOW   ❑ nerve fiber

BLUE   ❑ vein

VIOLET   ❑ goblet cell

PINK   ❑ epithelial cell of villus

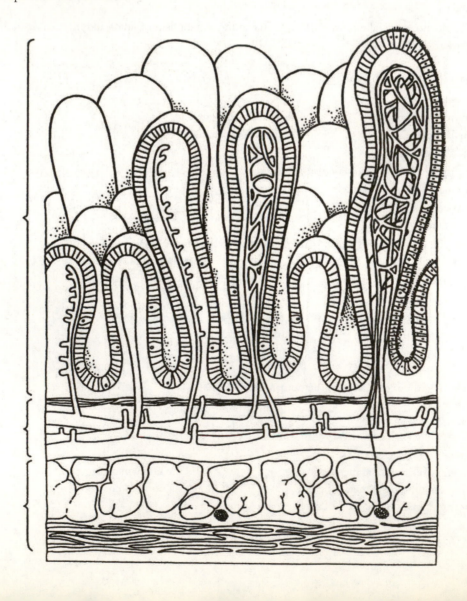

Color the parts of the illustration below as indicated.

RED ❑ liver
GREEN ❑ gall bladder
YELLOW ❑ esophagus
BLUE ❑ pancreas
ORANGE ❑ stomach
BROWN ❑ large intestine
TAN ❑ small intestine
PINK ❑ rectum and anus
VIOLET ❑ vermiform appendix

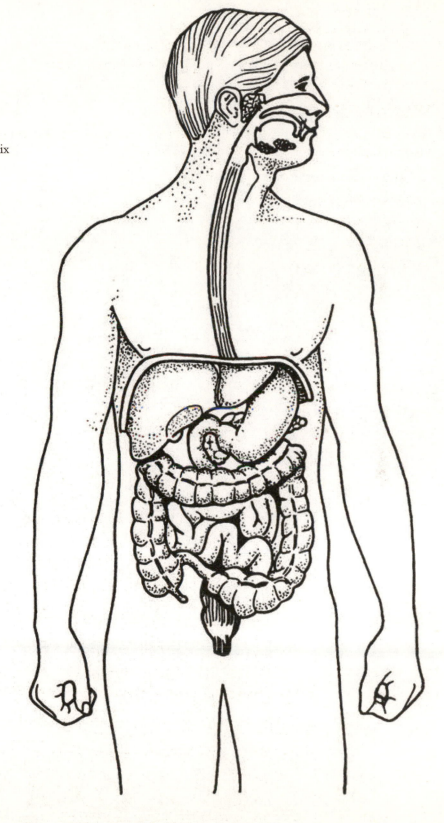

CHAPTER 47

Osmoregulation and Disposal of Metabolic Wastes

The water content of the animal body, as well as the concentration and distribution of ions in body fluids is carefully regulated. Most animals have excretory systems that function to rid the body of excess water, ions, and metabolic wastes. The excretory system collects fluid from the blood and interstitial fluid, adjusts its composition by reabsorbing from it the substances the body needs, and expels the adjusted excretory product, e.g., urine in humans. The principal metabolic wastes are water, carbon dioxide, and nitrogenous wastes. Excretory systems among invertebrates are diverse and adapted to the body plan and lifestyle of each species. The kidney, with the nephron as its functional unit, is the primary excretory organ in vertebrates.

REVIEWING CONCEPTS

Fill in the blanks.

INTRODUCTION

1. _____ is the medium in which most metabolic reactions take place.

2. Two processes that maintain homeostasis of fluids in the body are _____ and _____.

MAINTAINING FLUID AND ELECTROLYTE BALANCE

3. _____ controls the concentration of salt and water so that body fluids don't become too dilute or too concentrated.

METABOLIC WASTE PRODUCTS

4. Principal metabolic wastes in most animals are water, carbon dioxide, and nitrogenous wastes in the form of _____.

5. The production of _____ is an important water-conserving adaptation in many terrestrial animals, including insects, certain reptiles, and birds.

6. The urea cycle is a series of steps that produces urea from (a)_____ and (b)_____.

OSMOREGULATION AND EXCRETION IN INVERTEBRATES
Nephridial organs are specialized for osmoregulation and/or excretion

7. The nephridial organs of many invertebrates consist of tubes that open to the outside of the body through (a)_____. In flatworms, these excretory organs are called (b)_____, and in annelids and mollusks they are known as (c)_____.

Malpighian tubules conserve water

8. Malpighian tubules are slender extensions of the _____.

9. Malpighian tubules have blind ends that lie in the _____.

10. Wastes are transferred from blood to the Malpighian tubules by
    _____.

OSMOREGULATION AND EXCRETION IN VERTEBRATES

11. In most vertebrates, the _____
    and the kidneys all help in ridding the body of wastes and maintaining fluid
    balance.

Freshwater vertebrates must rid themselves of excess water

12. In freshwater fishes, (a)_____ is the main nitrogenous
    waste; about 10% of nitrogenous wastes are excreted as (b)_____.

Marine vertebrates must replace lost fluid

13. To compensate for fluid loss, marine fishes _____.

Terrestrial vertebrates must conserve water

14. Birds conserve water by excreting nitrogen as (a)_____;
    mammals excrete (b)_____.

THE URINARY SYSTEM

15. The urinary system is the principal excretory system in human beings and other
    vertebrates. Urine produced in the kidneys is transported to the urinary bladder
    through two tubes called (a)_____, then it passes from the bladder to the
    outside through the (b)_____.

The nephron is the functional unit of the kidney

16. The functional units of the kidneys are the nephrons. Each nephron consists of a
    cup-shaped (a)_____ and a (b)_____. The filtrate
    passes through structures in the nephron in this sequence:
    (c)_____ → proximal convoluted tubule →
    (d)_____ → distal convoluted tubule →
    (e)_____.

17. There are two types of nephrons in kidneys. The "ordinary variety" in the cortex
    are called (a)_____ nephrons, and those with an exceptionally long
    loop of Henle that extends deep into the medulla are called
    (b)_____ nephrons.

Urine is produced by filtration, reabsorption, and secretion

18. Urine is produced by a combination of three processes: _____,
    _____, and _____.

19. Plasma is filtered out of the glomerular capillaries and into
    _____.

20. Most of the filtrate is reabsorbed from the _____ back
    into the blood in order to return needed materials to the blood and adjust the
    composition of the filtrate.

Urine becomes concentrated as it passes through the renal tubule

21. Filtrate is concentrated as it moves downward through the (a)_____ loop and diluted as it moves upward through the (b)_____ loop. This movement of filtrate in opposite directions, called the (c)_____, helps maintain a hypertonic interstitial fluid that draws water out of the collecting ducts.

22. The _____ are capillaries that extend from the efferent arterioles of the juxtamedullary nephrons; they collect water from interstitial fluid.

Urine consists of water, nitrogenous wastes, and salts

23. _____ is the adjusted filtrate consisting of water, nitrogenous wastes, salts, and traces of other substances.

Hormones regulate kidney function

24. Urine volume is regulated by the hormone (a)_____. It is released by the (b)_____.

25. The "thirst center" that responds to dehydration is located in the
_____.

26. Aldosterone secreted by the (a)_____ stimulates distal tubules and collecting ducts to reabsorb more (b)_____.

BUILDING WORDS

Use combinations of prefixes and suffixes to build words for the definitions that follow.

Prefixes	The Meaning	Suffixes	The Meaning
juxta-	beside, near	-cyte	cell
podo-	foot		
proto-	first, earliest form of		

Prefix	Suffix	Definition
_____	-nephridium	1. The flame cell excretory organs of flatworms and nemerteans; the earliest form of specialized excretory organ.
_____	_____	2. A specialized epithelial cell possessing elongated foot processes, which cover the surfaces of most of the glomerular capillaries.
_____	-medullary	3. Pertains to nephrons situated nearest the medulla of the kidney.

MATCHING

Terms:

a. Aldosterone
b. Antidiuretic hormone
c. Bowman's capsule
d. Osmoconformer
e. Osmoregulator

f. Glomerulus
g. Malpighian tubule
h. Metanephridium
i. Nephron
j. Osmoregulation

k. Reabsorption
l. Urea
m. Ureter
n. Urethra
o. Uric acid

For each of these definitions, select the correct matching term from the list above.

_____ 1. An animal that maintains an optimal salt concentration of its body fluids despite changes in the salinity of its surroundings.

_____ 2. A hormone produced by the vertebrate adrenal cortex that functions to increase sodium reabsorption.

_____ 3. The knot of capillaries at the proximal end of a nephron; enclosed by the Bowman's capsule.

_____ 4. The tube that conducts urine from the bladder to the outside of the body.

_____ 5. The principal nitrogenous excretory product of many terrestrial animals, including insects and certain birds and reptiles.

_____ 6. The active regulation of the osmotic pressure of body fluids.

_____ 7. One of the paired ducts that conducts urine from the kidney to the bladder.

_____ 8. The principal nitrogenous excretory product of mammals; one of the water-soluble end products of protein metabolism.

_____ 9. The functional unit of the vertebrate kidney.

_____10. A hormone secreted by the posterior lobe of the pituitary that controls the rate of water reabsorption by the kidney.

MAKING COMPARISONS

Fill in the blanks.

Organism	Excretory Mechanism/Structure
Marine sponges	Diffusion
Vertebrates	#1
Insects	#2
Earthworms	#3
Flatworms	#4

MAKING CHOICES

Place your answer(s) in the space provided. Some questions may have more than one correct answer.

_____ 1. The amount of needed substances that can be reabsorbed from renal tubules is a function of
  a. Tm.
  b. blood pH.
  c. tubular transport maximum.
  d. concentration of urea in forming urine.
  e. maximum rate.

_____ 2. The principal functional unit(s) in the vertebrate kidneys is/are
  a. nephridia.
  b. nephrons.
  c. Bowman's capsules.
  d. antennal glands.
  e. Malpighian tubules.

_____ 3. Water is reabsorbed by interstitial fluids when osmotic concentration in interstitial fluid is increased by
  a. salts from filtrate.
  b. urea from filtrate.
  c. free ions.
  d. concentration of filtrate.
  e. deamination in kidney cells.

_____ 4. Urea is a principal nitrogenous waste product produced by
  a. amphibians.
  b. the kidneys.
  c. nephrons.
  d. mammals.
  e. the liver.

_____ 5. Most reabsorption of filtrate in the kidney takes place at the
  a. ureter.
  b. loop of Henle.
  c. proximal convoluted tubule.
  d. urethra.
  e. collecting duct.

_____ 6. Urea is synthesized
  a. from uric acid.
  b. in kidneys.
  c. from ammonia and carbon dioxide.
  d. in the urea cycle.
  e. in aquatic invertebrates.

_____ 7. Relative to sea water, fluids in the bodies of marine organisms
  a. are isotonic.
  b. are hypotonic.
  c. are hypertonic.
  d. lose water.
  e. gain water.

_____ 8. The duct in humans that leads from the urinary bladder to the outside is the
  a. ureter.
  b. loop of Henle.
  c. proximal convoluted tubule.
  d. urethra.
  e. collecting duct.

_____ 9. Which of the following statements most accurately describes changes in the concentration of filtrate in the two portions of the loop of Henle?
  a. decreases in both
  b. increases in both
  c. increases in descending/decreases in ascending
  d. decreases in descending/increases in ascending
  e. remains essentially the same in both

_____10. A potato bug would excrete wastes by means of
  a. nephridia.
  b. nephrons.
  c. a pair of kidney-like structures.
  d. green glands.
  e. Malpighian tubules.

_____11. Relative to fresh water, fluids in the bodies of aquatic organisms
    a. are isotonic.   d. lose water.
    b. are hypotonic.   e. gain water.
    c. are hypertonic.

_____12. Excretion is specifically defined as
    a. maintaining water balance.   d. the elimination of undigested wastes.
    b. homeostasis.   e. the concentration of nitrogenous products.
    c. removal of metabolic wastes from the body.

_____13. The main way(s) that _any_ excretory system maintains homeostasis in the body is/are to
    a. excrete metabolic wastes.   d. regulate salt and water.
    b. eliminate undigested food.   e. concentrate urea in urine.
    c. regulate body fluid constituents.

_____14. The group(s) of animals that has/have no specialized excretory systems include
    a. insects.   d. sponges.
    b. cnidarians.   e. sharks and rays.
    c. annelids.

_____15. The principal form(s) of nitrogenous waste products in various animal groups include(s)
    a. uric acid.   d. ammonia.
    b. carbon dioxide.   e. urea.
    c. amino acids.

_____16. The principal nitrogenous waste product(s) in human urine is/are
    a. uric acid.   d. ammonia.
    b. carbon dioxide.   e. urea.
    c. amino acids.

_____17. The first step in the catabolism of amino acids
    a. is conversion of ammonia to uric acid.   d. is removal of the amino group.
    b. is deamination.   e. is accomplished with peptidases.
    c. produces ammonia.

_____18. The type(s) of excretory organ(s) found in animals that collect wastes in flame cells is/are
    a. nephridia.   d. green glands.
    b. nephrons.   e. Malpighian tubules.
    c. branching tubes that open to the outside through pores.

VISUAL FOUNDATIONS

Color the parts of the illustration below as indicated.

RED	☐	abdominal aorta
GREEN	☐	adrenal gland
YELLOW	☐	ureter
BLUE	☐	inferior vena cava
ORANGE	☐	urethra
BROWN	☐	kidney
TAN	☐	urinary bladder
PINK	☐	renal artery
VIOLET	☐	renal vein

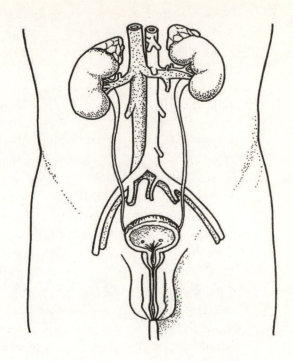

Color the parts of the illustration below as indicated. Label the juxtamedullary nephron and the cortical nephron.

RED	☐ artery	ORANGE	☐	loop of Henle
GREEN	☐ renal pelvis	BROWN	☐	medulla
YELLOW	☐ collecting duct	BLUE	☐	vein
PINK	☐ glomerulus	TAN	☐	cortex
VIOLET	☐ proximal convoluted tubule and distal convoluted tubule			

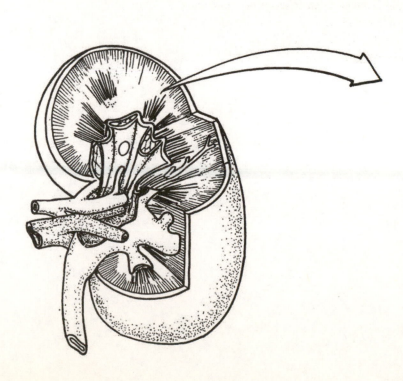

❏

CHAPTER 48

Endocrine Regulation

This chapter discusses the actions of a variety of hormones and examines how overproduction or deficiency of various hormones interferes with normal functioning. The endocrine system is a diverse collection of glands and tissues that secrete hormones — chemical messengers that are transported by the blood to target tissue where they stimulate a physiological change. The endocrine system works closely with the nervous system to maintain the steady state of the body. Hormones may be steroids, peptides, proteins, or derivatives of amino acids or fatty acids. They diffuse from the blood into the interstitial fluid and then combine with receptor molecules on or in the cells of the target tissue. Some hormones activate genes that lead to the synthesis of specific proteins. Others activate a second messenger that relays the hormonal message to the appropriate site within the cell. Most invertebrate hormones are secreted by neurons rather than endocrine glands. They regulate growth, metabolism, reproduction, molting, and pigmentation. In vertebrates, hormonal activity is controlled by the hypothalamus, which links the nervous and endocrine systems. Vertebrate hormones regulate growth, reproduction, salt and fluid balance, and many aspects of metabolism. Malfunction of any of the endocrine glands can lead to specific disorders.

REVIEWING CONCEPTS

Fill in the blanks.

INTRODUCTION

1. The tissues and organs of the endocrine system secrete chemical messengers, or _____, that signal other cells.

AN OVERVIEW OF ENDOCRINE REGULATION

2. Pheromones are generally not classified as hormones because they are generally produced by _____ and do not regulate metabolic activities with the animal that produces them.

3. Endocrine glands differ from exocrine glands in that they have no _____ and secrete their hormones into the surrounding interstitial fluid or blood.

The endocrine system and nervous system interact to regulate the body

4. Hormones secreted by neurons are (a)_____. They are an important link between the endocrine and (b)_____ systems.

Negative feedback systems regulate endocrine activity

5. Most endocrine action is regulated by (a)_____ _____ that act to restore (b)_____.

Hormones are assigned to four chemical groups

6. The four chemical groups of hormones are _____ _____.

7. In vertebrates, the adrenal cortex, testis, ovary, and placenta secrete steroids synthesized from _____.

TYPES OF ENDOCRINE SIGNALING

Neurohormones are transported in the blood

8. Neurohormones are transported down (a)_____ and released into the (b)_____.

Some local regulators are considered hormones

9. In _____ a hormone acts on the very cell that produces it.

MECHANISMS OF HORMONE ACTION

10. Receptors are continuously synthesized and degraded. Their numbers can be increased or decreased by (a)_____ and (b)_____.

Some hormones enter target cells and activate genes

11. Specific protein receptors in the cytoplasm or in the nucleus bind with the hormone to form a _____.

Many hormones bind to cell-surface receptors

12. Peptide hormones do not enter the target cell. Instead, they bind to a specific _____ in the plasma membrane.

13. Many protein hormones combine with receptors in the cell membrane of the target cell and act by way of a second messenger. Two common second messengers are (a)_____, which is derived from ATP, and (b)_____, which, for example, takes a role in disassembling microtubules.

INVERTEBRATE NEUROENDOCRINE SYSTEMS

14. Most invertebrate hormones are secreted by _____ rather than by endocrine glands. They help to regulate regeneration, molting, metamorphosis, reproduction, and metabolism.

15. Hormones control development in insects. Neurosecretory cells in the brain of insects produce the "brain hormone" called (a)_____, which stimulates (b)_____ glands.

16. The molting hormone in insects is also called _____.

THE VERTEBRATE ENDOCRINE SYSTEM

Homeostasis depends on normal concentrations of hormones

17. In hypersecretion, a target cell is (a)_____, and in hyposecretion, a target cell is (b)_____.

The hypothalamus regulates the pituitary gland

18. The pituitary gland secretes at least _____(#?) peptide hormones that exert influence over body activities.

The posterior lobe of the pituitary gland releases hormones produced by the hypothalamus

19. The hypothalamus produces vasopressin, also known as

_____.

20. The hypothalamus also produces _____, which stimulates milk production in nursing mothers.

The anterior lobe of the pituitary gland regulates growth and other endocrine glands

21. _____ hormones stimulate other endocrine glands.

22. The anterior lobe of the pituitary gland secretes _____, which stimulates mammary glands to produce milk.

23. Among hormones that influence growth are growth hormone (GH), also called (a)_____, secreted by the (b)_____ gland, and thyroid hormones, secreted by the thyroid gland.

24. Hypersecretion of GH during childhood may cause a disorder known as (a)_____. In adulthood, hypersecretion can cause "large extremities," or (b)_____.

Thyroid hormones increase metabolic rate

25. One thyroid hormone, (a)_____, is also known as T4 because each molecule contains four atoms of (b)_____.

26. Thyroid secretion is regulated by a negative feedback system between the thyroid gland and the (a)_____ gland, which releases less (b)_____ when thyroid hormone blood titers rise above normal.

27. Hyposecretion of thyroid hormones during infancy and childhood may result in retarded development, a condition known as _____.

The parathyroid glands regulate calcium concentration

28. Parathyroid hormone regulates calcium levels in body fluids by stimulating release of calcium from (a)_____ and calcium reabsorption by the (b)_____.

29. A negative feedback system induces the (a)_____ gland to secret (b)_____, which inhibits parathyroid hormone activity.

The islets of the pancreas regulate glucose concentration

30. Numerous clusters of cells in the pancreas, called (a)_____, secrete hormones that regulate glucose concentration in the blood. When blood glucose levels are high, (b)_____ cells release the hormone (c)_____; when it is low, (d)_____ cells release (e)_____.

31. Insulin lowers the concentration of glucose in the blood by stimulating uptake of glucose by _____.

32. In the endocrine metabolism disorder known as _____, cells are unable to utilize glucose properly and turn to fat and protein for fuel.

The adrenal glands help the body respond to stress

33. The adrenal medulla and the adrenal cortex secrete hormones that help the body cope with stress. The two secreted by the adrenal medulla are _____ and _____; they increase heart rate, metabolic rate, and strength of muscle contraction, and reroute the blood to organs that require more blood in time of stress.

34. The adrenal cortex produces three types of hormones in appreciable amounts.; these are (a)_____. The sex hormone precursors are converted to (b)_____, the principal male sex hormone, and (c)_____, the principal female sex hormone.

35. Kidneys reabsorb more sodium and excrete more potassium in response to the mineralocorticoid hormone _____.

36. The main function of _____ is to enhance gluconeogenesis in the liver.

37. Stress stimulates secretion of CRF, which is the abbreviation for (a)_____, from the hypothalamus. CRF stimulates the anterior pituitary to release (b)_____, which regulates the secretion of glucocorticoids and aldosterones.

Many other hormones are known

38. The (a)_____ gland in the brain produces the hormone (b)_____, which influences the onset of sexual maturation.

39. The thymus gland produces the hormone _____.

40. The heart secretion ANF, which stands for _____, lowers blood pressure.

BUILDING WORDS

Use combinations of prefixes and suffixes to build words for the definitions that follow.

Prefixes	The Meaning	Suffixes	The Meaning
hyper-	over	-megaly	enlargement
hypo-	under		
neuro-	nerve		

Prefix	Suffix	Definition
_____	-hormone	1. A hormone secreted by certain nerve cells.
_____	-endocrine	2. Refers to a nerve cell that secretes a neurohormone.
_____	-secretion	3. An excessive secretion; over secretion.
_____	-secretion	4. A diminished secretion; under secretion.
_____	-thyroidism	5. A condition resulting from an overactive thyroid gland.
_____	-glycemia	6. An abnormally low level of glucose in the blood.
_____	-glycemia	7. An abnormally high level of glucose in the blood.
acro-	_____	8. An abnormal condition characterized by enlargement of the head, and sometimes other structures.

MATCHING

Terms:

a. Adrenal gland
b. Aldosterone
c. Beta cells
d. Calcitonin
e. Endocrine gland

f. Glucagon
g. Hormone
h. Insulin
i. Oxytocin
j. Prostaglandin

k. Thymus
l. Thyroid gland
m. Thyroxine
n. Tropic hormone

For each of these definitions, select the correct matching term from the list above.

_____ 1. A hormone produced by the hypothalamus and released by the posterior lobe of the pituitary; causes the uterus to contract and stimulates the release of milk from the mammary glands.

_____ 2. A hormone secreted by the thyroid gland that rapidly lowers the calcium content in the blood.

_____ 3. An endocrine gland that lies anterior to the trachea and releases hormones that regulate the rate of metabolism.

_____ 4. Paired endocrine glands, each located just superior to each kidney.

_____ 5. General term for an organic chemical produced in one part of the body and transported to another part where it affects some aspect of metabolism.

_____ 6. Cells of the pancreas that secrete insulin.

_____ 7. A gland that secretes products directly into the blood or tissue fluid instead of into ducts.

_____ 8. General term for a hormone that helps regulate another endocrine gland.

_____ 9. One of the hormones produced by the thyroid gland.

_____10. An endocrine gland that produces thymosin; important in the development of the immune response mechanism.

MAKING COMPARISONS

Fill in the blanks.

Hormone	Chemical Group	Function
Testosterone	Steroid	Develops and maintains sex characteristics of males; promotes spermatogenesis
Aldosterone	#1	#2
Thyroxine	#3	#4
#5	Peptide	Stimulates reabsorption of water; conserves water
#6	#7	Helps body adapt to long-term stress; mobilizes fat; raises blood glucose levels
#8	#9	Affects wide range of body processes; may interact with other hormones to regulate metabolic activities
ACTH	Peptide	#10
Epinephrine	#11	#12

Hormone	Chemical Group	Function
#13	#14	Stimulates uterine contraction

MAKING CHOICES

Place your answer(s) in the space provided. Some questions may have more than one correct answer.

_____ 1. GH is referred to as an anabolic hormone because it
  - a. suppresses appetite .
  - b. stimulates other endocrine glands.
  - c. promotes tissue growth.
  - d. regulates the anterior pituitary gland.
  - e. affects neural membrane potentials.

_____ 2. The thyroid gland is located
  - a. in the neck.
  - b. behind the trachea.
  - c. in front of the trachea.
  - d. above the larynx.
  - e. below the larynx.

_____ 3. The hormone and target tissue that are involved in raising glucose concentration in blood by glycogenolysis and gluconeogenesis are
  - a. insulin and pancreas.
  - b. glucagon and liver.
  - c. thyroid-stimulating hormone and thyroid gland.
  - d. adrenocorticotropic hormone and adrenal cortex.
  - e. thyroxine and various metabolically active cells.

_____ 4. Activity of the anterior lobe of the pituitary gland is controlled by
  - a. the hypothalamus.
  - b. ADCH.
  - c. epinephrine and norepinephrine.
  - d. releasing hormones.
  - e. inhibiting hormones.

_____ 5. A chemical produced by one cell that has a specific regulatory effect on another cell is the definition of a
  - a. neurohormone.
  - b. hormone.
  - c. exocrine gland secretion.
  - d. pheromone.
  - e. prohormone.

_____ 6. A three-year old cretin and an adult suffering from myxedema most likely have
  - a. a low metabolic rate.
  - b. Cushing's disease.
  - c. Addison's disease.
  - d. a low level of thyroid hormones.
  - e. a high level of thyroid hormones.

_____ 7. A person with a fasting level of 750 mg glucose per 100 ml of blood
  - a. is hyperglycemic.
  - b. is hypoglycemic.
  - c. is about normal.
  - d. probably has cells that are not using enough glucose.
  - e. has too much insulin.

_____ 8. Increased skeletal growth results directly and/or indirectly from activity of
  - a. aldosterone.
  - b. growth hormone.
  - c. hormones from the hypothalamus.
  - d. andosterides.
  - e. epinephrine and/or norepinephrine.

_____ 9. Diabetes in an adult with ample numbers of adequately functioning beta cells and normal concentrations of insulin
  - a. is type I diabetes.
  - b. is type II diabetes.
  - c. indicates that target cells are not using insulin.
  - d. indicates that not enough insulin is being produced.
  - e. is unusual since this condition usually occurs in children.

_____10. Small, hydrophobic hormones that form a hormone-receptor complex with intranuclear receptors include
a. steroids.
b. cyclic AMP.
c. chemicals that activate genes.
d. thyroid hormones.
e. prostaglandins.

_____11. A possible negative consequence of hyposecretion of insulin is
a. Addison's disease.
b. Cushing's disease.
c. myxedema.
d. diabetes mellitus.
e. dwarfism.

VISUAL FOUNDATIONS

Color the parts of the illustration below as indicated. Also label transcription, translation, and the target cell.

RED ☐ receptor molecule

YELLOW ☐ hormone molecules

BLUE ☐ DNA

ORANGE ☐ mRNA

BROWN ☐ endocrine gland cell

TAN ☐ nucleus

PINK ☐ blood vessel

VIOLET ☐ protein molecule

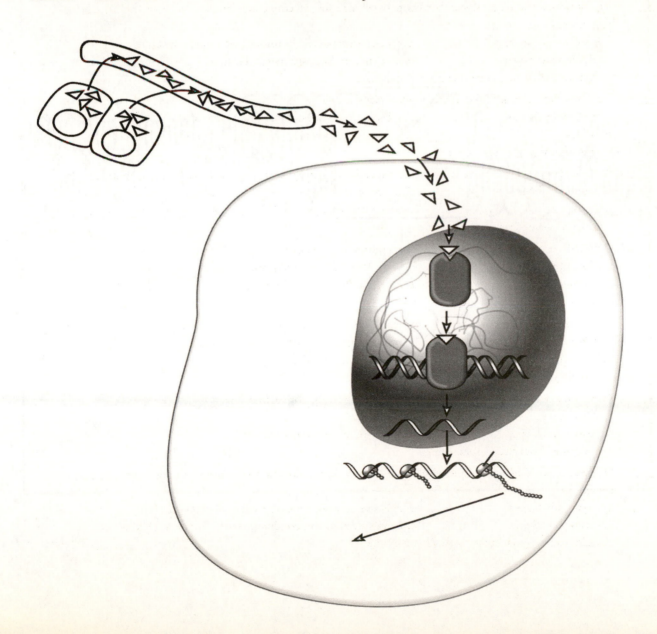

Color the parts of the illustration below as indicated.

RED   ☐   receptor

GREEN   ☐   hormone

YELLOW   ☐   cytosol

BLUE   ☐   second messengers

ORANGE   ☐   extracellular fluid

BROWN   ☐   plasma membrane

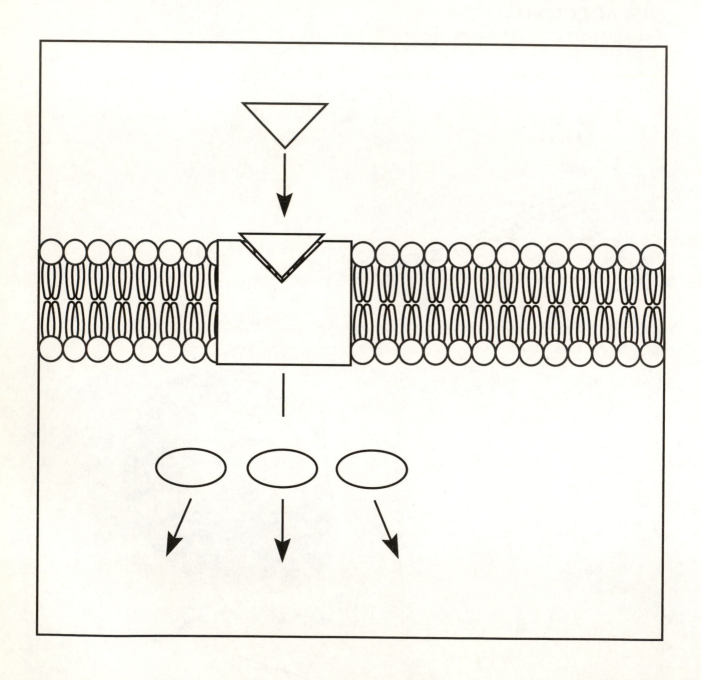

Color the parts of the illustration below as indicated.

RED	❑	pituitary gland
GREEN	❑	pineal gland
YELLOW	❑	thyroid gland
BLUE	❑	parathyroid gland
ORANGE	❑	hypothalamus
BROWN	❑	thymus gland
TAN	❑	pancreas
PINK	❑	testis and ovary
VIOLET	❑	adrenal gland

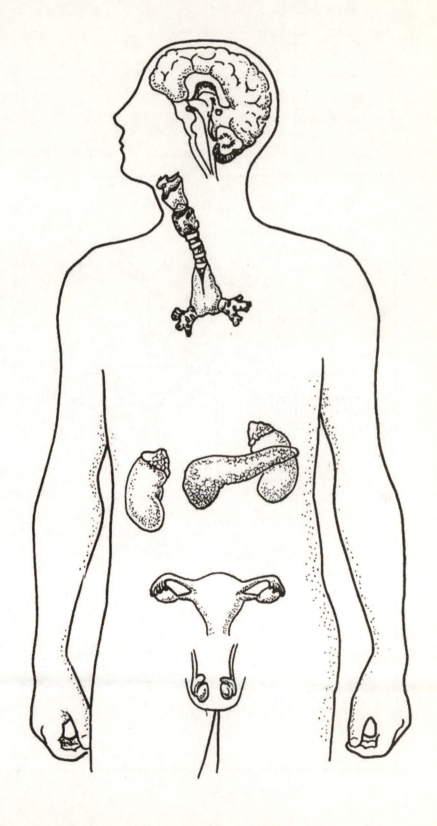

CHAPTER 49

Reproduction

For a species to survive, it is essential that it replace individuals that die. Animals do this either asexually or sexually. In asexual reproduction, a single parent splits, buds, or fragments, giving rise to two or more offspring that are genetically identical to the parent. In sexual reproduction, sperm contributed by a male parent, and an egg contributed by a female parent, unite, forming a zygote that develops into a new organism. This method of reproduction promotes genetic variety among the offspring, and therefore, gives rise to individuals that may be better able to survive than either parent. Depending on the species, fertilization occurs either outside or inside the body. Some animals have both asexual and sexual stages. In some, the unfertilized egg may develop into an adult. In still others, both male and female reproductive organs may occur in the same individual. More typically, animals have only a sexual stage, fertilization is required for development to occur, and male and female reproductive organs occur only in separate individuals. The complex structural, functional, and behavioral processes involved in reproduction in vertebrates are regulated by hormones secreted by the brain and gonads. Several hormones regulate the birth process in humans. There are a variety of effective methods of birth control. Next to the common cold, sexually transmitted diseases are the most prevalent communicable diseases in the world.

REVIEWING CONCEPTS
Fill in the blanks.

INTRODUCTION

1. Sexual reproduction in animals involves the production and fusion of two types of gametes _____.

ASEXUAL AND SEXUAL REPRODUCTION
Asexual reproduction is an efficient strategy

2. In asexual reproduction, a single parent endows its offspring with a set of genes identical to its own. Sponges and cnidarians can reproduce by _____, involving the production of a new individual from a group of expendable cells that separate from the "parent's" body.

3. _____ is a form of asexual reproduction whereby a "parent" breaks into several pieces, each piece giving rise to a new individual.

4. Parthenogenesis refers to the production of a complete organism from a(n) _____.

Most animals reproduce sexually

5. Many aquatic animals practice _____ in which the gametes meet outside the body.

6. _____ is a form of sexual reproduction in which a single individual produces both eggs and sperm.

Sexual reproduction increases genetic variability

7. _____ reproduction removes harmful mutations from a population.

HUMAN REPRODUCTION: THE MALE

The testes produce gametes and hormones

8. The initial development of a sperm begins with an undifferentiated stem cell called a (a)_____, which enlarges to form the (b)_____ that goes through meiosis.

9. The _____ at the front of a sperm contains enzymes that assist the sperm in penetrating an egg.

10. The testes, housed in an external sac called the (a)_____, contain the (b)_____, where sperm are produced.

11. The _____ are the passage ways that testes take as they descend through the abdominal wall into the scrotum.

A series of ducts store and transport sperm

12. Sperm complete their maturation and are stored in the (a)_____, from which they are moved up into the body through the (b)_____.

13. During ejaculation, sperm pass through the penis though the _____.

The accessory glands produce the fluid portion of semen

14. Human semen contains an average of about_____(#?) sperm.

15. Sperm are suspended in secretions that are mostly produced by two glands called the _____.

The penis transfers sperm to the female

16. The penis consists of an elongated (a)_____ that terminates in an expanded portion called the (b)_____, which is partially covered by a fold of skin called the (c)_____. An operation that removes this "extra" skin is known as (d)_____.

17. The penis contains three columns of erectile tissue, two (a)_____ and one _____. When they become engorged with (b)_____, the penis erects.

Testosterone has multiple effects

18. _____ is the principal male sex hormone.

The hypothalamus, pituitary gland, and testes regulate male reproduction

19. The two gonadotropic hormones, (a)_____, along with testosterone, stimulate sperm production. LH stimulates secretion of the hormone (b)_____, which is responsible for establishing and maintaining primary and secondary sex characteristics in the male.

20. The male hormone (a)_____ is produced by the (b)_____ cells in the testes.

HUMAN REPRODUCTION: THE FEMALE

The ovaries produce gametes and sex hormones

21. Formation of ova is called (a)_____. In this process, stem cells in the fetus, the (b)_____, enlarge, forming the (c)_____, which begin meiosis before birth.

22. A developing ovum and the cells immediately around it comprise the (a)_____. At puberty, the hormone (b)_____ stimulates development of these cells, causing some of them to complete meiosis. One of the resulting cells, the (c)_____, continues development to form the ovum.

23. After ovulation, the portion of the follicle that remains in the ovary develops into a temporary endocrine gland called the _____.

The oviducts transport the secondary oocyte

24. In a tubal pregnancy, the embryo begins to develop in the _____ _____ rather than the uterus.

The uterus incubates the embryo

25. Embryos normally implant in the (a)_____, which is the inner layer of cells of the uterus. This signals the beginning of pregnancy. If there is no embryo, or the embryo fails to implant, the bleeding phase known as (b)_____ commences.

26. The lower portion of the uterus, the _____, extends slightly into the vagina.

The vagina receives sperm

27. The vagina serves as a _____ for sperm.

The vulva are external genital structures

28. The female external genitalia are collectively known as the _____.

29. The most sensitive part in the vulva is the _____.

The breasts function in lactation

30. _____ is the production of milk.

31. Milk production begins as a result of stimulation by the hormone _____.

The hypothalamus, pituitary gland, and ovaries interact to regulate female reproduction

32. The first day of menstrual bleeding marks the first day of the menstrual cycle. _____ occurs at about day 14 in a 28-day menstrual cycle.

33. FSH stimulates follicle development. The developing follicles release (a)_____, which stimulates development of the endometrium. The two hormones (b)_____ stimulate ovulation, and (c)_____ also promotes development of the corpus luteum.

34. The corpus luteum secretes _____, which stimulates final preparation of the uterus for possible pregnancy.

Menstrual cycles stop at menopause

35. A change in the (a)_____ triggers menopause, a period when (b)_____ are no longer produced and a woman becomes (c)_____.

Most mammals have estrous cycles

36. In an estrous cycle, the endometrium is _____ if conception does not occur.

SEXUAL RESPONSE

37. _____ are two basic physiological responses to sexual stimulation.

38. _____ are the four phases of the sexual response cycle.

FERTILIZATION AND EARLY DEVELOPMENT

39. Membranes that develop around an embryo secrete a peptide hormone called _____ that signals the corpus luteum to continue to function.

THE BIRTH PROCESS

40. The stretching of the uterine muscle by the growing fetus combined with the effects of increased _____ produce strong uterine contractions.

BIRTH CONTROL METHODS

Most hormone contraceptives prevent ovulation

41. The most common contraceptives are combinations of _____ _____.

Intrauterine devices are widely used

42. The IUD is used in an estimated _____(#?) women around the world.

Other common contraceptive methods include the diaphragm and condom

43. The condom is the only contraceptive device that affords some protection against _____.

Emergency contraception is available

44. The most common types of emergency contraception are _____ _____, which can decrease the probability of pregnancy by about 89%.

Sterilization renders an individual incapable of producing offspring

45. Male sterilization is accomplished by a _____.

46. Female sterilization is accomplished by _____.

Abortions can be spontaneous or induced

47. _____ can be administered to induce uterine contractions and expel an embryo.

SEXUALLY TRANSMITTED DISEASES

48. (a)_____ is the most
common STD in the United States. About (b)_____(#?)
people are currently infected and about (c)_____(%?) have been
infected at some point in their lives.

BUILDING WORDS

Use combinations of prefixes and suffixes to build words for the definitions that follow.

Prefixes	The Meaning	Suffixes	The Meaning
acro-	extremity, height	-ectomy	excision
circum-	around, about	-gen(esis)	production of
contra-	against, opposite, opposing	-some	body
endo-	within		
intra-	within		
oo-	egg		
partheno-	virgin		
post-	behind, after		
pre-	before, prior to, in advance of, early		
spermato-	seed, "sperm"		
vaso-	vessel		

Prefix	Suffix	Definition
_____	_____	1. The production of an adult organism from an egg in the absence of fertilization; virgin development.
_____	_____	2. The production of sperm.
_____	_____	3. The organelle (body) at the extreme tip of the sperm head that helps the sperm penetrate the egg.
_____	-cision	4. The removal of all or part of the foreskin by cutting all the way around the penis.
_____	_____	5. The production of eggs or ova.
_____	-metrium	6. The lining within the uterus.
_____	-ovulatory	7. Pertains to a period of time after ovulation.
_____	-ovulatory	8. Pertains to a period of time prior to ovulation.
vas-	_____	9. Male sterilization in which the vas deferentia are partially excised.
_____	-congestion	10. The engorgement of erectile tissues with blood.
_____	-uterine	11. Located or occurring within the uterus.
_____	-ception	12. Birth control; techniques or devices opposing conception.

MATCHING

Terms:

a. Budding
b. Clitoris
c. Corpus luteum
d. Ejaculatory duct
e. Fertilization

f. Fragmentation
g. Ovary
h. Oviduct
i. Ovulation
j. Progesterone

k. Prostate gland
l. Scrotum
m. Semen
n. Testosterone
o. Zygote

For each of these definitions, select the correct matching term from the list above.

_____ 1. Fluid composed of sperm suspended in various glandular secretions that is ejaculated from the penis during orgasm.

_____ 2. A hormone produced by the corpus luteum of the ovary and by the placenta; acts with estradiol to regulate menstrual cycles and to maintain pregnancy.

_____ 3. A gland in male mammals that secretes an alkaline fluid that is part of the seminal fluid.

_____ 4. A small, erectile structure at the anterior part of the vulva in female mammals; homologous to the male penis.

_____ 5. One of the paired female gonads; responsible for producing eggs and sex hormones.

_____ 6. Pocket of endocrine tissue in the ovary that is derived from the follicle cells; secretes the hormone progesterone and estrogen.

_____ 7. The fusion of the male and female gametes; results in the formation of a zygote.

_____ 8. A short duct that passes through the prostate gland and connects the vas deferens to the urethra.

_____ 9. The release of a mature "egg" from the ovary.

_____10. The external sac of skin found in most male mammals that contains the testes and their accessory organs.

MAKING COMPARISONS

Fill in the blanks.

Endocrine Gland	Hormone	Target Tissue	Action in Male	Action in Female
Hypothalamus	GnRH	Anterior pituitary	Stimulates release of FSH and LH	Stimulates release of FSH and LH
#1	FSH	#2	#3	#4
Anterior pituitary	LH	Gonad	#5	#6
#7	#8	General	Establishes and maintains primary & secondary sex characteristics	Not produced
#9	#10	General	Not produced	Establishes and maintains primary & secondary sex characteristics

MAKING CHOICES

Place your answer(s) in the space provided. Some questions may have more than one correct answer.

_____ 1. The hormone in urine or blood that indicates early pregnancy is
    a. GnRH.   d. inhibin.
    b. FSH.   e. ABP.
    c. hCG.

_____ 2. The tube(s) that transport(s) gametes away from gonads is/are the
    a. vas deferens.   d. vagina.
    b. seminal vesicle.   e. oviduct.
    c. sperm cord.

_____ 3. A human female at birth already has
    a. primary oocytes in the first prophase.   d. some secondary oocytes in meiosis II.
    b. preovulatory Graafian follicles.   e. zona pellucida.
    c. all the formative oogonia she will ever have.

_____ 4. The _basic_ physiological responses that result from effective sexual stimulation are
    a. vasocongestion.   d. muscle tension.
    b. erection.   e. orgasm.
    c. vaginal lubrication.

_____ 5. Human chorionic gonadotropin (hCG)
    a. is produced by embryonic membranes.   d. is originally produced in the pituitary.
    b. is secreted by the endometrium.   e. causes menstrual flow to begin.
    c. affects functioning of the corpus luteum.

_____ 6. The hormone(s) primarily responsible for secondary sexual characteristics is/are
    a. luteinizing hormone.   d. testosterone.
    b. estrogen.   e. human chorionic gonadotropin.
    c. follicle-stimulating hormone.

_____ 7. The corpus luteum is directly or indirectly involved in
    a. secreting estrogen.   d. responding to signals from embryonic membrane secretions.
    b. secreting progesterone.   e. maintaining a newly implanted embryo.
    c. stimulating glands to produce nutrients for an implanted embryo.

_____ 8. Follicle-stimulating hormone (FSH) stimulates development of
    a. seminiferous tubules.   d. the corpus luteum.
    b. interstitial cells.   e. ovarian follicles.
    c. sperm cells.

_____ 9. When used properly, the most effective form of birth control in this list for a fertile person is
    a. condoms.   d. the rhythm method.
    b. douching.   e. spermicidal sponges.
    c. oral contraception.

_____10. The first measurable response to effective sexual stimulation
    a. is orgasm.   d. is penile erection.
    b. is ejaculation.   e. is vaginal lubrication.
    c. occurs in the excitement phase.

_____11. The form(s) of birth control that can definitely deter contraction of sexually transmitted diseases is/are
    a. condoms.
    b. douching.
    c. oral contraception.
    d. diaphragms.
    e. spermicides.

_____12. Your authors consider a count of 200 million sperm in a single ejaculation as
    a. clinical sterility.
    b. above average.
    c. below average.
    d. about average.
    e. impossible.

_____13. A Papanicolaou test (Pap smear)
    a. can detect early cancer.
    b. requires an abdominal incision.
    c. involves examination of cells taken from the lower portion of the uterus.
    d. looks at cells from the menstrual flow.
    e. examines cells scraped from the cervix.

_____14. Primitive, undifferentiated stem cells that line the seminiferous tubules are called
    a. sperm.
    b. spermatids.
    c. primary spermatocytes.
    d. secondary spermatocytes.
    e. spermatogonia.

_____15. The external genitalia of the human female are collectively known as the
    a. vagina.
    b. clitoris.
    c. mons pubis.
    d. major and minor lips.
    e. vulva.

_____16. In gametogenesis in the human male, the cell that develops after the first meiotic division is the
    a. sperm.
    b. spermatid.
    c. primary spermatocyte.
    d. secondary spermatocyte.
    e. spermatogonium.

_____17. The portion of the ovarian follicle that remains behind after ovulation
    a. develops into the corpus luteum.
    b. becomes a gland.
    c. is called the Graafian follicle.
    d. is surrounded by the zona pellucida.
    e. transforms into the endometrium.

_____18. The term gonad refers to the
    a. penis.
    b. testes.
    c. vagina.
    d. ovaries.
    e. gametes.

_____19. Human sperm cell production occurs in the
    a. epididymis.
    b. sperm tube.
    c. testes.
    d. accessory glands.
    e. seminiferous tubules.

_____20. The normal, periodic flow of blood from the vagina is associated with
    a. ovulation.
    b. orgasm.
    c. sloughing of the endometrium.
    d. removal of the inner lining of the vagina.
    e. miscarriages.

_____21. Human female breast size is primarily a function of the amount of
    a. alveoli.
    b. milk glands.
    c. adipose tissue (fat).
    d. milk the woman can produce after giving birth.
    e. colostrum.

_____22. Which of the following is/are true of the human penis?

a. It has three erectile tissue bodies.   d. It contains a bone.

b. Circumcision involves removal of the prepuce.   e. Erection results from increased blood pressure.

c. The glans is an expanded portion of the spongy body.

VISUAL FOUNDATIONS

Color the parts of the illustrations below as indicated. Label the first meiotic division and second meiotic division.

RED	☐	sperm
PINK	☐	large chromosome
BLUE	☐	small chromosome
ORANGE	☐	primary spermatocyte
BROWN	☐	secondary spermatocyte
YELLOW	☐	spermatids

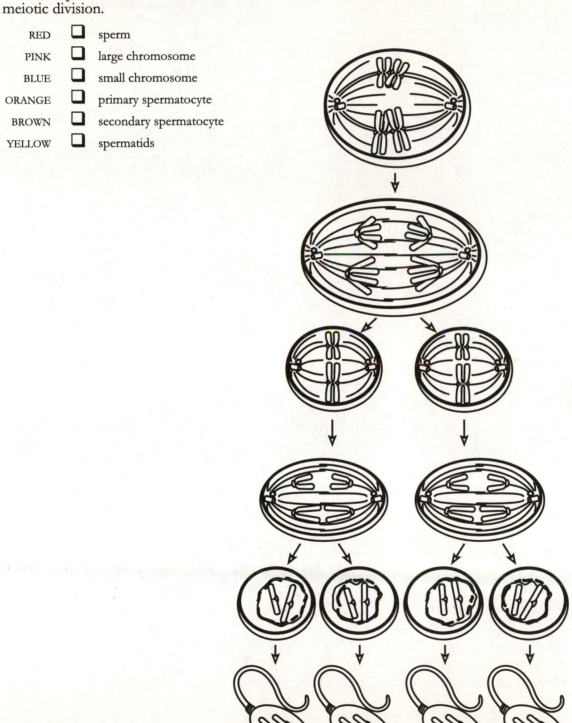

Color the parts of the illustrations below as indicated. Label first and second meiotic divisions.

RED ▢ ovum

PINK ▢ large chromosome

BLUE ▢ small chromosome

ORANGE ▢ primary oocyte

BROWN ▢ secondary oocyte

YELLOW ▢ polar body

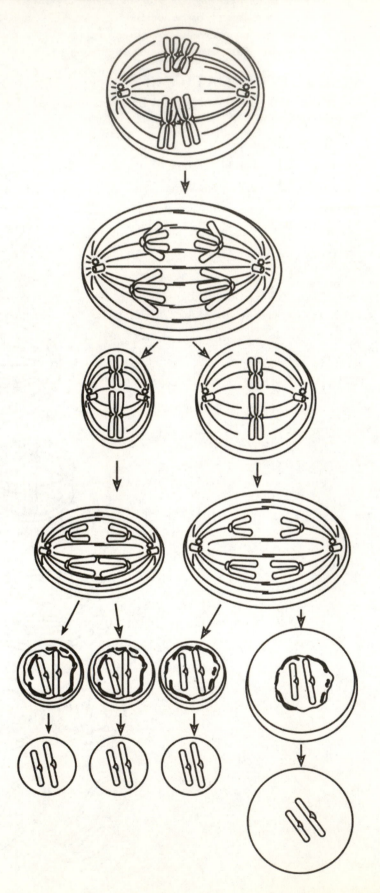

◻

CHAPTER 50

Animal Development

Development encompasses all of the changes that take place during the life of an organism from fertilization to death. Guided by instructions encoded in the DNA of the genes, the organism develops from one cell to billions, from a formless mass of cells to an intricate, highly specialized and organized organism. Development is a balanced combination of several interrelated processes — cell division, increase in the number and size of cells, cellular movements that arrange cells into specific structures and appropriate body forms, and biochemical and structural specialization of cells to perform specific tasks. The stages of early development, which are basically similar for all animals, include fertilization, cleavage, gastrulation, and organogenesis. All terrestrial vertebrates have four extraembryonic membranes that function to protect and nourish the embryo. Environmental factors, such as nutrition, vitamin and drug intake, cigarette smoke, and disease-causing organisms, can adversely influence developmental processes. Aging is a developmental process that results in decreased functional capacities of the mature organism.

REVIEWING CONCEPTS

Fill in the blanks.

INTRODUCTION

1. During development, the number of cells _____, resulting in growth.

DEVELOPMENT OF FORM

2. The process by which cells become specialized is called _____.

3. Cells undergo _____, a process during which they become increasingly organized, shaping the intricate pattern of tissues and organs that characterizes a multicellular animal.

FERTILIZATION

4. Fertilization restores the (a)_____ number of chromosomes, during which time the (b)_____ of the individual is determined.

The first step in fertilization involves contact and recognition

5. _____, a species-specific protein on the acrosome, adheres to receptors on the vitelline envelope.

6. Before a mammalian sperm can participate in fertilization, it must first undergo (a)_____, a maturation process that occurs in the female reproductive tract. During this process, sperm become increasingly motile and capable of undergoing an (b)_____ when they encounter an egg.

Sperm entry is regulated

7. Fertilization of the egg by more than one sperm is _____.

8. The block that prevents multiple sperm from entering an egg is called the
_____.

Fertilization activates the egg

9. During activation, aerobic respiration (a)_____, enzymes
and proteins become (b)_____, and a burst of
(c)_____ synthesis occurs.

Sperm and egg pronuclei fuse, restoring the diploid state.

10. It is believed that after the sperm enters the egg, it is guided toward the egg
nucleus by a system of _____.

CLEAVAGE

11. The inherent potential of the egg to form all of the different cell types of the fully
differentiated individual is referred to as _____.

12. The zygote undergoes cleavage, first forming a morula and then a
_____.

The pattern of cleavage is affected by yolk

13. The isolecithal eggs of (a)_____
undergo (b)_____ cleavage, that is, the entire egg divides into
cells of about the same size. (c)_____ cleavage of isolecithal eggs is
common in echinoderms, while some annelids and mollusks undergo
(d)_____ cleavage.

14. The (a)_____ cleavage that occurs in the telolecithal eggs of
reptiles and birds is restricted to the (b)_____.

Cleavage may distribute developmental determinants

15. Mosaic development is a consequence of the _____
of important materials in the cytoplasm of the zygote.

Cleavage provides building blocks for development

16. _____ are important in helping cells recognize
one another and determining which ones adhere to form tissues.

GASTRULATION

17. The blastula becomes a three-layered embryo during _____.

18. Embryonic layers are collectively called (a)_____; they include
the (b)_____.

The amount of yolk affects the pattern of gastrulation

19. In the amphibian, cells from the animal pole move down over the yolk-rich cells
and invaginate, forming the dorsal lip of the _____.

20. In the bird, invagination occurs at the _____ and
no archenteron forms.

ORGANOGENESIS

21. The (a)_____ induces the formation of the neural plate, from
which cells migrate downward thus forming the (b)_____,
flanked on each side by neural folds. When neural folds fuse, the

(c)_____ is formed. This structure gives rise to the
(d)_____.

22. (a)_____ meet the
(b)_____ thereby forming branchial arches that give
rise to elements of the face, jaws, and neck.

EXTRAEMBRYONIC MEMBRANES

23. Terrestrial vertebrates have four extraembryonic membranes:
_____.

24. The _____ is a fluid-filled sac that surrounds the embryo and keeps it
moist.

25. In humans, the _____, an outgrowth of the digestive tract, is
small and nonfunctional, except that its blood vessels contribute to the formation
of umbilical vessels joining the embryo to the placenta.

HUMAN DEVELOPMENT

26. The (a)_____ dissolves when the embryo enters the uterus.
At this time, the embryo is in the (b)_____ stage, but now begins to
differentiate into a (c)_____.

27. The outer layer of embryonic cells, or _____, forms the
chorion and amnion.

28. The trophoblast cells of the blastocyst secrete enzymes that digest an opening
into the uterine wall in a process called (a)_____. The
process begins at about the (b)_____ day of development.

The placenta is an organ of exchange

29. In placental mammals, the _____
tissues give rise to the placenta, the organ of exchange between mother and
developing child.

Organ development begins during the first trimester

30. Gastrulation occurs during the _____ weeks of human development.

31. Sex (gender) can be determined by external observation by the
_____.

Development continues during the second and third trimesters

32. During the (a)_____ month of pregnancy the cerebrum grows
rapidly and develops (b)_____.

More than one mechanism can lead to a multiple birth

33. A sonogram is a picture taken by the technique known as _____.

Environmental factors affect the embryo

34. Ultrasound imaging techniques and _____, the analysis of
intrauterine fluids, are helpful in anticipating potential birth defects.

The neonate must adapt to its new environment

35. A neonate's first breath may be initiated by _____
in the blood after the umbilical cord is cut.

Aging is not a uniform process

36. Women live an average of _____ (#?) years longer than men.

Homeostatic response to stress decreases during aging

37. All body systems _____ with age, but not at the same rate.

BUILDING WORDS

Use combinations of prefixes and suffixes to build words for the definitions that follow.

<u>Prefixes</u>	<u>The Meaning</u>		<u>Suffixes</u>	<u>The Meaning</u>
arch-	primitive		-blast	embryo
blasto-	embryo		-mere	part
holo-	whole, entire			
iso-	equal			
mero-	part, partial			
neo-	new, recent			
telo-	end			
tri-	three			
tropho-	nourishment			

<u>Prefix</u>	<u>Suffix</u>	<u>Definition</u>
_____	-lecithal	1. Having an accumulation of yolk at one end (vegetal pole) of the egg.
_____	-lecithal	2. Having a fairly equal distribution of yolk in the egg.
_____ _____		3. Any cell that is a part of the early embryo (specifically, during cleavage).
_____	-cyst	4. The blastula of the mammalian embryo.
_____	-pore	5. The opening into the archenteron of an early embryo.
_____	-blastic	6. A type of cleavage in which the entire egg divides.
_____	-blastic	7. A type of cleavage in which only the blastodisc divides; partial cleavage.
_____	-enteron	8. The central cavity resulting from gastrulation; the primitive digestive system.
_____ _____		9. The extraembryonic part of the blastocyst that chiefly nourishes the embryo or that develops into fetal membranes with nutritive functions.
_____	-mester	10. A term or period of three months.
_____	-nate	11. A newborn child.

MATCHING

Terms:

a. Amnion
b. Animal pole
c. Blastula
d. Chorion
e. Cleavage

f. Ectoderm
g. Endoderm
h. Fertilization
i. Gastrula
j. Germ layer

k. Implantation
l. Morphogenesis
m. Placenta
n. Vegetal pole

For each of these definitions, select the correct matching term from the list above.

_____ 1. A membrane that forms a fluid-filled sac for the protection of the developing embryo.

_____ 2. The first of several cell divisions in early embryonic development that converts the zygote into a multicellular blastula.

_____ 3. Any of the three embryonic tissue layers.

_____ 4. The yolk pole of a vertebrate or echinoderm egg.

_____ 5. Usually a spherical structure produced by cleavage of a fertilized ovum; consists of a single layer of cells surrounding a fluid-filled cavity.

_____ 6. The outer germ layer that gives rise to the skin and nervous system.

_____ 7. The attachment of the developing embryo to the uterus of the mother.

_____ 8. The development of the form and structures of an organism and its parts.

_____ 9. Early stage of embryonic development during which the embryo has three layers and is cup-shaped.

_____10. The fusion of the egg and the sperm, resulting in the formation of a zygote.

MAKING COMPARISONS

Fill in the blanks.

Human Developmental Process	Approximate Time	Event
Fertilization	0 hour	Restores the diploid number of chromosomes
#1	2, 3 weeks	Blastocyst becomes a three-layered embryo
Early cleavage	#2	Formation of the two-cell stage
#3	#4	Activates the egg
Implantation	#5	#6
#7	0 hour	Establishes the sex of the offspring
#8	2.5 weeks	Neural plate begins to form
#9	#10	Morula is formed, reaches uterus

MAKING CHOICES

Place your answer(s) in the space provided. Some questions may have more than one correct answer.

_____ 1. The liver, pancreas, trachea, and pharynx
     a. arise from Hensen's node.   d. are derived from ectoderm.
     b. are formed prior to the primitive streak.   e. are derived from mesoderm.
     c. are derived from endoderm.

_____ 2. A hollow ball of several hundred cells
     a. is a morula.   d. is the gastrula.
     b. is a blastula.   e. contains a blastocoel.
     c. surrounds a fluid-filled cavity.

_____ 3. Which of the following organ systems is the first to form during organogenesis?
     a. circulatory system   d. nervous system
     b. reproductive system   e. muscular system
     c. excretory system

_____ 4. The duration of pregnancy is referred to as the
     a. parturition.   d. neonatal period.
     b. induction cycle.   e. prenatal period.
     c. gestation period.

_____ 5. The fertilization envelope (membrane)
     a. promotes binding of sperm to egg.   d. expedites entry of sperm into egg.
     b. facilitates egg-sperm recognition.   e. blocks entrance of sperm.
     c. prevents polyspermy.

_____ 6. The cavity within the embryo that is formed during gastrulation is the
     a. archenteron.   d. blastopore.
     b. gastrulacoel.   e. neural tube.
     c. blastocoel.

_____ 7. An embryo is called a fetus
     a. when the brain forms.   d. after two months of development.
     b. when the limbs forms.   e. once it is firmly implanted in the endometrium.
     c. when its heart begins to beat.

_____ 8. Embryonic cells are arranged in three distinct germ layers (endoderm, mesoderm, ectoderm) during
     a. cleavage.   d. organogenesis.
     b. gastrulation.   e. blastocoel formation.
     c. blastulation.

_____ 9. Amniotic fluid
     a. replaces the yolk sac in mammals.   d. fills the space between the amnion and chorion.
     b. is secreted by the allantois.   e. fills the space between the amnion and embryo.
     c. is secreted by an embryonic membrane.

_____10. Recognition of species compatibility between a sperm and an egg is due to the recognition protein
     a. acrosin.   d. bindin.
     b. actin.   e. pellucidin.
     c. vitelline.

_____11. The organ of exchange between a placental mammalian embryo and its mother develops from

    a. chorion and amnion.

    b. chorion and allantois.

    c. chorion and uterine tissue.

    d. amnion and uterine tissue.

    e. amnion and allantois.

_____12. Eggs that have a large amount of yolk concentrated at one pole

    a. are isolecithal.

    b. are telolecithal.

    c. have a metabolically active animal pole.

    d. have a vegetal pole that is metabolically active.

    e. do not have food for the embryo in the egg.

_____13. _____ gives rise to skeletal tissues, muscle, and the circulatory system.

    a. Ectoderm

    b. Mesoderm

    c. Endoderm

    d. One of the germ layers

    e. The blastocoel

VISUAL FOUNDATIONS

Color the parts of the illustration below as indicated. Also label the blastocyst and umbilical cord.

RED ☐ umbilical artery

GREEN ☐ inner cell mass

YELLOW ☐ yolk sac

BLUE ☐ amniotic cavity and amnion

ORANGE ☐ chorionic cavity and chorion

BROWN ☐ trophoblast cells

TAN ☐ uterine epithelium

PINK ☐ embryo

VIOLET ☐ maternal blood and blood vessel

GREY ☐ placenta

CHAPTER 51

❑

Animal Behavior

Animal behavior consists of those movements or responses an animal makes in response to signals from its environment; it tends to be adaptive and homeostatic. The behavior of an organism is as unique and characteristic as its structure and biochemistry. An organism adapts to its environment by synchronizing its behavior with cyclic change in its environment. Although behaviors are inherited, they can still be modified by experience. Behavioral ecology focuses on the interactions between animals and their environments and on the survival value of their behavior. Social behaviors, that is, interactions between two or more animals of the same species, have advantages and disadvantages. Advantages include confusing predators, repelling predators, and finding food. Disadvantages include competition for food and habitats, and increased ease of disease transmission. Some species that engage in social behaviors form societies. Some societies are loosely organized, whereas others have a complex structure. A system of communication reinforces the organization of the society. Animals communicate in a wide variety of ways; some use sound, some use scent, and some use pheromones. Some invertebrate societies exhibit elaborate and complex patterns of social interactions, such as the bees, ants, wasps, and termites. Vertebrate societies are far less rigid.

REVIEWING CONCEPTS
Fill in the blanks.

INTRODUCTION

1. Behavior refers to the responses an organism makes to _____ from its environment.

2. Persistent changes in behavior brought about by experience is _____.

3. _____ is the study of behavior in natural environments from an evolutionary perspective.

BEHAVIOR AND ADAPTATION

4. _____ is an individual's reproductive success as measured by the number of viable offspring it produces.

INTERACTION OF GENES AND ENVIRONMENT

5. (a)_____ is inborn, genetically programmed behavior, whereas (b)_____ is behavior that has been modified in response to environmental experience.

Behavior depends of physiological readiness

6. Behavior is influenced primarily by two systems of the body, namely, the _____ systems.

Many behavior patterns depend on motor programs

7. Fixed action pattern (FAP) behaviors are elicited by a _____.

LEARNING FROM EXPERIENCE

8. Learning is defined as a persistent change in behavior due to
_____.

An animal habituates to irrelevant stimuli

9. _____ is a type of learning in which an animal learns to ignore a repeated, irrelevant stimulus.

Imprinting occurs during an early critical period

10. Imprinting establishes a parent-offspring bond during a critical period early in development, ensuring that the offspring _____ the mother.

In classical conditioning, a reflex becomes associated with a new stimulus

11. In classical conditioning, an animal makes an association between an irrelevant stimulus and a normal body response. It occurs when a
(a)_____ becomes a substitute for an
(b)_____ stimulus.

In operant conditioning spontaneous behavior is reinforced

12. In operant conditioning, the behavior of the animal is rewarded or punished after it performs a behavior discovered by chance. Repetitive rewards bring about
(a)_____, and consistent punishment induces
(b)_____.

Animal cognition is controversial

13. Some animals have a form of cognition called _____
_____, which is the ability to adapt past experiences that may involve different stimuli to solve a new problem.

Play may be practice behavior

14. Hypotheses for the ultimate causes of play behavior include _____
_____.

BIOLOGICAL RHYTHMS AND MIGRATION

Biological rhythms affect behavior

15. Daily cycles of activity are known as _____.

16. In mammals, the master clock is located in the
_____ in the hypothalamus.

Migration involves interactions among biological rhythms, physiology, and environment

17. _____ refers to travel by migrating animals in a specific direction.

FORAGING BEHAVIOR

18. _____ refers to an animal's ability to obtain food in the most efficient manner.

COMMUNICATION AND LIVING IN GROUPS

19. A _____ is an actively cooperating group of individuals belonging to the same species and often closely related.

Communication is necessary for social behavior

20. _____ are chemical signals that convey information between members of a species.

21. Chemoreceptor cells in the nose of animals comprise the _____ organ.

Animals benefit from social organization

22. _____ refers to a linear "pecking order" into which animals in a population may organize according to status; those lower in the hierarchy are subordinate to those above them.

23. In some animals, the hormone _____ is the apparent cause of their aggressive behaviors.

Many animals defend a territory

24. Most animals have a _____, a geographical area that they seldom leave.

SEXUAL SELECTION

Animals of the same sex compete for mates

25. In _____, individuals of the same sex compete for mates.

Animals choose quality mates

26. A _____ is a small display area in which the males of some species of insects, birds, and bats compete for females.

27. _____ ensure that the male is a member of the same species, and it provides the female with a means of evaluating the male.

Sexual selection favors polygynous mating systems

28. (a)_____ is a mating system in which males fertilize the eggs of many females during a breeding season, whereas (b)_____ is a mating system in which a female mates with several males.

Some animals care for their young

29. _____ is the contribution each parent makes in producing and rearing offspring.

HELPING BEHAVIOR

Altruistic behavior can be explained by inclusive fitness

30. Indirect selection, also called _____, is a form of natural selection that increases inclusive fitness through the breeding success of close relatives

Helping behavior may have alternative explanations

31. One example of kin selection includes _____, keeping watch for predators and warning of threats.

Some animals help nonrelatives

32. _____ is a type of behavior in which one animal helps another with no immediate benefit; however, at some later time the animal that was helped repays the debt.

HIGHLY ORGANIZED SOCIETIES

Social insects form elaborate societies

33. The most sophisticated known mode of communication among non-mammals is a stereotyped series of body movements performed by bees referred to as a _____.

Vertebrate societies tend to be relatively flexible

34. (a)_____, or behavior common to a population, is maintained by (b)_____.

Sociobiology explains human social behavior in terms of adaptation

35. Sociobiology focuses on the evolution of social behavior through _____.

BUILDING WORDS

Use combinations of prefixes and suffixes to build words for the definitions that follow.

Prefixes	The Meaning	Suffixes	The Meaning
mono-	one	-andr(y)	male
poly-	many, much, multiple, complex	-gyn(y)	female
socio-	social, sociological, society		

Prefix	Suffix	Definition
_____	-biology	1. The school of ethology that focuses on the evolution of social behavior through natural selection.
_____	_____	2. A mating system in which a female mates with many males.
_____	_____	3. A mating system in which a male mates with many females.
_____	-gamy	4. Mating with a single partner during a breeding season.

MATCHING

Terms:

a. Altruistic behavior
b. Biological clock
c. Circadian rhythm
d. Crepuscular animal
e. Diurnal animal
f. Ethology
g. Habituation
h. Imprinting
i. Innate behaviors
j. Migration
k. Nocturnal animal
l. Optimal foraging
m. Pheromone
n. Sign stimulus
o. Territoriality

For each of these definitions, select the correct matching term from the list above.

_____ 1. Means by which activities of plants or animals are adapted to regularly-recurring changes.

_____ 2. An animal that is most active during the day.

_____ 3. Behaviors that are inherited and typical of the species.

_____ 4. A substance secreted by organisms into the external environment that influences the development or behavior of other members of the same species.

_____ 5. The theory that animals feed in a manner that maximizes benefits and/or minimizes costs.

_____ 6. The process by which organisms become accustomed to a stimulus and cease to respond to it.

_____ 7. A form of rapid learning by which a young bird or mammal forms a strong social attachment to an individual or object within a few hours after hatching or birth.

_____ 8. Behavior in which one individual appears to act in such a way as to benefit others rather than itself.

_____ 9. A daily cycle of activity around which the behavior of many organisms is organized.

_____10. The study of animal behavior.

MAKING COMPARISONS

Fill in the blanks.

Classification of Behavior	Example of the Behavior
Innate (FAP)	Egg rolling in graylag goose
#1	Wasp responding to cone arrangement to locate nest
#2	Dog salivating at sound of bell
#3	Child sitting quietly for praise
#4	Birds tolerant of human presence
#5	Chicks learning the appearance of the parent
#6	Primate stacking boxes to reach food

MAKING CHOICES

Place your answer(s) in the space provided. Some questions may have more than one correct answer.

_____ 1. A behavior that resembles a simple reflex in some ways and a volitional behavior in others, and is triggered by a sign stimulus is best described as a/an

    a. releaser.
    d. behavioral pattern (formerly FAP).
    b. conditioned reflex.
    e. unconditioned reflex.
    c. conditioned stimulus.

_____ 2. If your dog perks up its ears when you clap your hands, but stops perking its ears after you have repeatedly clapped your hands for a period of time, your dog has displayed a form of

    a. learning.
    d. sensory adaptation.
    b. operant conditioning.
    e. habituation.
    c. classical conditioning.

_____ 3. A change in your behavior derived from experience in your environment is the result of

    a. a FAP.
    d. redirected behavior.
    b. learning.
    e. habituation and/or sensitization.
    c. inherited behavioral characteristics.

_____ 4. Suppose you decide to perform an experiment with a type of lizard that instantly attacks any lizard with green scales on its sides. You place a live lizard with green scales on its sides in a cage next to a styrofoam block with green paint on its sides. Your experimental lizard ignores the spotted lizard, preferentially attacking the styrofoam block. The styrofoam-attacking lizard is responding to a/an

a. inducer.

b. sign stimulus.

c. conditioned reflex.

d. displacement syndrome.

e. redirected behavioral syndrome.

_____ 5. Which of the following is/are used by a worker honey bee to indicate the distance of a food source?

a. round dance

b. waggle dance

c. angle relative to a north–south axis

d. angle relative to gravity

e. angle relative to sun

_____ 6. The study of the behavior of animals in their natural environments from an evolutionary perspective is

a. behavioral ecology (formerly ethology).

b. behavioral genetics.

c. social behaviorist.

d. behavioral ecology.

e. neurobiology.

_____ 7. Genetically programmed behavior has been variously termed

a. innate behavior.

b. inborn behavior.

c. determinism.

d. instinct.

e. patterned behavior.

_____ 8. A simple signal that triggers a specific behavioral response is known as a

a. initiator.

b. patterned stimulus.

c. releaser.

d. sign stimulus.

e. motor stimulus.

PART VIII

The Interactions of Life: Ecology

CHAPTER 52

❑

Introduction to Ecology:
Population Ecology

This is the first of five chapters that cover the foundations of ecology. It focuses on the study of populations as functioning systems. A population is all the members of a particular species that live together in the same area. Populations have properties that do not exist in their component individuals or in the community of which the population is a part. These properties include birth rates, death rates, growth rates, population density, population dispersion, age structure, and survivorship. Population ecology deals with the number of individuals of a particular species that are found in an area and how and why those numbers change over time. Populations must be understood in order to manage forests, field crops, game, fishes, and other populations of economic importance. Population growth is determined, in part, by the rate of arrival and the rate of departure of organisms. Population growth is limited by density-dependent and density-independent factors. Each species has its own survival strategy that enables it to survive. Most organisms fall into one of two survival strategies: One emphasizes a high rate of natural increase and the other emphasizes maintenance of a population near the carrying capacity of the environment. The principles of population ecology apply to humans as well as other organisms. Environmental degradation is related to population growth and resource consumption.

REVIEWING CONCEPTS
Fill in the blanks.

INTRODUCTION

1. Interactions among organisms are called (a)_____,
   while interactions between organisms and their nonliving, physical environment
   are called (b)_____.

FEATURES OF POPULATIONS

2. Populations of organisms have properties that individual organisms do not have.
   Some properties of importance are

   _____

   _____.

3. The study of changes in populations is known as _____.

Density and dispersion are important features of populations

4. _____ is the number of individuals of a species
   per unit of habitat area or volume at a given time.

5. Individuals within a population may exhibit a characteristic spacing. For example,
   (a)_____ exists when individuals in a population are
   spaced throughout an area in a manner that is unrelated to the presence of others;

(b)_____occurs when individuals are more evenly spaced than would be expected from a random occupation of a given habitat; and (c)_____ occurs when individuals are concentrated in specific parts of the habitat.

CHANGES IN POPULATION SIZE

6. Changes in the size of a population result from the difference between _____.

7. The rate of change in a population is expressed as:

$$\Delta N / \Delta t = N(b - d), \text{ where}$$

ΔN is the (a)_____, Δt is the (b)_____, b is the birth rate or (c)_____, and d is the death rate or (d)_____.

Dispersal affects the growth rate in some populations

8. (a)_____ occurs when individuals enter a population and thus increase in size; (b)_____ occurs when individuals leave a population and thus decrease its size.

Each population has a characteristic intrinsic rate of increase

9. The maximum rate at which a population of a given species could increase under ideal conditions when resources are abundant and its population density is low is known as its _____.

10. _____ is the accelerating population growth rate that occurs when optimal conditions allow a constant per capita growth rate.

No population can increase exponentially indefinitely

11. The largest population that can be sustained for an indefinite period by a particular environment is that environment's _____.

12. Population growth curves take a variety of forms, their shapes primarily a function of biotic potential and limiting factors. The _____ growth curve shows the population's initial exponential increase, followed by a leveling out as the carrying capacity of the environment is approached.

FACTORS INFLUENCING POPULATION SIZE

Density-dependent factors regulate population size

13. Examples of density-dependent factors are _____.

14. Density-dependent limiting factors are most effective at _____ (high or low?) population densities, and therefore they tend to stabilize populations.

15. _____ is an effective density-dependent limiting factor because transmission of infectious organisms is more likely among members of dense populations.

16. In (a)_____, also called contest competition, certain dominant individuals obtain an adequate supply of the limited resource at the expense of other individuals in the population. In

(b)_____, also called scramble competition, all the individuals in a population "share" the limited resource more or less equally.

Density-independent factors are generally abiotic

17. Random weather events that reduce population size serve as

_____.

LIFE HISTORY TRAITS

18. (a)_____ species expend their energy in a single, immense reproductive effort. (b)_____ species reproduce during several breeding seasons throughout their lifetimes.

19. Each species has a life history strategy. (a)_____ are usually opportunists found in variable, temporary, or unpredictable environments where the probability of long-term survival is low. (b)_____ tend to be found in relatively constant or stable environments, where they have a high competitive ability.

Life tables and survivorship curves indicate mortality and survival

20. _____ is the probability that a given individual in a population or cohort will survive to a particular age.

METAPOPULATIONS

21. Good habitats, called (a)_____, are areas where local reproductive success is greater than local mortality. Lower-quality habitats, called (b)_____, are areas where local reproductive success is less than local mortality.

HUMAN POPULATIONS

22. Zero population growth is the point at which the (a)_____ equals the (b)_____.

Not all countries have the same growth rate

23. The amount of time that it takes for a population to double in size is its (a)_____. In general, the shorter this time, the (b)_____(more or less?) developed the country.

24. The number of children that a couple must produce to replace themselves is known as the (a)_____, a value that is generally relatively (b)_____(higher or lower?) in developed countries and (c)_____(higher or lower?) in undeveloped and developing countries.

25. The average number of children born to a woman during her lifetime is called the (a)_____. On a worldwide basis, this value is currently (b)_____(higher or lower?) than the replacement value.

The age structure of a country helps predict future population growth

26. To predict the future growth of a population, it is important to know its age structure, which is the _____ of people at each age in a population.

Environmental degradation is related to population growth and resource consumption

27. People overpopulation occurs when the environment is worsening from too many people. Consumption overpopulation occurs when each individual in a population consumes too large a share of resources. The (a)_____ (former or latter?) is the current problem in many developing nations, and the (b)_____ (former or latter?) is the current problem in most affluent, highly developed nations.

BUILDING WORDS

Use combinations of prefixes and suffixes to build words for the definitions that follow.

Prefixes	The Meaning
bio-	life
inter-	between, among
intra-	within

Prefix	Suffix	Definition
_____	-specific competition	1. Competition for resources within a population.
_____	-specific competition	2. Competition for resources among populations.
_____	-sphere	3. The entire zone of air, land, and water at the surface of the earth that is occupied by living things.

MATCHING

Terms:

a. Intrinsic rate of increase
b. Carrying capacity
c. Contest competition
d. Density-dependent factor
e. Density-independent factor
f. Doubling time
g. Ecology
h. Emigration
i. Exponential growth
j. Immigration
k. Mortality
l. Natality
m. Population
n. Population crash
o. Random dispersion
p. Replacement-level fertility
q. Uniform dispersion

For each of these definitions, select the correct matching term from the list above.

_____ 1. A pattern of spacing in which individuals in a population are spaced unpredictably.

_____ 2. Growth that occurs at a constant rate of increase over a period of time.

_____ 3. The rate at which organisms produce offspring.

_____ 4. The maximum number of organisms that an environment can support.

_____ 5. Any factor, such as climate, that does not depend on the density of populations.

_____ 6. The number of offspring a couple must produce in order to "replace" themselves.

_____ 7. The movement of individuals out of a population.

_____ 8. The maximum rate of increase of a species that occurs when all environmental conditions are optimal.

_____ 9. The amount of time it takes for a population to double in size, assuming that its current rate of increase does not change.

_____10. An abrupt decline in a population from a high to very low population density.

_____11. The study of the interactions between living things and their environment, both physical and biotic.

_____12. A group of organisms of the same species that live in the same geographical area at the same time.

MAKING COMPARISONS

Fill in the blanks.

Examples of Environmental Factors That Affect Population Size	Is Factor Density-Dependent or Density-Independent?	Description of Factor
Predation	Density-dependent	As the density of a prey species increases, predators are more likely to encounter an individual of the prey species
Killing frost	#1	#2
Contest competition	#3	#4
Disease	#5	#6
Hurricane	#7	#8

MAKING CHOICES

Place your answer(s) in the space provided. Some questions may have more than one correct answer.

Questions 1-5 pertain to the following equation. Use the list of choices below to answer them.

$$\frac{\Delta N}{\Delta t} = N(b - d)$$

a. growth only
b. rate of growth
c. change only

d. ΔN
e. Δt
f. $b - d$

g. value "b"
h. value "d"
i. $\Delta N / \Delta t$

_____ 1. Natality.

_____ 2. Mortality.

_____ 3. Population change.

_____ 4. Rate of population change.

_____ 5. The parameter derived from the equation.

_____ 6. The equation $dN/dt = rN$ is used to derive the
a. biotic potential (r_m).
b. instantaneous growth rate.
c. decline in growth due to limiting factors.
d. per capita growth rate (r).
e. growth rate at carrying capacity.

_____ 7. During the exponential growth phase of bacterial growth, the population
   a. increases moderately.
   b. increases dramatically.
   c. decreased moderately.
   d. decreases dramatically.
   e. does not change substantially.

_____ 8. A cultivated field of wheat would most likely display
   a. uniform dispersion.
   b. random dispersion.
   c. clumped dispersion.
   d. no dispersion pattern.
   e. a form of dispersion not found in nature.

_____ 9. In the logistic equation, if environmental limits or resistance are sustained, "K" is the
   a. growth rate.
   b. birth rate.
   c. mortality rate.
   d. carrying capacity.
   e. equivalent of the lag phase.

_____10. A grove of trees that originated from one seed displays
   a. uniform dispersion.
   b. random dispersion.
   c. clumped dispersion.
   d. no dispersion pattern.
   e. a form of dispersion not found in nature.

_____11. Population growth is most frequently controlled by which _one_ of the following?
   a. environmental limits
   b. predation
   c. disease
   d. behavioral modifications
   e. resource depletion

_____12. Organisms referred to as r-strategists usually
   a. are mobile animals.
   b. have a high "r" value.
   c. inhabit variable environments.
   d. survive well in the long term.
   e. develop slowly.

_____13. The redwood stands in California are examples of organisms that
   a. use the K-strategy.
   b. use the r-strategy.
   c. often experience local extinction.
   d. live in a relatively stable environment.
   e. pioneer new habitats.

_____14. The number of individuals of a species per unit of habitat area is the _____ for that species.
   a. distribution curve
   b. distribution pattern
   c. population density
   d. ΔN
   e. b − d

Community Ecology

A community consists of an association of populations of different species that live and interact in the same place at the same time. A community and its abiotic environment together compose an ecosystem. The interactions among species in a community include competition, predation, and symbiosis. Every organism has its own ecological niche, that is, its own role within the structure and function of the community. Its niche reflects the totality of its adaptations, its use of resources, and its lifestyle. Communities vary greatly in the number of species they contain. The number is often high where the number of potential ecological niches is great, where a community is not isolated or severely stressed, at the edges of adjacent communities, and in communities with long histories. A community develops gradually over time, through a series of stages, until it reaches a state of maturity; species in one stage are replaced by different species in the next stage. Most ecologists believe that the species composition of a community is due primarily to abiotic factors.

REVIEWING CONCEPTS

Fill in the blanks.

INTRODUCTION

1. An association of populations of different species living and interacting in the same place at the same time comprises a _____.

2. A biological community and its abiotic environment together comprise an _____.

COMMUNITY STRUCTURE AND FUNCTIONING

Community interactions are often complex and not readily apparent

3. The three main types of interactions that occur among species in a community are _____.

The niche is a species' ecological role in the community

4. The totality of adaptations by a species to its environment, its use of resources, and the lifestyle to which it is suited comprises its _____.

5. The potential ecological niche of a species is its (a)_____ niche, while the lifestyle that a species actually pursues and the resources that it actually uses make up its (b)_____ niche.

6. Any environmental resource that, because it is scarce or unfavorable, tends to restrict the ecological niche of a species is called a _____.

Biotic and abiotic factors influence a species' ecological niche

Competition is intraspecific or interspecific

7. Competition among individuals within a population is called _____.

8. According to the _____, two species cannot indefinitely occupy the same niche in the same community.

9. The reduction in competition for environmental resources among coexisting species as a result of each species' niche differing from the others in one or more ways is called _____.

10. _____ is the divergence in traits in two similar species living in the same geographic area.

Natural selection shapes the body forms and behaviors of both predator and prey

11. _____ refers to the independent evolution of two interacting species.

12. Conspicuous colors or patterns that advertise a species' unpalatability to potential predators are known as (a)_____ or (b)_____.

13. _____ is a defense strategy in which a defenseless species is protected from predation by its resemblance to a species that is dangerous in some way.

Symbiosis involves a close association between species

14. The three forms of symbiosis are
_____.

15. When a parasite causes disease and sometimes the death of a host, it is called a _____.

Keystone species and dominant species affect the character of a community

16. (a)_____ species determine the nature of an entire community, having an impact that is out of proportion to their abundance; (b)_____ species influence a community as a result of their greater size or abundance.

COMMUNITY BIODIVERSITY

Ecologists seek to explain why some communities have more species than others

17. Isolated island communities are generally much _____ (more? less?) diverse than are communities in similar environments found on continents.

18. The _____ is a phrase used to describe the relatively greater species diversity near the margins of distinct communities as compared to their centers.

Species richness may promote community stability

19. Community complexity is expressed in terms of (a)_____ _____, the number of species within a community, and (b)_____, a measure of the relative importance of each species within a community based on abundance, productivity, or size.

COMMUNITY DEVELOPMENT

20. _____ succession is the change in species composition over time in a soilless habitat that was not previously inhabited by organisms.

21. _____ succession is the change in species composition that takes place after some disturbance removes the existing vegetation; soil is already present.

Disturbance influences succession and species richness

22. The relatively stable, often long term stage at the end of ecological succession is the _____.

Ecologists continue to study community structure

23. The cooperative view of community that stresses the interaction of the members is known as the _____ model of a community.

24. The view that biological interactions are less important than abiotic factors in producing communities, that, in fact, a community may be a classification category with no reality, is the _____ model of communities.

BUILDING WORDS

Use combinations of prefixes and suffixes to build words for the definitions that follow.

Prefixes	The Meaning		Suffixes	The Meaning
eco-	home		-gen	production of
inter-	between, among		-vore	eating
intra-	within			
patho-	suffering, disease, feeling			

Prefix	Suffix	Definition
carni-	_____	1. Any animal that eats flesh.
_____	-system	2. A community along with its nonliving environment.
_____ _____		3. Any disease-producing organism.
_____ -specific competition		4. Competition for resources within a population.
herbi-	_____	5. Any animal that eats plants.
_____ -specific competition		6. Competition for resources among populations.

MATCHING

Terms:

a. Batesian mimicry
b. Commensalism
c. Detritus
d. Ecological niche
e. Ecotone

f. Edge effect
g. Mullerian mimicry
h. Mutualism
i. Parasitism
j. Pathogen

k. Primary consumer
l. Secondary succession
m. Succession
n. Symbiosis
o. Community

For each of these definitions, select the correct matching term from the list above.

_____ 1. The sequence of changes in a plant community over time.

_____ 2. An ecological succession that occurs after some disturbance destroys the existing vegetation.

_____ 3. An organism that is capable of producing disease.

_____ 4. Resemblance of different species, each of which is equally obnoxious to predators.

_____ 5. An intimate relationship between two or more organisms of different species.

_____ 6. A symbiotic relationship between individuals of different species in which one benefits from the relationship and one is harmed.

_____ 7. A transition region between adjacent communities.

_____ 8. A type of symbiosis in which one organism benefits and the other one is neither harmed nor helped.

_____ 9. An assemblage of different species of organisms that live and interact in a defined area or habitat.

MAKING COMPARISONS

Fill in the blanks.

Interaction	Species 1	Effect on Species 1	Species 2	Effect on Species 2
Mutualism of Species 1 and Species 2	Coral	Beneficial	Zooxanthellae	Beneficial
#1	Silverfish	#2	Army ants	#3
#4	Orcas	#5	Salmon	#6
#7	Tracheal mites	#8	Honeybees	#9
#10	_Entamoeba histolytica_	#11	Humans	#12
Mutualism of Species 1 and Species 2	_Rhizobium_	#13	Beans	#14
Competition between Species 1 and Species 2	_Balanus balanoides_	#15	_Chthamalus stellatus_	#16
#17	Orchids	#18	Host tree	#19

MAKING CHOICES

Place your answer(s) in the space provided. Some questions may have more than one correct answer.

_____ 1. The local environment in which a species lives is its
a. ecosystem.
b. habitat.
c. fundamental niche.
d. realized niche.
e. niche.

_____ 2. The actual lifestyle of a species and the resources it uses are known as the
a. ecosystem.
b. habitat.
c. fundamental niche.
d. realized niche.
e. niche.

_____ 3. Resource partitioning results in
    a. extinction of the least adapted species.   d. increased competition among coexisting species.
    b. interspecific breeding.   e. competition for niches among similar species.
    c. reduced competition among coexisting species.

_____ 4. An association of different species of organisms interacting and living together is a/an
    a. ecosystem.   d. pyramid.
    b. community.   e. niche.
    c. food web.

_____ 5. The difference between an organism's potential niche and the niche it comes to occupy may result from
    a. competition.   d. overabundance of resources.
    b. an inverted pyramid.   e. limited space or resources.
    c. factors that exclude it from part of its fundamental niche.

_____ 6. The totality of an organism's adaptations, its use of resources, and the lifestyle to which it is fitted are all embodied in the term
    a. ecosystem.   d. pyramid.
    b. community.   e. niche.
    c. food web.

_____ 7. All food energy in the biosphere is ultimately provided by
    a. producers.   d. consumers.
    b. nitrogen fixers.   e. decomposers.
    c. phosphorus fixers.

_____ 8. The major roles of organisms in a community are
    a. producers.   d. consumers.
    b. nitrogen fixers.   e. decomposers.
    c. phosphorus fixers.

_____ 9. Critically important minerals and other essential substances would be permanently lost to organisms if it were not for the
    a. producers.   d. consumers.
    b. nitrogen fixers.   e. decomposers.
    c. phosphorus fixers.

_____10. The nonliving environment together with different interacting species is a/an
    a. ecosystem.   d. pyramid.
    b. community.   e. niche.
    c. food web.

_____11. Divergence of traits of two similar species living in the same geographical area is known as
    a. coevolution.   d. Batesian mimicry.
    b. aposematic displacement.   e. Mullerian mimicry.
    c. character displacement.

CHAPTER 54

Ecosystems and the Biosphere

This chapter examines the dynamic exchanges between communities and their physical environments. Energy cannot be recycled and reused. Energy moves through ecosystems in a linear, one-way direction from the sun to producer to consumer to decomposer. Much of this energy is converted into less useful heat as the energy moves from one organism to another. Matter, the material of which living things are composed, cycles from the living world to the abiotic physical environment and back again. All materials vital to life are continually recycled through ecosystems and so become available to new generations of organisms. Key cycles include the carbon cycle, the nitrogen cycle, the phosphorus cycle, and the water, or hydrologic, cycle. Life is possible on earth because of its unique environment. The atmosphere protects organisms from most of the harmful radiation from the sun and space, while allowing life supporting visible light and infrared radiation to penetrate. An area's climate is determined largely by temperature and precipitation. Many ecosystems contain fire-adapted organisms.

REVIEWING CONCEPTS

Fill in the blanks.

INTRODUCTION

1. An individual community *and* its abiotic components together make up an
   _____.

2. All of earth's communities combined comprise the Earth's
   _____.

ENERGY FLOW THROUGH ECOSYSTEMS

3. Energy flows through ecosystems. It enters in the form of (a)_____
   energy, which plants can convert to useful (b)_____ energy that is
   stored in the bonds of molecules. When cellular respiration breaks these
   molecules apart, energy becomes available to do work. As the work is
   accomplished, energy escapes the organisms and dissipates into the environment
   as (c)_____ energy where organisms cannot use it.

4. A complex of interconnected food chains in an ecosystem is a
   _____.

5. Each "link" in a food chain is a trophic level, the first and most basic of which is
   comprised of the photosynthesizers, or (a)_____. Once
   "fixed," materials and energy pass linearly through the chain from
   photosynthesizers to the (b)_____, and then to the
   (c)_____.

Ecological pyramids illustrate how ecosystems work

6. The ecological pyramid of (a)_____ illustrates how many organisms there are in each trophic level; while the pyramid of (b)_____ is a quantitative estimate of the amount of living material at each level.

7. The pyramid of _____ is a somewhat unique representation since this pyramid, unlike the others, and because of the second law of thermodynamics, can never be inverted.

Ecosystems vary in productivity

8. The rate at which energy is captured during photosynthesis is known as the _____ of an ecosystem.

9. The energy that remains in plant tissues after cellular respiration has occurred is called _____; it represents the rate at which organic matter is actually incorporated into plant tissues to produce growth.

Food chains and poisons in the environment

10. The increase in concentration as a toxin passes through successive levels of the food web is known as _____.

CYCLES OF MATTER IN ECOSYSTEMS

11. The cycling of matter from one organism to another and from living organisms to the abiotic environment and back again is referred to as _____.

12. Four cycles that are of particular importance to organisms are the cycles for _____.

Carbon dioxide is the pivotal molecule in the carbon cycle

13. Carbon is incorporated into the biota, or "fixed," primarily through the process of (a)_____, and in the short term much of it is returned to the abiotic environment by (b)_____. Some carbon is tied up for long periods of time in wood and fossil fuels, which may eventually be released through combustion.

Bacteria are essential to the nitrogen cycle

14. The nitrogen cycle has five steps: (a)_____, which is the conversion of nitrogen gas to ammonia; the conversion of ammonia or ammonium to nitrate, called (b)_____; absorption of nitrates, ammonia, or ammonium by plant roots, called (c)_____; the release of organic nitrogen compounds back into the abiotic environment in the form of ammonia and ammonium ions in waste products and by decomposition, a process known as (d)_____; and finally, to complete the cycle, (e)_____ reduces nitrates to gaseous nitrogen.

The phosphorus cycle lacks a gaseous component

15. Erosion of (a)_____ releases inorganic phosphorus to the (b)_____ where it is taken up by plant roots.

Water moves among the ocean, land, and atmosphere in the hydrologic cycle

16. The hydrological cycle results from the evaporation of water from both land and sea and its subsequent precipitation when air is cooled. The movement of surface water from land to ocean is called _____.

17. Water seeps down through the soil to become _____, which supplies much of the water to streams, rivers, ocean, soil, and plants.

ECOSYSTEM REGULATION FROM THE BOTTOM UP AND THE TOP DOWN

18. Bottom-up processes predominate in aquatic ecosystems with (a)_____ _____ while top-down processes predominate in ecosystems with (b)_____.

ABIOTIC FACTORS IN ECOSYSTEMS

19. The two abiotic factors that probably most affect organisms in ecosystems are _____.

The sun warms Earth

20. Ultimately, all of the sun's energy that is absorbed by the earth's surface and atmosphere is returned to space in the form of _____.

The atmosphere contains several gases essential to organisms

21. Winds generally blow from (a)_____(high or low?) pressure areas to (b)_____(high or low?) pressure areas, but they are deflected from these paths by Earth's rotation, a phenomenon known as the (c)_____.

The global ocean covers most of Earth's surface

22. The _____ Ocean covers one third of the Earth's surface and contains more than half of Earth's water.

23. Earth's rotation from west to east causes surface ocean currents to swerve to the (a)_____ (left or right?) in the Northern Hemisphere, producing a (b)_____ (clockwise or counterclockwise?) gyre of water currents, and to the (c)_____ (left or right?) in the Southern Hemisphere, producing a (d)_____ (clockwise or counterclockwise?) gyre.

Climate profoundly affects organisms

24. _____ is the average weather conditions, plus extremes, that occur in a given location over an extended period of time.

25. _____ occurs when air heavily saturated with water vapor cools and loses its moisture-holding ability.

26. The dry lands on the sides of the mountains away from the prevailing wind are called _____.

Fires are a common disturbance in some ecosystems

27. Fires have several effects on organisms. First, they (a)_____; secondly, they (b)_____ _____; and thirdly, they (c)_____.

STUDYING ECOSYSTEM PROCESSES

28. _____ is the clearance of large expanses of forest for agriculture or other uses.

BUILDING WORDS

Use combinations of prefixes and suffixes to build words for the definitions that follow.

Prefixes	The Meaning
atmo-	air
bio-	life
hydro-	water
micro-	small

Prefix	Suffix	Definition
_____	-mass	1. The total weight of all living things in a particular habitat.
_____	-sphere	2. The gaseous envelope surrounding the earth; the air.
_____	-logic	3. The water cycle, which includes evaporation, precipitation, and flow to the seas.
_____	-climate	4. A localized variation in climate; the climate of a small area.

MATCHING

Terms:

a. Assimilation
b. Coriolis effect
c. Ecological pyramid
d. Biosphere
e. Ecosystem

f. Food chain
g. Food web
h. Fossil fuel
i. Heterocyst
j. Nitrification

k. Nitrogen fixation
l. Rain shadow
m. Trophic level
n. Upwelling

For each of these definitions, select the correct matching term from the list above.

_____ 1. The distance of an organism in a food chain from the primary producers of a community.

_____ 2. Cells of certain cyanobacteria that are the site of nitrogen fixation.

_____ 3. The interacting system that encompasses a community and its nonliving, physical environment.

_____ 4. The system of interconnected food chains in a community.

_____ 5. The ability of certain microorganisms to bring atmospheric nitrogen into chemical combination.

_____ 6. An area on the downwind side of a mountain range with very little precipitation.

_____ 7. All of the Earth's communities.

_____ 8. A sequence of organisms through which energy is transferred from its ultimate source in a plant; each organism eats the preceding member and is eaten by the following member of the sequence.

_____ 9. A graphical representation of the relative energy value at each trophic level.

_____10. The oxidation of ammonium salts or ammonia into nitrites and nitrates by soil bacteria.

MAKING COMPARISONS

Fill in the blanks.

Biogeochemical Cycle	Material Cycled	Description
Phosphorus	Phosphorus	No gas formed; dissolves in water, is taken into plants
#1	#2	As a gas, is taken into plants, converted into glucose; returned to atmosphere by respiration
#3	#4	Converted from a gas into form usable by plants, then fixed in plant tissue, then consumed by animals
#5	Water	#6

MAKING CHOICES

Place your answer(s) in the space provided. Some questions may have more than one correct answer.

_____ 1. The energy that remains as biomass in plants after they have carried out cellular respiration is
   a. significant only in a changing ecosystem.
   b. called gross primary productivity.
   c. called net primary productivity.
   d. equal to the primary producer's metabolic rate.
   e. evident only in a climax community.

_____ 2. Compared to the overall regional climate, a microclimate
   a. may be significantly different.
   b. is insignificant to organisms.
   c. is more important to resident organisms.
   d. covers a larger area.
   e. exists for a brief period.

_____ 3. When it is warmest in the United States, it is coolest in
   a. the Southern Hemisphere.
   b. the Northern Hemisphere.
   c. Rio de Janiero.
   d. Europe.
   e. Mexico.

_____ 4. Activities performed by microorganisms involved in the nitrogen cycle include
   a. liberation of nitrogen from nitrate.
   b. oxidization of ammonia to nitrates.
   c. production of ammonia from proteins, urea, or uric acid.
   d. fixation of molecular nitrogen.
   e. continuous cycling of nitrogen.

_____ 5. An animal burrow in a hostile desert environment is an example of a/an
   a. subterranean biome.
   b. O-horizon.
   c. macroclimate.
   d. microclimate.
   e. ecosystem.

_____ 6. Which of the following would be positioned at the top of a typical pyramid?
   a. producers
   b. consumers
   c. the trophic level with the greatest biomass
   d. trophic level with the least numbers
   e. trophic level with the least energy

_____ 7. The main collector(s) of solar energy used to power life processes is/are the
   a. producers.
   b. atmosphere.
   c. hydrosphere.
   d. green leaves.
   e. photosynthetic organisms.

_____ 8. Most carbon is fixed _____ and liberated _____
   a. in proteins/as organic compounds.
   b. in CO_2/as complex compounds.
   c. in complex compounds/as CO_2.
   d. in humus/by cyanobacteria.
   e. by plants/by plants.

_____ 9. Phosphorus enters aquatic communities by means of
   a. decomposers.
   b. producers.
   c. primary consumers.
   d. secondary consumers.
   e. erosion.

_____ 10. The totality of the Earth's living inhabitants is the
   a. ecosphere.
   b. lithosphere.
   c. biosphere.
   d. biome.
   e. hydrosphere.

VISUAL FOUNDATIONS

Color the parts of the illustration below as indicated. Label N_2 in the atmosphere, NO_3^- in the soil, and NH_3 in the soil.

RED	☐ nitrogen fixation	BLUE	☐ atmosphere	TAN	☐ animal
GREEN	☐ plants	ORANGE	☐ assimilation	PINK	☐ nitrification
YELLOW	☐ ammonification	BROWN	☐ soil	VIOLET	☐ denitrification

Color the parts of the illustration below as indicated.

RED	❑	equator
GREEN	❑	trade winds
YELLOW	❑	warm air
BLUE	❑	cool air
ORANGE	❑	westerlies
VIOLET	❑	polar easterlies

CHAPTER 55

Ecology and the Geography of Life

This chapter examines the characteristics of Earth's major biomes, aquatic ecosystems, and biogeographical realms. Major terrestrial life zones (biomes) are large, relatively distinct ecosystems characterized by similar climate, soil, plants, and animals, regardless of where they occur on earth. Their boundaries are determined primarily by climate. Temperature and precipitation are the major determinants of plant and animal inhabitants. Examples of biomes include tundra; the taiga, or boreal forests; temperate forests; temperate grasslands; chaparral; deserts; tropical grasslands, or savanna; and tropical forests. In aquatic life zones, salinity, dissolved oxygen, nutrient minerals, and light, are the major determinants of plant and animal inhabitants. Freshwater ecosystems include flowing water (rivers and streams), standing water (lakes and ponds), and freshwater wetlands. Marine ecosystems include estuaries, the intertidal zone, the ocean floor, the neritic province, and the oceanic province. All terrestrial and aquatic life zones interact with one another. The study of biogeography helps biologists relate ancient communities to modern communities that exist in a given area. Biologists believe that each species originated only once, and each has a characteristic geographic distribution. The ranges of different species do not include everywhere that they could survive. The world's land areas are divided into six biogeographical realms.

REVIEWING CONCEPTS

Fill in the blanks.

INTRODUCTION

BIOMES

1. Biomes are terrestrial regions that have similar climate, soil, plants, and animals . _____ is the most influential factor in determining the characteristics and boundaries of a biome.

Tundra is the cold, boggy plains of the far north

2. Tundra is the northernmost biome. It is characterized by low-growing vegetation, a short growing season, and a permanently frozen layer of subsoil called _____ .

Boreal forest is the evergreen forest of the north

3. Most of the boreal forest is not suitable for agriculture because of its _____ .

4. Boreal forest trees are currently the primary source of the world's _____ .

Temperate rain forest has cool weather, dense fog, and high precipitation

5. Temperate rain forest is one of the richest wood producers in the world, supplying us with _____ .

Temperate deciduous forest has a canopy of broad-leaf trees

6. The temperate deciduous forest receives approximately
_____ inches (cm) of rain and is dominated by broad-leaf
hardwood trees.

Temperate grasslands occur in areas of moderate precipitation

7. Temperate grasslands are also known as _____.

Chaparral is a thicket of evergreen shrubs and small trees

8. Chaparral is an assemblage of small-leaved shrubs and trees in areas of modest
rainfall and mild winters, climatic conditions referred to as
_____.

Deserts are arid ecosystems

9. During the driest months of the year, many desert animals tunnel underground,
where they remain inactive; this period of dormancy is known as
_____.

Savanna is a tropical grassland with scattered trees

10. The best known savanna, with its large herds of grazing, hoofed animals and
predators, is the _____ savanna.

11. Smaller savannas also occur on the continents of _____
_____.

There are two basic types of tropical forests

12. Tropical rain forests have about _____ inches (cm) of
rainfall annually. They have many diverse species and are noted for tall foliage,
many epiphytes, and leached soil.

13. Most of the nutrients in tropical rain forests are tied up in the
_____, not in the soil.

AQUATIC ECOSYSTEMS

14. Aquatic life zones differ from terrestrial largely due to the different properties of
water and air. The principal limiting factors in water habitats include
_____.

15. Aquatic life is ecologically divided into the free-floating (a)_____, the
strong swimming (b)_____, and the bottom dwelling (c)_____
organisms.

16. The base of the aquatic food chain is composed of the free-floating
(a)_____, while the free-floating
nonphotosynthetic protozoa and other tiny animals comprise the
(b)_____.

Freshwater ecosystems are closely linked to land and marine ecosystems

17. Rivers and streams are flowing-water ecosystems. The kinds of organisms they
contain depends primarily on the _____.

18. Lakes and ponds are standing-water ecosystems. Large lakes have three zones:
the (a)_____ zone is the shallow water area along the shore; the
(b)_____ zone is the open water area away from shore that is illuminated

by sunlight; and the (c)_____ zone, which is the deeper, open water area beneath the illuminated portion. Of these, the (d)_____ zone is the most productive.

19. Vertical thermal layering of standing water is called
(a)_____. It is usually marked in the summer by an abrupt temperature transition at some depth called the
(b)_____. When atmospheric temperatures drop in fall, vertical layers mix, a phenomenon known as (c)_____. In spring, surface waters sink and bottom waters return to the surface. This is called
(d)_____.

20. Freshwater wetlands may be (a)_____, where grasslike plants dominate, or (b)_____, where woody trees or shrubs dominate.

21. Freshwater wetlands may be small, shallow ponds called
(a)_____, or (b)_____ that are dominated by mosses.

Estuaries occur where fresh water and salt water meet

22. Temperate estuaries usually contain _____, shallow wetlands in which salt-tolerant grasses dominate.

Marine ecosystems dominate earth's surface

23. The main marine habitats are similar to those of fresh water but are modified by tides and currents. The marine habitat near the shoreline between low and high tide is the _____.

24. The _____ is open ocean that overlies the continental shelves.

25. The portion of the neritic province that contains sufficient light to support photosynthesis is called the _____ region.

26. The deeper region of the open ocean, beyond the reach of light, is known as the
_____.

ECOTONES

27. Within landscapes, ecosystems intergrade with one another at their
_____.

BIOGEOGRAPHY

28. One of the basic tenets of biogeography is that each species originated only once. The particular place where this occurred is known as the species'
_____.

29. Some species have a nearly worldwide distribution and occur on more than one continent or throughout much of the ocean. Such species are said to be
_____.

Land areas are divided into six biogeographic realms

30. The six biogeographical realms are the _____
_____.

BUILDING WORDS

Use combinations of prefixes and suffixes to build words for the definitions that follow.

Prefixes	The Meaning
inter-	between, among
phyto-	plant
thermo-	heat, hot
zoo-	animal

Prefix	Suffix	Definition
_____	-plankton	1. Planktonic plants; primary producers.
_____	-plankton	2. Planktonic protists and animals.
_____	-tidal	3. Pertains to the zone of a shoreline between high tide and low tide.
_____	-cline	4. A layer of water marked by an abrupt temperature transition.

MATCHING

Terms:

a.	Benthos	f.	Ecotone	k.	Neritic province
b.	Biome	g.	Endemic species	l.	Permafrost
c.	Chaparral	h.	Estuary	m.	Savanna
d.	Desert	i.	Euphotic zone	n.	Taiga
e.	Desertification	j.	Limnetic zone	o.	Tundra

For each of these definitions, select the correct matching term from the list above.

_____ 1. Northern coniferous forest biome that stretches across North America and Eurasia.

_____ 2. Upper reaches of the neritic province; region of aquatic habitats lying near enough the surface that sufficient light is present for photosynthesis to take place.

_____ 3. The open water area away from the shore of a lake or pond, extending down as far as sun light penetrates.

_____ 4. Organisms that live on the bottom of oceans or lakes.

_____ 5. Distinctive vegetation type characteristic of certain areas with cool moist winters and long dry summers.

_____ 6. Permanently frozen ground.

_____ 7. Dry areas found in temperate and subtropical or tropical regions.

_____ 8. The conversion of marginal savanna into desert due to severe overgrazing by domestic animals..

_____ 9. A coastal body of water, partly surrounded by land, with access to the open ocean and a large supply of fresh water from rivers.

_____ 10. Localized, native species that are not found anywhere else in the world.

MAKING COMPARISONS

Fill in the blanks.

Biome	Climate	Soil	Characteristic Plants and Animals
Temperate rain forest	Cool weather, dense fog, high rainfall	Relatively nutrient-poor	Large conifers, epiphytes, squirrels, deer
Tundra	#1	#2	#3
#4	Not as harsh as tundra	Acidic, mineral poor, decomposed needles	Coniferous and deciduous trees, medium to small animals and a few larger species
Temperate deciduous forest	#5	#6	#7
#8	Moderate, uncertain rainfall	#9	Grasses, grazing animals and predators such as wolves
Chaparral	#10	Thin, infertile	#11
#12	Temperate and tropical, low rainfall	#13	Organisms adapted to water conservation, small animals
#14	Tropical areas with low or seasonal rainfall	#15	Grassy areas with scattered trees; hoofed mammals, large predators
Tropical rain forest	#16	#17	#18

MAKING CHOICES

Place your answer(s) in the space provided. Some questions may have more than one correct answer.

_____ 1. One expects to find organisms with adaptations for retaining moisture and clinging securely to the substrate in the

a. salt marsh.

b. intertidal zone.

c. profundal zone.

d. littoral zone.

e. estuary.

_____ 2. The communities containing the most permafrost are called

a. polar.

b. steppes.

c. boreal.

d. tundra.

e. muskegs.

_____ 3. A community with two distinct soil layers abounding in trees, reptiles, and amphibians is

a. grassland.

b. deciduous forest.

c. temperate rain forest.

d. temperate.

e. chaparral.

_____ 4. Insect larvae, water striders, and crayfish are found mainly in
    a. the littoral zone.   d. shallow lakes.
    b. the limnetic zone.   e. the area just below the compensation point.
    c. the profundal zone.

_____ 5. Small or microscopic organisms carried about by currents are
    a. littoral.   d. benthic.
    b. all producers.   e. planktonic.
    c. nektonic.

_____ 6. Most of midwestern North America is
    a. grassland.   d. temperate.
    b. deciduous forest.   e. chaparral.
    c. temperate rain forest.

_____ 7. The area(s) essentially corresponding to the waters above the continental shelf is/are the
    a. littoral zone.   d. neritic province.
    b. limnetic zone.   e. benthic zone.
    c. profundal zone.

_____ 8. The two most diverse of communities are the _____ and _____
    a. deciduous forest.   d. coral reef.
    b. tropical rain forest .   e. grassland.
    c. coniferous forest.

_____ 9. An aquatic zone at the juncture of a river and the sea is a/an
    a. salt marsh.   d. littoral zone.
    b. intertidal zone.   e. estuary.
    c. profundal zone.

_____10. Communities dominated by evergreens and containing acidic, mineral-poor soil is/are
    a. in South America.   d. typically impregnated with a layer of permafrost.
    b. called taiga.   e. boreal forest.
    c. temperate coniferous forests.

_____11. The factor that determines the nature of the biogeographic realms more than any other single factor is
    a. soil.   d. types of animals.
    b. climate.   e. altitude.
    c. latitude.

VISUAL FOUNDATIONS

Color the parts of the illustration below as indicated. Label intertidal zone, benthic environment, abyssal zone, hadal zone, and pelagic environment.

GREEN ☐ neritic province

BLUE ☐ oceanic province

TAN ☐ land

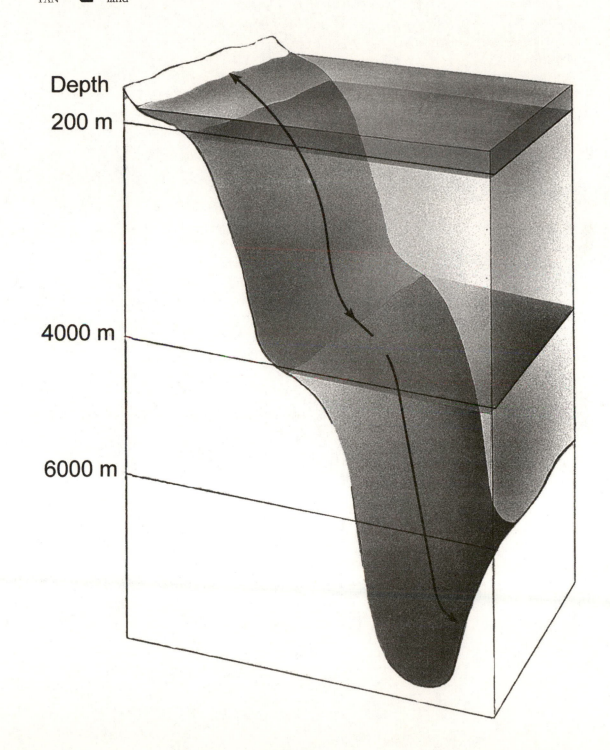

Global Environmental Issues

Chapter 52 introduced you to the principles of human population growth. This chapter examines some of the consequences of that growth on the biosphere. Although humans have been on earth a comparatively brief span of time, our biological success has been unparalleled. We have expanded our biological range into almost every habitat on earth, and, in the process, have placed great strain on the Earth's resources and resilience, and profoundly affected other life forms. As a direct result of our actions, many environmental concerns exist today. Species are disappearing from earth at an alarming rate. Although forests provide us with many ecological benefits, deforestation is occurring at an unprecedented rate. The production of atmospheric pollutants that trap solar heat in the atmosphere threaten to alter the earth's climate. Global warming may cause a rise in sea level, changes in precipitation patterns, death of forests, extinction of animals and plants, and problems for agriculture. It could result in the displacement of thousands or even millions of people. Ozone is disappearing from the stratosphere at an alarming rate. The ozone layer helps to shield Earth from damaging ultraviolet radiation. If the ozone were to disappear, Earth would become uninhabitable for most forms of life.

REVIEWING CONCEPTS
Fill in the blanks.

INTRODUCTION

1. Four serious environmental issues that affect the biosphere are: _____
_____
_____.

THE BIODIVERSITY CRISIS

2. Many people are concerned with _____, the ability to meet humanity's current needs without compromising the ability of future generations to meet their needs.

3. Biological diversity is the variety of living organisms considered at three levels: _____, _____, and _____.

4. When the number of individuals in a species is reduced to the point where extinction seems imminent, the species is said to be (a)_____, and when numbers are seriously reduced so that extinction is feared but thought to be less imminent, the species is said to be (b)_____.

Endangered species have certain characteristics in common

5. Although a variety of human activities threaten the continued existence of many organisms, most species facing extinction reach this perilous state as the result of _____.

Human activities contribute to declining biological diversity

6. The breakup of large areas of habitat into small, isolated patches is known as

_____.

7. _____, the introduction of a foreign species into an area where it is not native, often upsets the balance among the organisms living in that area.

CONSERVATION BIOLOGY

8. *In situ* conservation attempts to preserve species in _____ by establishing reserves (e.g., parks).

9. *Ex situ* conservation tries to preserve species in _____ such as zoos and seed banks.

In situ conservation is the best way to preserve biological diversity

10. The single best way to protect biological diversity is to protect _____
_____.

Landscape ecology considers ecosystem types on a regional scale

11. _____ is a subdivision of ecology studying the connections in a heterogeneous landscape consisting of multiple interacting ecosystems.

Restoring damaged or destroyed habitats is the goal of restoration ecology

12. (a)_____, in which the principles of ecology are used to return a degraded environment to one that is more functional and sustainable, is an important part of (b)_____.

Ex situ conservation attempts to save species on the brink of extinction

13. In (a)_____, sperm is collected from a male and used to impregnate a female, while in (b)_____ a female is treated with fertility drugs so she will produce more eggs.

The Endangered Species Act provides some legal protection for species and habitats

14. The Endangered Species Act authorizes the U.S. Fish and Wildlife Service to
_____
_____.

International agreements provide some protection of species and habitats

15. The Convention on International Trade in Endangered Species of Wild Flora and Fauna protects _____
_____.

DEFORESTATION

Why are tropical rain forests disappearing?

16. The most immediate causes of deforestation in tropical rains forests include
_____.

Why are boreal forests disappearing?

17. Boreal forests are currently the primary source of the world's _____
_____.

GLOBAL WARMING

Greenhouse gases cause global warming

18. Gases that are accumulating in the atmosphere as a result of human activities include _____.

19. Because CO2 and other gases trap the sun's radiation somewhat like glass does in a greenhouse, the natural trapping of heat is referred to as the _____.

What are the probable effects of global warming?

20. The probable effects of global warming include

_____.

DECLINING STRATOSPHERIC OZONE

21. According to the National Center for Atmospheric Research, ozone levels over Europe and North America have dropped almost _____% since the 1970s.

22. Ozone shields Earth from much of the harmful _____ _____ that comes from the sun.

Certain chemicals destroy stratospheric ozone

23. The primary chemicals responsible for ozone loss in the stratosphere are a group of chlorine compounds called _____.

24. The thinning of the ozone layer that was discovered over Antarctica occurs annually between _____ and _____.

Ozone depletion harms organisms

25. Excessive exposure to UV radiation is linked to human health problems, such as

_____.

International cooperation can prevent significant depletion of the ozone layer

26. In 1987, the Montreal Protocol was signed by many countries stipulating a (a)_____ % reduction in (b)_____ by 1998.

CONNECTIONS AMONG ENVIRONMENTAL PROBLEMS

27. All environmental problems are connected one way or another with human _____.

28. _____ is the disproportionately large consumption of resources by people in highly developed countries.

BUILDING WORDS

Use combinations of prefixes and suffixes to build words for the definitions that follow.

Prefixes	The Meaning
de-	to remove
strat(o)-	layer

Prefix	Suffix	Definition
_____	-forestation	1. The removal or destruction of all tree cover in an area.
_____	-sphere	2. The layer of the atmosphere between the troposphere and the mesosphere, containing a layer of ozone that protects life by filtering out much of the sun's ultraviolet radiation.

MATCHING

Terms:

a. Artificial insemination
b. Biological diversity
c. Endangered species
d. Ex situ conservation
e. Extinction

f. Genetic diversity
g. Greenhouse effect
h. Greenhouse gases
i. Host mothering
j. In situ conservations

k. Keystone species
l. Ozone
m. Slash-and-burn agriculture
n. Subsistence agriculture
o. Threatened species

For each of these definitions, select the correct matching term from the list above.

_____ 1. The warming of the Earth resulting from the retention of atmospheric heat caused by the build-up of certain gases, especially carbon dioxide.

_____ 2. A species whose population is low enough for it to be at risk of becoming extinct.

_____ 3. The number and variety of living organisms.

_____ 4. The disappearance of a species from a given habitat.

_____ 5. Carbon dioxide, methane, nitrous oxide, CFCs, and ozone.

_____ 6. A method of controlled breeding in zoos.

_____ 7. Efforts to preserve biological diversity in the wild.

_____ 8. The production of enough food to feed oneself and one's family.

_____ 9. A species whose numbers are so severely reduced that it is in imminent danger of becoming extinct.

_____10. The layer in the upper atmosphere that helps shield the Earth from damaging ultraviolet radiation.

MAKING COMPARISONS

Fill in the blanks.

Issue	Why It Is An Issue	Cause of Problem	Solution
Declining biological diversity	Diversity contributes to a sustainable environment; lost opportunities, lost solutions to future problems	Commercial hunting, habitat destruction, efforts to eradicate or control species, commercial harvesting	*In situ* conservation, *ex situ* conservation, habitat protection
#1	Forests are needed for habitat, watershed protection, prevention of erosion	#2	#3
#4	Threatens food production, forests, biological diversity; could cause submergence of coastal areas, change precipitation patterns	#5	#6
Ozone depletion	Ozone layer protects living organisms from exposure to harmful amounts of UV radiation	#7	#8

REFLECTIONS

The subject matter covered in this chapter is of critical importance to the continuance of life on earth as we know it. This chapter points out some of the dangers humanity faces if we continue some of our habits. Biologists and other responsible citizens can make a difference. One might even say that we have a responsibility to make a difference. A few genuinely concerned individuals will assume leadership roles in bringing about the social, cultural, and political changes required to protect the environment. All of us can participate in assuring that our children and grandchildren will inherit a clean, healthy, and productive environment.

We encourage you to reflect on what your personal role could be to bring about needed reforms. What might you contribute to the welfare of the planet? What changes are needed to ensure that future generations can enjoy a reasonably good quality of life? Think about your lifestyle and aspirations. Are they compatible with a long-term investment in the future of the planet?

VISUAL FOUNDATIONS

Color the parts of the illustration below as indicated. Also label greenhouse gases in atmosphere.

RED   ☐   radiated heat redirected back to Earth

GREEN   ☐   Earth

YELLOW   ☐   solar energy

BLUE   ☐   atmosphere

ORANGE   ☐   heat escaping to space

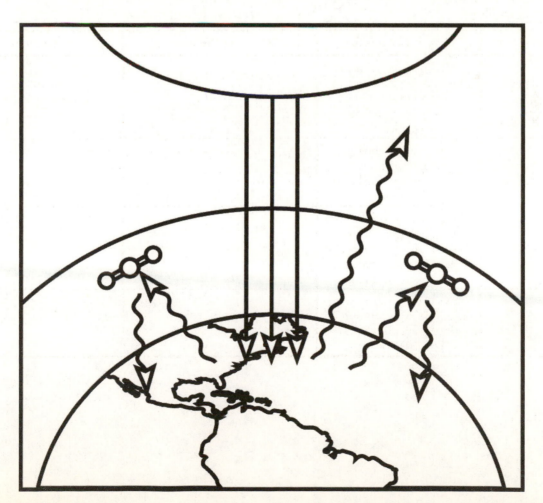

Color the parts of the illustration below as indicated. Also label normal ozone levels, and reduced ozone levels.

RED ☐ ozone molecules
GREEN ☐ oxygen molecules
BLUE ☐ troposphere
YELLOW ☐ stratosphere
ORANGE ☐ UV solar radiation

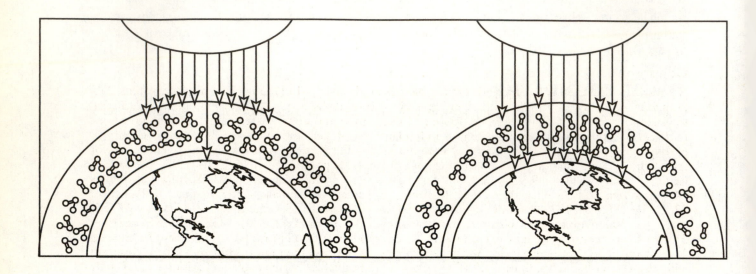

ANSWERS

□

Chapter 1

Reviewing Concepts: **1.** (a)decreasing biological diversity, (b)diminishing natural resources, (c) expanding human population, (d)prevention and cure of diseases **2.** (a)evolution, (b)information transmission, (c)energy flow **3.** composed of cells, grow and develop, regulate metabolic processes, respond to stimuli, reproduce, evolve, and adapt **4.** cell theory **5.** size and/or number **6.** development **7.** chemical activities **8.** homeostasis **9.** stimuli **10.** Louis Pasteur **11.** sexually or asexually **12.** adaptations **13.** cell **14.** biosphere **15.** population **16.** genes **17.** Watson and Crick **18.** proteins **19.** genus and species **20.** species **21.** taxonomy **22.** domain **23.** orders **24.** (a)Archaea and Bacteria, (b)Protista, (c)Fungi, (d)Plantae, (e)Animalia **25.** Plantae **26.** extinct **27.** *On the Origin of Species By Means of Natural Selection* **28.** (a)variation, (b)survive, (c)survive **29** gene pool **30.** environmental changes (or pressures) **31.** metabolism **32.** cellular respiration **33.** producers, consumers, and decomposers **34.** producers or autotrophs **35.** scientific method **36.** deductive **37.** inductive **38.** Alexander Fleming **39.** phenomenon or problem **40.** (a)consistent with well-established facts, (b)capable of being tested, (c)falsifiable **41.** variable **42.** double-blind study **43.** not all cases of what is being studied can be observed **44.** theory **45.** the Big Bang and evolution of major groups of organisms **46.** paradigm **47.** systems biologists **48.** genetic research, stem cell research, cloning, human and animal experimentation

Building Words: **1.** biology **2.** ecology **3.** asexual **4.** homeostasis **5.** ecosystem **6.** biosphere **7.** autotroph **8.** heterotroph **9.** protozoa **10.** photosynthesis

Matching: **1.** j **2.** f **3.** m **4.** c **5.** l **6.** n **7.** a **8.** e **9.** b **10.** k

Making Comparisons: **1.** Cellular level **2.** Tissue level **3.** Organ level **4.** Coordinated group of tissues and organs **5.** Coordinated group of organ systems

Making Choices: **1.** e **2.** c,e **3.** b **4.** d **5.** c **6.** a **7.** a,b,e **8.** b **9.** a,c,d **10.** d **11.** c **12.** c,d **13.** c **14.** e **15.** a-e

Visual Foundations: **1.** d **2.** b,c,d **3.** a

Chapter 2

Reviewing Concepts: **1.** water, many simple acids and bases, and simple salts **2.** chemical reactions (ordinary chemical means) **3.** carbon, hydrogen, oxygen, nitrogen **4.** (a)protons, neutrons, electrons, (b)protons and neutrons, (c)electrons **5.** (a)protons, (b)neutrons, (c)electrons **6.** atomic number **7.** eight **8.** atomic mass **9.** dalton **10.** (a)atomic mass (or numbers of neutrons), (b)atomic number (or numbers of protons) **11.** radioisotopes **12.** electron shell (or energy level) **13.** valence **14.** chemical properties **15.** two or more different elements **16.** molecule **17.** chemical **18.** structural **19.** atomic masses of its constituent atoms **20.** mole **21.** (a)reactants, (b)products **22.** (a)left, (b)right **23.** (a)nonpolar bonds, (b)polar bonds **24.** (a)cations, (b)anions **25.** hydration **26.** (a)weak, (b)strong **27.** electric charges **28.** redox **29.** (a)oxidized, (b)reduced **30.** 70 **31.** hydrogen bonding **32.** capillary action **33.** (a)hydrophilic, (b)hydrophobic **34.** specific heat

35. calorie (cal)   **36.** (a)OH— and a cation, (b)H$^+$ and an anion   **37.** (a)0-14, (b)7   **38.** (a)base, (b)acid
39. buffering system   **40.** electrolytes

Building Words: **1.** neutron   **2.** equilibrium   **3.** tetrahedron   **4.** hydration   **5.** hydrophobic
6. nonelectrolyte   **7.** hydrophilic

Matching: **1.** l   **2.** c   **3.** g   **4.** b   **5.** f   **6.** o   **7.** a   **8.** j   **9.** e   **10.** d

Making Comparisons: **1.** carbon dioxide   **2.** Ca   **3.** iron (III) oxide   **4.** CH$_4$ **5.** chloride ion
6. potassium ion   **7.** H$^+$ **8.** magnesium   **9.** NH$_4^+$ **10.** carbon dioxide   **11.** H$_2$O   **12.** glucose
13. HCO$_3^-$ **14.** carbonic acid   **15.** NaCl   **16.** hydroxide ion   **17.** NaOH

Making Choices: **1.** e   **2.** b   **3.** b   **4.** b   **5.** e   **6.** a,b   **7.** b   **8.** c,d   **9.** a   **10.** c,d,e   **11.** b,d
12. b   **13.** b   **14.** e   **15.** a   **16.** a,b,d   **17.** e   **18.** a,b,d,e   **19.** a,b

Visual Foundations: 1a. Carbon-12   **1b.** Carbon-14   **2.** e

Chapter 3

Reviewing Concepts: **1.** carbohydrates, lipids, proteins, nucleic acids   **2.** amino acids   **3.** four
4. single, double, triple   **5.** structural   **6.** geometric   **7.** enantiomers   **8.** functional   **9.** hydroxyl
10. aldehyde   **11.** carboxyl   **12.** monomers   **13.** hydrolysis   **14.** carbon, hydrogen, oxygen   **15.** glucose
16. glycosidic   **17.** simple sugars (monosaccharides)   **18.** starch   **19.** amylose   **20.** amyloplasts
21. glycogen   **22.** cellulose   **23.** chitin   **24.** (a)glycoprotein, (b)glycolipids   **25.** carbon, hydrogen,
oxygen   **26.** (a)glycerol, (b)fatty acid   **27.** van der Waals interactions   **28.** amphipathic   **29.** isoprene
units   **30.** (a)carbon, (b)six, (c)five   **31.** fatty acids   **32.** (a)amino and carboxyl, (b)R group (radical)
33. peptide   **34.** peptide   **35.** amino acids   **36.** primary, secondary, tertiary, quaternary
37. polypeptide chains   **38.** hydrogen   **39.** side chains   **40.** polypeptides   **41.** function   **42.** nucleotide
43. (a)purine or pyrimidine, (b)sugar, (c)phosphate   **44.** ATP

Building Words: **1.** isomer   **2.** macromolecule   **3.** monomer   **4.** hydrolysis   **5.** monosaccharide
6. hexose   **7.** disaccharide   **8.** polysaccharide   **9.** amyloplast   **10.** monoacylglycerol   **11.** diacylglycerol
12. triacylglycerol   **13.** amphipathic   **14.** dipeptide   **15.** polypeptide

Matching: **1.** i   **2.** c   **3.** l   **4.** n   **5.** a   **6.** o   **7.** g   **8.** k   **9.** m   **10.** e

Making Comparisons: **1.** phosphate   **2.** organic phosphates   **3.** R—CH$_3$ **4.** hydrocarbon;
nonpolar   **5.** R—NH$_2$ **6.** amines   **7.** thiols   **8.** R—OH   **9.** alcohols   **10.** polar because
electronegative oxygen attracts covalent electrons

Making Choices: **1.** b   **2.** a,c,e   **3.** c,d,e   **4.** b,d   **5.** c,e   **6.** b,c,d   **7.** a,c,d   **8.** a,d   **9.** a,c,d,e
10. b,c,e   **11.** a,b,d   **12.** b,c,d,e   **13.** a,b,c,d   **14.** a,b,d   **15.** a,c

Visual Foundations: **1.** b,d   **2.** b   **3.** b   **4.** c

Chapter 4

Reviewing Concepts: **1.** cells   **2.** cells are the basic living units of organization and function in all
organisms, and all cells come from other cells   **3.** plasma membrane   **4.** organelles   **5.** surface area-to-
volume ratio   **6.** 1/1000   **7.** sizes and shapes   **8.** (a)magnification, (b)resolution   **9.** ultrastructure
10. cell fractionation   **11.** prokaryotic   **12.** eukaryotic   **13.** cell nucleus   **14.** compartments
15. endomembrane system   **16.** nuclear envelope   **17.** chromatin   **18.** (a)RNA and proteins,
(b)nucleolus   **19.** endoplasmic reticulum   **20.** ribosomes   **21.** transport vesicles   **22.** cisternae
23. digestive enzymes   **24.** Golgi complex   **25.** tonoplast   **26.** hydrogen peroxide   **27.** mitochondria
28. free radicals   **29.** chloroplasts   **30.** thylakoid   **31.** (a)stroma, (b)thylakoids   **32.** strength, shape, and
movement   **33.** microfilaments and microtubules   **34.** centrioles   **35.** alpha and beta tubulin
36. myosin   **37.** intermediate filaments   **38.** glycocalyx   **39.** cellulose

Building Words: **1.** chlorophyll   **2.** chromoplasts   **3.** cytoplasm   **4.** cytoskeleton   **5.** glyoxysome
6. leukoplast   **7.** lysosome   **8.** microfilaments   **9.** microtubules   **10.** microvilli   **11.** myosin
12. peroxisome   **13.** proplastid   **14.** prokaryotes   **15.** cytosol   **16.** eukaryote

Matching: **1.** b   **2.** h   **3.** f   **4.** m   **5.** n   **6.** c   **7.** e   **8.** l   **9.** d   **10.** j

Making Comparisons: **1.** ribosomes **2.** prokaryotic, most eukaryotic **3.** cytoplasm **4.** modifies, sorts, and packages proteins **5.** cytoplasm **6.** ATP synthesis (aerobic respiration) **7.** eukaryotic **8.** nucleus (and cytoplasm during some stages of mitosis and meiosis) **9.** inheritance, control center **10.** eukaryotic **11.** cytoplasm **12.** cell division, microtubule formation **13.** eukaryotic (animals primarily) **14.** encloses cell, including the plasma membrane **15.** prokaryotic, eukaryotic plants **16.** plasma membrane **17.** boundary of cell cytoplasm **18.** prokaryotic, eukaryotic **19.** cytoplasm **20.** captures light energy (site of photosynthesis)

Making Choices: **1.** a,c,e **2.** d,e **3.** a-e **4.** c,e **5.** d,e **6.** a,b,d,e **7.** a,b,c,d **8.** a-e **9.** a,c,d,e **10.** b,c,e **11.** a,b,d,e **12.** a,d **13.** a,b,c,e **14.** e **15.** a,d

Visual Foundations: **1.** prokaryotic **2.** nuclear area **3.** eukaryotic plant cell **4.** prominent vacuole, chloroplast, mitochondrion, internal membrane system (excluding the nucleus) **5.** prominent vacuole, chloroplast **6.** eukaryotic animal cell **7.** centriole, lysosome, mitochondrion, internal membrane system (excluding the nucleus) **8.** centriole

Chapter 5
Reviewing Concepts: **1.** plasma membrane **2.** cadherins **3.** phospholipids **4.** (a)hydrophilic, (b)hydrophobic **5.** amphipathic **6.** (a)fluid mosaic, (b)lipid bilayer, (c)proteins **7.** (a)head groups, (b)fatty acid chains **8.** hydrocarbon chains **9.** plasma membrane **10.** integral membrane proteins **11.** peripheral membrane proteins **12.** (a)free ribosomes, (b)cytoplasm **13.** protein **14.** nonpolar **15.** aquaporins **16.** size, shape, electrical charge **17.** osmosis **18.** osmotic pressure **19.** (a)higher, (b)lower **20.** fewer **21.** negatively **22.** (a)uniporters, (b)symporters **23.** indirect active transport **24.** vesicle **25.** (a)phagocytosis, (b)pinocytosis **26.** desmosomes, tight junctions, gap junctions **27.** (a)adhering junctions, (b)desmosomes **28.** blood-brain barrier **29.** (a)proteins or connexin, (b)cylinders **30.** gap junctions **31.** desmotubule

Building Words: **1.** endocytosis **2.** exocytosis **3.** hypertonic **4.** hypotonic **5.** isotonic **6.** phagocytosis **7.** pinocytosis **8.** desmosome **9.** plasmodesma

Matching: **1.** g **2.** i **3.** f **4.** b **5.** a **6.** k **7.** e **8.** m **9.** l **10.** h

Making Comparisons: **1.** phagocytosis **2.** facilitated diffusion **3.** endocytosis **4.** carrier-mediated transport **5.** carrier-mediated active transport **6.** the active transport of materials out of the cell by fusion of cytoplasmic vesicles with the plasma membrane **7.** osmosis **8.** active transport **9.** cell takes in materials dissolved in droplets of water; literally means "cell drinking" **10.** desmosomes

Making Choices: **1.** d,e **2.** a,c,e **3.** c **4.** c,e **5.** b,c,d **6.** b,c **7.** a **8.** b,e **9.** d **10.** a,b,d,e **11.** b,e **12.** b,c,d,e **13.** b,d **14.** a **15.** c,d

Visual Foundations: **1.** fluid mosaic model **2.** transmembrane proteins

Chapter 6
Reviewing Concepts: **1.** the outside environment **2.** secreting chemical signals **3.** biofilm **4.** cell signaling **5.** target cells **6.** signal transduction **7.** paracrine regulation **8.** prostaglandins **9.** neurotransmitters **10.** receptor **11.** ligand **12.** cryptochromes **13.** (a)high, (b)decreasing **14.** (a)low, (b)increasing **15.** (a)ion channel-linked receptors or ligand-gated channels, (b)G protein-linked receptors or G protein-coupled receptors (c)enzyme-linked receptors **16.** enzyme-linked receptors **17.** transcription factors **18.** plasma membrane **19.** ligand-gated chloride ion channels **20.** signal transduction **21.** signaling cascade **22.** cyclic AMP (cAMP) **23.** cytoplasmic **24.** (a)cAMP, (b)protein kinase A, (c)proteins **25.** low **26.** (a)kinases, (b)phosphorylate **27.** intracellular receptors **28.** (a)guanylyl cyclase, (b)cGMP, (c)blood vessel walls **29.** (a)intracellular signaling molecules, (b)signaling complexes **30.** integrins **31.** steroid hormones **32.** abscisic acid **33.** signal amplification **34.** (a)signal transduction, (b)inactive **35.** prokaryotes

Building Words: **1.** paracrine **2.** neurotransmitter **3.** phytochrome **4.** intracellular **5.** triphosphate **6.** diphosphate **7.** diacylglycerol

Matching: 1. h 2. n 3. j 4. i 5. f 6. d 7. c 8. g 9. e 10. p 11. k 12. o 13. a 14. b 15. l 16. m

Making Comparisons: 1. phosphate group 2. cAMP 3. protein 4. ATP 5. phosphorylated protein 6. G protein 7. signaling molecule 8. GDP 9. GTP split from protein 10. plasma membrane

Making Choices: 1. c 2. e 3. b 4. a 5. c 6. a,b,d 7. e 8. a,b,c,e 9. a 10. b,e 11. c,d 12. a,c,d,e 13. a,b 14. a-e 15. d,e 16. a,c,d 17. a,c,e 18. a-e

Chapter 7

Reviewing Concepts: 1. photosynthetic organisms (producers) 2. chemical 3. work 4. (a)potential energy, (b)kinetic energy 5. thermodynamics 6. (a)created nor destroyed, (b)transferred 7. heat 8. randomness or disorder 9. bond energy 10. total bond energies 11. entropy 12. (a)enthalpy, (b)entropy 13. exergonic reactions 14. $\triangle G$ 15. positive 16. (a)high, (b)lower 17. dynamic equilibrium 18. (a)release, (b)require input of 19. more 20. adenine, ribose, three phosphates 21. phosphorylation 22. (a)catabolic, (b)anabolic 23. large 24. acceptor molecule 25. break existing bonds 26. increase 27. enzyme-substrate complex 28. induced fit 29. -ase 30. active site 31. coenzyme 32. apoenzyme 33. temperature, pH, ion concentration 34. (a)product, (b)substrate 35. metabolic pathway 36. inhibits 37. reversible 38. competitive 39. noncompetitive 40. PABA

Building Words: 1. kilocalorie 2. anabolism 3. catabolism 4. exergonic 5. endergonic 6. allosteric

Matching: 1. e 2. b 3. k 4. i 5. l 6. a 7. f 8. d 9. j 10. c

Making Comparisons: 1. catabolic 2. exergonic 3. anabolic 4. endergonic 5. catabolic 6. exergonic 7. catabolic 8. exergonic 9. anabolic 10. endergonic 11. anabolic 12. endergonic

Making Choices: 1. b,c,e 2. a–e 3. a,c 4. a,b,e 5. a,b,d 6. a,d 7. b,d,e 8. b,d 9. c 10. a–e 11. c 12. d

Chapter 8

Reviewing Concepts: 1. catabolism 2. anabolism 3. adenosine triphosphate (ATP) 4. aerobic respiration 5. (a)carbon dioxide and water, (b)energy 6. glycolysis, formation of acetyl coenzyme A, citric acid cycle, and electron transport chain and chemiosmosis 7. cytosol (or cytoplasm) 8. glyceraldehyde-3-phosphate (PGAL) 9. pyruvate 10. two 11. (a)two, (b)two 12. (a)three, (b)two, (c)eight 13. two 14. (a)NAD^+ and FAD, (b)the electron transport chain and chemiosmotic phosphorylation 15. (a)electron transport chain, (b)molecular oxygen 16. NADH 17. oxygen 18. fatty acids 19. deamination 20. glycerol and fatty acid 21. beta oxidation 22. anaerobic respiration 23. organic molecule 24. ethyl alcohol 25. lactate 26. (a)two, (b)36-38

Building Words: 1. aerobe 2. anaerobe 3. dehydrogenation 4. decarboxylation 5. deamination 6. glycolysis

Matching: 1. c 2. d 3. h 4. i 5. a 6. e 7. k 8. l 9. f 10. n

Making Comparisons: 1. fructose-6-phosphate 2. phosphor-fructokinase 3. –1 ATP 4. glyceraldehydes-3-phosphate (G3P) 5. zero 6. zero 7. 3-phosphoglycerate 8. +2 ATP 9. 2-phosphoglycerate 10. zero 11. pyruvate 12. +2 ATP

Making Choices: 1. a,b,c 2. b,c,d,e 3. a,b,c,d 4. c,d,e 5. b,e 6. b 7. a,d 8. c,d 9. a,b,e 10. b 11. a-e 12. a,b,c 13. b,c,d,e 14. a,b,c,d

Visual Foundations: 1. citrate synthase 2. aconitase 3. isocitrate dehydrogenase 4. α-ketoglutarate dehydrogenase 5. succinyl CoA synthetase 6. succinate dehydrogenase 7. fumarase 8. malate dehydrogenase

Chapter 9

Reviewing Concepts: **1.** (a)carbon dioxide and water, (b)light (solar), (c) chemical **2.** (a)fluorescence, (b)electron acceptor **3.** (a)stroma, (b)mesophyll **4.** thylakoids **5.** blue and red **6.** (a)chlorophyll *a*, (b)chlorophyll *b*, (c)carotenoids **7.** (a)absorption spectrum, (b)spectrophotometer **8.** (a)action spectrum, (b)accessory pigments **9.** synthesis **10.** (a)NADPH and ATP, (b)carbohydrate **11.** chlorophyll *a* **12.** antenna complexes **13.** reaction center **14.** electron transfer reactions **15.** P680 **16.** reaction center **17.** ferredoxin **18.** antenna complex **19.** photolysis **20.** thylakoids membrane **21.** cyclic electron transport **22.** electron transport **23.** carbon fixation **24.** (a) ribulose biphosphate (RuBP), (b)phosphoglycerate (PGA) **25.** (a)CO_2, (b)NADPH, (c)ATP **26.** (a)RuBP carboxylate, (b)oxygen **27.** (a)C_4, (b)C_3 **28.** CO_2 **29.** CO_2 **30.** oxaloacetate **31.** stomata **32.** (a)heterotrophs, (b)autotrophs **33.** photolysis of water

Building Words: **1.** mesophyll **2.** photosynthesis **3.** photolysis **4.** photophosphorylation **5.** chloroplast **6.** chlorophyll **7.** autotroph **8.** heterotroph

Matching: **1.** m **2.** n **3.** h **4.** e **5.** a **6.** c **7.** k **8.** f **9.** j **10.** d

Making Comparisons: **1.** anabolic **2.** catabolic **3.** thylakoid membrane **4.** cristae of mitochondria **5.** H_2O **6.** glucose or other carbohydrate **7.** $NADP^+$ **8.** O_2

Making Choices: **1.** a **2.** b **3.** e **4.** a,b,d **5.** c **6.** a,c,d,e **7.** a,c,d **8.** a **9.** c **10.** c,e **11.** a,c,e **12.** c **13.** b **14.** a,c,d **15.** c

Chapter 10

Reviewing Concepts: **1.** genome **2.** chromatin **3.** 25,000 **4.** nucleosomes **5.** 46 **6.** cell cycle **7.** (a)first gap, (b)synthesis, (c)second gap **8.** centromere **9.** kinetochore **10.** mitotic spindle **11.** spindle microtubules **12.** metaphase plate **13.** depolymerize **14.** arrival of the chromosomes at the poles **15.** cell plate **16.** chromosomes **17.** binary fission **18.** cytokinins **19.** asexual **20.** (a)sexual, (b)diploid, (c)zygote **21.** size, shape, and the position of their centromeres **22.** meiosis I and meiosis II **23.** synapsis **24.** (a)prophase, (b)genetic variability **25.** anaphase **26.** (a)four, (b)two **27.** (a)mitosis, (b)meiosis **28.** (a)gametes, (b)gametogenesis **29.** spermatogenesis **30.** oogenesis

Building Words: **1.** chromosome **2.** interphase **3.** haploid **4.** diploid **5.** centromere **6.**gametogenesis **7.** spermatogenesis **8.** oogenesis **9.** gametophyte **10.** sporophyte

Matching: **1.** n **2.** l **3.** f **4.** c **5.** h **6.** b **7.** g **8.** j **9.** i **10.** d

Making Comparisons: **1.** M phase **2.** Telophase I and II **3.** does not occur **4.** anaphase I **5.** does not occur **6.** prophase I **7.** telophase **8.** telophase I, telophase II **9.** anaphase **10.** anaphase II **11.** metaphase **12.** metaphase I, metaphase II **13.** S phase of interphase **14.** S phase of interphase

Making Choices: **1.** c,d,e **2.** c,d **3.** b **4.** b,c,e **5.** a,d **6.** b,f **7.** b **8.** h **9.** d **10** b **11.** c **12.** d **13.** b **14.** d **15.** g **16.** h **17.** b **18.** e **19.** a,i **20.** a,g,h **21.** c **22.** j **23.** b **24.** g **25.** d

Chapter 11

Reviewing Concepts: **1.** Gregor Mendel **2.** (a)first filial, (b)F_2 or second filial **3.** (a)dominant, (b)recessive **4.** segregate **5.** locus **6.** homologous **7.** heterozygous **8.** Punnett square **9.** genotype **10.** homozygous recessive **11.** two loci **12.** meiosis **13.** linearly arranged **14.** product rule **15.** (a)zero, (b)one **16.** product **17.** sum rule **18.** (a)past events, (b)future events **19.** linked **20.** crossing over **21.** recombination **22.** autosomes **23.** (a)XY male, (b)XX female **24.** (a)hyperactive, (b)inactivation **25.** Barr body **26.** incomplete dominance **27.** (a)three or more, (b)locus **28.** pleiotropy **29.** epistasis **30.** polygenic inheritance **31.** norm of reaction

Building Words: **1.** polygene **2.** homozygous **3.** heterozygous **4.** dihybrid **5.** monohybrid **6.** hemizygous

Matching: **1.** d **2.** a **3.** k **4.** l **5.** b **6.** m **7.** i **8.** g **9.** c

Making Comparisons: **1.** tall plant with yellow seeds (TT Yy)) **2.** tall plant with green seeds (TT yy) **3.** tall plant with yellow seeds (Tt Yy) **4.** tall plant with green seeds (Tt yy) **5.** tall plant with yellow seeds (Tt YY) **6.** tall plant with yellow seeds (Tt Yy) **7.** short plant with yellow seeds (tt YY) **8.** short plant with yellow seeds (tt Yy) **9.** tall plant with yellow seeds (Tt Yy) **10.** tall plant with green seeds (Tt yy) **11.** short plant with yellow seeds (tt Yy) **12.** short plant with green seeds (tt yy)

Making Choices: **1.** b,c,d,e,f **2.** j **3.** both j **4.** RrTT **5.** l,m,n,r,s,t **6.** g **7.** c/n and d/r **8.** g **9.** b/m and e/s **10.** a,c,d **11.** b,c,e **12.** d **13.** e **14.** b **15.** b or d **16.** a,c,d **17.** d **18.** a,b,e **19.** a,e **20.** c

Chapter 12

Reviewing Concepts: **1.** inheritance **2.** transformation **3.** bacteriophages **4.** pentose sugar (deoxyribose), phosphate, nitrogenous base **5.** (a)adenine and guanine, (b)cytosine and thymine **6.** X-ray diffraction **7.** antiparallel nucleotide strands **8.** (a)two, (b)three **9.** template **10.** density **11.** mutation **12.** DNA polymerases **13.** (a)5, (b)3, (c)Okazaki **14.** nuclease, DNA polymerase, and DNA ligase **15.** apoptosis **16.** cell aging and apoptosis

Building Words: **1.** avirulent **2.** antiparallel **3.** semiconservative **4.** telomere

Matching: **1.** l **2.** d **3.** k **4.** i **5.** g **6.** h **7.** b **8.** e

Making Comparisons: **1.** Frederick Griffith **2.** Hershey-Chase **3.** found that the ratios of guanine to cytosine, purines to pyrimidines, and adenine to thymine were very close to 1 **4.** Franklin and Wilkins **5.** Avery, MacLeod, and McCarty

Making Choices: **1.** a,c,e **2.** a,c **3.** c,d **4.** b **5.** d **6.** a,c **7.** a,c,d **8.** d **9.** b,c,e **10.** a,c **11** a **12.** b **13.** e **14.** d **15.** a-e **16.** d **17.** d

Chapter 13

Reviewing Concepts: **1.** DNA bases **2.** recessive mutant allele **3.** mRNA, tRNA, and rRNA **4.** amino acid **5.** (a)transcription, (b)translation **6.** codon **7.** methionine and tryptophan **8.** (a)64, (b)61, (c)3 **9.** reading frame **10.** RNA polymerases **11.** (a)upstream, (b)downstream **12.** promoter **13.** leader sequence **14.** trailing sequences **15.** peptide bonds **16.** aminoacyl-tRNA synthetases **17.** initiation **18.** AUG **19.** elongation **20.** peptidyl transferase **21.** stop codons **22.** polyribosome **23.** precursor mRNA **24.** (a)intervening sequences, (b)expressed sequences **25.** protein domains **26.** posttranscriptional **27.** a specific RNA or polypeptide **28.** reverse transcriptase **29.** nucleotide sequence **30.** (a)point or base-substitution, (b)missense, (c)nonsense **31.** inserted or deleted **32.** movable **33.** hot spots **34.** mutagens

Building Words: **1.** anticodon **2.** ribosome **3.** polysome **4.** mutagen **5.** carcinogen

Matching: **1.** g **2.** j **3.** h **4.** m **5.** b **6.** c **7.** k **8.** d **9.** f

Making Comparisons: **1.** 5'—UGG—3' **2.** 3'—ACC —5' **3.** 3'—AAA—5' **4.** 3'—AAA—5' **5.** 3'—CCT—5' **6.** 5'—UAA—3', 5'—UAG—3', 5'—UGA—3' **7.** no tRNA molecule binds to a stop codon **8.** 3'—TAC—5' **9.** 5'—UTC—3'

Making Choices: **1.** d **2.** c **3.** a,c **4.** c,e **5.** a,c **6.** a,c,d **7.** b,c **8.** a,c,d **9.** a,c,e **10.** d **11.** a **12.** b **13.** b,c **14.** b,d **15.** a,b

Chapter 14

Reviewing Concepts: **1.** control of the amount or mRNA transcribed, the rate of translation of mRNA, and the activity of the protein product **2.** transcriptional-level control **3.** constitutive genes **4.** operon **5.** promoter region **6.** operator **7.** inducer **8.** corepressor **9.** posttranscriptional controls **10.** feedback inhibition **11.** TATA box **12.** upstream promoter elements (UPEs) **13.** (a)heterochromatin, (b)euchromatin **14.** differential mRNA processing **15.** chemical modification **16.** kinases **17.** phosphatases

Building Words: **1.** euchromatin **2.** heterochromatin **3.** corepressor **4.** heterodimer

Matching: **1.** f **2.** a **3.** j **4.** k **5.** d **6.** b **7.** g **8.** c **9.** e

Making Comparisons: **1.** temporal regulation **2.** assist newly synthesized proteins to fold properly **3.** genomic imprinting **4.** switches that activate or inactivate existing enzymes **5.** feedback inhibition **6.** enhancers **7.** translational controls

Making Choices: **1.** b,e **2.** a **3.** a,d **4.** c **5.** e **6.** e **7.** b **8.** b,d **9.** a,b,c,e **10.** a,d,e **11.** a,d,e **12.** b,c,d **13.** a,e

Chapter 15

Reviewing Concepts: **1.** genetic engineering **2.** restriction enzymes **3.** plasmids **4.** palindromic **5.** cosmid cloning vectors **6.** genomic DNA library **7.** genetic probes **8.** reverse transcriptase **9.** (a)DNA polymerase, (b)heat **10.** Taq **11.** positive pole **12.** Southern blot **13.** polymorphism **14.** chain termination method **15.** expressed sequence tags **16.** gene targeting **17.** DNA microarrays **18.** bioinformatics, pharmacogenetics, and proteomics **19.** gene therapy **20.** polymorphic **21.** fertilized egg or stem cell **22.** gene targeting **23.** genetically modified crops **24.** risk assessment

Building Words: **1.** transgenic **2.** retrovirus **3.** bacteriophage

Matching: **1.** i **2.** e **3.** a **4.** h **5.** f **6.** j **7.** g **8.** c **9.** d

Making Comparisons: **1.** restriction enzymes **2.** genetic probe **3.** cloning **4.** polymerase chain reaction (PCR) **5.** recombinant DNA vectors derived from bacterial DNA **6.** DNA fragments comprising the total DNA of an organism **7.** proteomics

Making Choices: **1.** a **2.** d **3.** c **4.** b **5.** c,e **6.** d **7.** a **8.** c,e **9.** b,e **10.** b **11.** d,e **12.** c,d **13.** b,c **14.** c,d,e **15.** a,c **16.** b

Chapter 16

Reviewing Concepts: **1.** nucleotides in DNA **2.** cytogenetics **3.** karyotype **4.** pedigree analysis **5.** 2.9 billion **6.** 500 **7.** gene targeting **8.** polyploidy **9.** (a)aneuploidy, (b)trisomic, (c)monosomic **10.** Down syndrome **11.** Klinefelter syndrome **12.** Turner syndrome **13.** Translocation **14.** deletion **15.** Fragile sites **16.** inborn errors of metabolism **17.** enzyme **18.** hemoglobin **19.** ion transport **20.** mucous **21.** (a) lipid, (b) brain **22.** autosomal dominant allele **23.** factor VIII **24.** (a)mutant, (b)normal **25.** amniocentesis **26.** chorionic villus sampling (CVS) **27.** preventive medicine **28.** probability **29.** normal

Building Words: **1.** cytogenetics **2.** polyploidy **3.** trisomy **4.** translocation

Matching: **1.** a **2.** j **3.** f **4.** h **5.** l **6.** e **7.** i **8.** c **9.** b **10.** g

Making Comparisons: **1.** sickle cell anemia **2.** autosomal recessive trait **3.** Klinefelter syndrome **4.** sex chromosome nondisjunction during meiosis **5.** absence of an enzyme that converts phenylalanine to tyrosine; instead phenylalanine is converted into toxic phenylketones that accumulate **6.** Hemophilia A **7.** X-linked recessive trait **8.** Tay-Sachs **9.** autosomal recessive trait **10.** Huntington's disease **11.** a mutated nucleotide triplet (CAG) that is repeated many times **12.** Down syndrome **13.** 47 chromosomes because they have extra copies of chromosome 21

Making Choices: **1.** b **2.** c **3.** a,c,e **4.** b,d **5.** d,e **6.** b,c,d **7.** a,b,c **8.** b,e **9.** b,c,e **10.** e **11.** c,e **12.** a,d

Visual Foundations: **1.** c **2.** a **3.** d

Chapter 17

Reviewing Concepts: 1. immunoflourescence   2. (a)cell determination, (b)cell differentiation   3. (a)morphogenesis, (b)pattern formation   4. nuclear equivalence   5. transcriptional   6. totipotent   7. transgenic organisms   8. pluripotent   9. characteristics that allow for the efficient analysis of biological processes   10. egg, larval, pupal, and adult   11. maternal effect   12. (a)segmentation genes, (b)homeotic genes   13. homeobox   14. founder cells   15. mosaic   16. chimera   17. regulative   18. SEPALLATA   19. malignant   20. oncogenes   21. tumor suppressor gene or antioncogene

Building Words: 1. morphogenesis   2. oncogene   3. totipotent

Matching: 1. j   2. k   3. b   4. i   5. e   6. d   7. c   8. a   9. f   10. g

Making Comparisons: 1. ectoderm   2. epidermis   3. ectoderm   4. nervous system   5. none   6. lungs   7. endoderm   8. mesoderm

Making Choices: 1. c,e   2. c,e   3. a,c,d   4. c   5. a,d   6. c,e   7. b   8. c   9. d   10. c,e   11. a   12. b

Chapter 18

Reviewing Concepts: 1. populations   2. links all fields of the life sciences into a unified body of knowledge   3. Leonardo da Vinci   4. Jean Baptiste de Lamarck (Lamarck)   5. (a)H.M.S. Beagle, (b)Galapagos Islands   6. artificial selection   7. Thomas Malthus   8. Alfred Russell Wallace   9. *Origin of Species by Means of Natural Selection*   10. overproduction, variation, limits on population growth (a struggle for existence), and differential reproductive success ("survival of the fittest")   11. the modern synthesis   12. mutations   13. natural selection   14. 100,000   15. index fossils   16. homologous   17. homoplastic features   18. vestigial structures (organs)   19. biogeography   20. continental drift   21. developmental genes   22. amino acid   23. DNA sequencing

Building Words: 1. homologous   2. biogeography   3. homoplastic

Matching: 1. a   2. d   3. f   4. h   5. j   6. l   7. i   8. e   9. b   10. c

Making Comparisons: 1. homologous features   2. homoplastic features   3. the presence of useless structures is to be expected as a species adapts to a changing mode of life   4. areas of the world that have been separated from the rest of the world for a long time have organisms unique to that area   5. evidence that all life is related

Making Choices: 1. a,c,d   2. a,d   3. c   4. b   5. a   6. b,e   7. a   8. e   9. c   10. c   11. d   12. c,d   13. c

Chapter 19

Reviewing Concepts: 1. genetic variability   2. gene pool,   3. genotype frequency   4. (a)$p^2 + 2pq + q^2$, (b)genetic equilibrium   5. large population, isolation, no mutations, no selection, random mating   6. codominant   7. inbreeding   8. genetic fitness   9. assortative mating   10. mutation   11. genetic drift   12. bottleneck   13. gene flow   14. stabilizing selection   15. directional selection   16. disruptive selection   17. phenotypes   18. heterozygote advantage   19. frequency-dependent selection   20. neutral variation   21. geographic variation

Building Words: 1. microevolution   2. phenotype

Matching: 1. l   2. i   3. g   4. e   5. b   6. a   7. d   8. f   9. h   10. k

Making Comparisons: 1. yes   2. yes   3. no   4. yes   5. yes   6. yes   7. no   8. yes   9. no

Making Choices: 1. b,c   2. c,e   3. c   4. c,d   5. d   6. c   7. c   8. c   9. j   10. g   11. y   12. k   13. h

Chapter 20

Reviewing Concepts: **1.** 10-100 million **2.** Linnaeus **3.** (a)reproductively, (b)gene pool **4.** fertilization **5.** (a)temporal isolation, (b) behavioral isolation, (c)mechanical isolation, (d)gametic isolation **6.** hybrid inviability **7.** hybrid sterility **8.** Lysin **9.** speciation **10.** geographically isolated **11.** change in ploidy and a change in ecology **12.** (a)polyploidy, (b)allopolyploidy **13.** hybrid zone **14.** punctuated equilibrium **15.** gradualism **16.** (a)allometric growth, (b)paedomorphosis **17.** (a)adaptive zones, (b)adaptive radiation **18.** background extinction **19.** mass **20.** (a)microevolutionary, (b)macroevolutionary

Building Words: **1.** polyploidy **2.** allopatric **3.** allopolyploidy **4.** macroevolution **5.** allometric

Matching: **1.** g **2.** k **3.** n **4.** a **5.** d **6.** e **7.** i **8.** m **9.** b **10.** c

Making Comparisons: **1.** hybrid breakdown **2.** gametes of similar species are incompatible **3.** hybrid sterility **4.** similar species have structural differences in their reproductive organs **5.** hybrid inviability **6.** temporal isolation

Making Choices: **1.** a **2.** e **3.** a,b,e **4.** c **5.** d **6.** e **7.** b,d **8.** a,c,e **9.** c,e **10.** e **11.** a,c,e **12.** c,d **13.** b,d

Chapter 21

Reviewing Concepts: **1.** chemical evolution **2.** absence of free oxygen, energy, chemical building blocks, time **3.** (a)Oparin, (b)Haldane **4.** (a)Miller and Urey, (b)hydrogen (H_2), methane (CH_4), ammonia (NH_3) **5.** protobionts **6.** (a)RNA, (b)ribosomes **7.** (a)microfossils, (b)3.5 **8.** (a)anaerobic, (b)prokaryotic **9.** (a)heterotrophs, (b)autotrophs **10.** oxygen **11.** endosymbiont **12.** Ediacaran fossils **13.** 542 million **14.** (a)Ordovician, (b)ostracoderm **15.** 251 million **16.** (a)Triassic, Jurassic, Cretaceous, (b)Triassic, (c)Jurassic **17.** (a)Paleogene period, (b)Neogene period

Building Words: **1.** autotroph **2.** protobiont **3.** heterotroph **4.** precambrian **5.** aerobe **6.** anaerobe

Matching: **1.** j **2.** e **3.** a **4.** l **5.** h **6.** d **7.** c **8.** g **9.** k

Making Comparisons: **1.** Ordovician **2.** Paleozoic **3.** Triassic **4.** Mesozoic **5.** Cretaceous **6.** Mesozoic **7.** Cambrian **8.** Paleozoic **9.** Neogene (formerly Quaternary) **10.** Cenozoic **11.** Devonian **12.** Paleozoic **13.** Paleogene (formerly Tertiary) **14.** Cenozoic **15.** Carboniferous **16.** Paleozoic

Making Choices: **1.** d **2.** b,c **3.** d **4.** c **5.** b,c **6.** a,d **7.** a-e **8.** c,d,e **9.** c,e **10.** a,b,c,d **11.** d **12.** b,d **13.** a-e

Chapter 22

Reviewing Concepts: **1.** placental mammals **2.** 56 million **3.** (a)prosimii, (b)tarsiiformes, (c)anthropoidea **4.** prehensile tails **5.** hominoids **6.** gibbons (*Hylobates*), orangutans (*Pongo*), gorillas (*Gorilla*), chimpanzees (*Pan*), humans (*Homo*) **7.** 6-7 million **8.** Africa **9.** sexual dimorphism **10.** *Australopithecus afarensis* **11.** Tanzania **12.** 2.3 million **13.** *Homo habilis* **14.** 1.7 million **15.** tools **16.** Germany **17.** "out of Africa" **18.** "multiregional" **19.** knowledge **20.** development of hunter/gatherer societies, development of agriculture and the Industrial Revolution **21.** 10,000 **22.** environment

Building Words: **1.** anthropoid **2.** quadrupedal **3.** bipedal **4.** hominoid **5.** supraorbital

Matching: **1.** e **2.** i **3.** a **4.** b **5.** k **6.** j **7.** g **8.** c **9.** d

Making Comparisons: **1.** *Australopithecus afarensis* **2.** Neandertal **3.** *Ardipithecus ramidus* **4.** *Australopithecus africanus* **5.** 2.3 mya **6.** more modern hominid features than *Australopithecus*, fashioned primitive tools; **7.** 6-7 mya **8.** 1.7 mya **9.** larger brain than *H. habilis*, made more sophisticated tools

Making Choices: 1. c,d,e 2. a,b,e 3. a,e 4. a,d,e 5. c 6. c 7. b 8. b 9. a 10. d,e 11. d 12. a,b,c 13. c 14. d

Chapter 23

Reviewing Concepts: 1. biological diversity or biodiversity 2. taxonomy 3. binomial system 4. (a)genus, (b)specific epithet 5. (a)family, (b)class, (c)phylum 6. taxon 7. clade 8. horizontal gene transfer 9. evolutionary relationships 10. homoplasy 11. ancient 12. recent 13. combination 14. (a)amino acid sequences, (b)nucleotide sequences 15. molecular systematics 16. polyphyletic group 17. monophyletic taxon 18. phenotypic similarities 19. evolutionary systematics and cladistics 20. outgroup 21. ancestral characters 22. mutations 23. parsimony

Building Words: 1. subphylum 2. monophyletic 3. polyphyletic

Matching: 1. i 2. g 3. l 4. a 5. b 6. k 7. j 8. d 9. e 10. h

Making Comparisons: 1. Archaea 2. Archaea 3. Eukarya 4. Protista 5. Eukarya 6. Fungi 7. Eukarya 8. Animalia 9. Eukarya 10. Plantae

Making Choices: 1. c 2. a 3. e 4. c,e 5. d 6. a,c,d 7. a 8. b,d 9. b 10. e

Chapter 24

Reviewing Concepts: 1. disease 2. nucleic acid (DNA or RNA) 3. capsid 4. escaped gene hypothesis 5. host range 6. polyhedral head 7. lytic and lysogenic cycles 8. virulent 9. attachment (absorption), penetration, replication and synthesis, assembly, release 10. temperate 11. (a)prophage, (b)lysogenic cells 12. lysogenic conversion 13. retroviruses 14. reverse transcriptase 15. single-stranded RNA 16. RNA 17. transmissible spongiform encephalopathies or TSEs 18. (a)cocci, (b)diplococci, (c)streptococci, (d)staphylococci 19. bacilli 20. (a)spirillum, (b)spirochete 21. vibrio 22. nucleoid 23. peptidoglycan 24. Gram-positive 25. (a)rotating flagella, (b)chemotaxis 26. plasmids 27. binary fission, budding, and fragmentation 28. transformation 29. transduction 30. conjugation 31. mutations 32. endospores 33. biofilm 34. saprotrophs 35. photoheterotrophs 36. (a)aerobic, (b)facultative anaerobes, (c)obligate anaerobes 37. methanogens, extreme halophiles, extreme thermophiles 38. 20 39. Koch's postulates 40. bioremediation

Building Words: 1. bacteriophage 2. exotoxin 3. endotoxin 4. methanogen 5. endospore 6. pathogen 7. anaerobe 8. prokaryote 9. eukaryote

Matching: 1. l 2. a 3. g 4. e 5. b 6. c 7. h 8. d 9. k 10. i

Making Comparisons: 1. absent 2. present 3. present 4. absent (except in mitochondria and chloroplasts) 5. absent 6. absent 7. present 8. present 9. absent 10. absent 11. absent 12. absent 13. present

Making Choices: 1. a-e 2. a,c,d 3. b,d 4. a 5. c 6. a,b,e 7. b,d,e 8. b 9. b,c,d,e 10. a 11. c 12. a,b 13. a,b 14. c 15. d,e

Chapter 25

Reviewing Concepts: 1. plankton 2. (a)Eukaryotic, (b)Eukarya 3. coenocytic 4. plankton 5. symbiotic relationships 6. monophyletic 7. mitochondria 8. chlorophyll *a* and *b*, yellow and orange carotenoids 9. paramylon 10. *Trypanosoma* 11. (a)micronuclei, (b)macronucleus 12. conjugation 13. cellulose 14. Zooxanthellae 15. animals 16. sporozoites 17. *Plasmodium* 18. mycelium 19. (a)zoospores, (b)oospores 20. two 21. silica 22. diatomaceous earth 23. unicellular 24. seaweeds 25. holdfast 26. (a)phycoerythrin, (b)phycocyanin 27. (a)isogamous, (b)anisogamous, (c)oogamous 28. lichen 29. tests 30. axopods 31. pseudopodia 32. *Entamoeba histolytica* 33. sporangia 34. (a)swarm cell, (b)myxamoeba 35. pseudoplasmodium (slug) 36. microvilli

Building Words: 1. protozoa 2. cytopharynx 3. pseudopodium 4. micronucleus 5. macronucleus 6. pseudoplasmodium 7. isogamous

Matching: **1.** b **2.** k **3.** g **4.** i **5.** d **6.** c **7.** f **8.** e **9.** j **10.** l

Making Choices: **1.** a-d **2.** c,e **3.** d,e **4.** a,b **5.** b **6.** c **7.** b **8.** a,c,e **9.** b **10.** a

Chapter 26

Reviewing Concepts: **1.** decomposers **2.** less **3.** lipid droplets or glycogen **4.** chitin **5.** yeasts and molds **6.** (a)dikaryotic, (b)monokaryotic **7.** (a)plant, (b)animals **8.** Chytridiomycota, Zygomycota, Ascomycota, Basidiomycota, Glomeromycota **9.** polyphyletic **10.** sexual stage **11.** (a)thallus, (b)rhizoids **12.** *Rhizopus stolonifer* **13.** heterothallic **14.** mycorrhizae **15.** (a)asci, (b)conidia, (c)conidiophores **16.** ascocarps **17.** (a)basidium, (b)basidiospores **18.** (a)button, (b)basidiocarp **19.** (a) carbon dioxide, (b)minerals **20.** cellulose and lignin **21.** mycorrhizae (fungus-roots) **22.** (a)green alga, cyanobacterium, or both (b)ascomycete **23.** (a)crustose, (b)foliose, (c)fruticose **24.** soredia **25.** (a)yeasts, (b)fruit sugars, (c)grains, (d)carbon dioxide **26.** (a)*Penicillium*, (b)*Aspergillus tamarii* **27.** *Amanita* **28.** psilocybin **29.** (a)*Penicillium notatum*, (b)penicillin **30.** ergot **31.** aflatoxins **32.** cutinase **33.** haustoria

Building Words: **1.** coenocytic **2.** monokaryotic **3.** heterothallic **4.** conidiophore **5.** homothallic

Matching: **1.** j **2.** k **3.** b **4.** c **5.** m **6.** l **7.** e **8.** d **9.** h **10.** f

Making Comparisons: **1.** Zygomycota **2.** Ascomycota **3.** Basidiomycota **4.** Glomeromycota

Making Choices: **1.** a,c,d,e **2.** c,e **3.** a-e **4.** e **5.** a-e **6.** a,c,e **7.** c,d,e **8.** a,d **9.** b **10.** b,d **11.** a-e **12.** c **13.** a,c,d

Chapter 27

Reviewing Concepts: **1.** green algae **2.** (a)chlorophylls *a* and *b* and accessory pigments, yellow and orange carotenoids (b)starch **3.** (a)waxy cuticle, (b)stomata **4.** gametangia **5.** (a)antheridia, (b)archegonia **6.** (a)zygote (fused gametes), (b)spores **7.** (a)mosses and other bryophytes, (b)bryophytes and seedless vascular plants, (c)gymnosperms, (d)angiosperms (flowering plants) **8.** lignin **9.** mosses, liverworts, hornworts **10.** rhizoids **11.** (a)*Sphagnum*, (b)peat mosses **12.** archegonia and antheridia **13.** gemmae **14.** Anthocerophyta **15.** photoperiodism **16.** bryophytes **17.** (a)microphyll, (b)megaphyll **18.** coal deposits **19.** (a)rhizome, (b)fronds **20.** (a)sporangia, (b)sori, (c)prothallus **21.** (a)upright stems, (b) dichotomous **22.** fungus **23.** (a)homospory, (b)heterospory **24.** (a)microsporangia, (b)microsporocytes, (c)microspores, (d)megaspores **25.** apical meristem **26.** megafossils

Building Words: **1.** xanthophyll **2.** gametangium **3.** archegonium **4.** gametophyte **5.** sporophyte **6.** microphyll **7.** megaphyll **8.** sporangium **9.** homospory **10.** heterospory **11.** megaspore **12.** microspore **13.** bryophyte

Matching: **1.** d **2.** c **3.** g **4.** i **5.** b **6.** e **7.** a **8.** k **9.** m **10.** h

Making Comparisons: **1.** vascular **2.** sporophyte **3.** vascular **4.** sporophyte **5.** vascular **6.** sporophyte **7.** vascular **8.** naked seeds **9.** nonvascular **10.** seedless, reproduce by spores **11.** gametophyte **12.** seedless, reproduce by spores **13.** seedless, reproduce by spores **14.** sporophyte

Making Choices: **1.** a,c,e **2.** a-e **3.** a,c,d **4.** b,d **5.** b,c,e **6.** c,d **7.** b,d **8.** a,e **9.** a,c **10.** a,c,d **11.** a,e **12.** b,d

Chapter 28

Reviewing Concepts: **1.** seeds **2.** (a)gymnosperms, (b)angiosperms **3.** (a)xylem, (b)phloem **4.** Coniferophyta, Ginkgophyta, Cycadophyta, Gnetophyta **5.** monoecious **6.** pines, spruces, firs, hemlocks, etc. **7.** (a)sporophylls, (b)microsporocytes, (c)male gametophytes (pollen grains) **8.** (a)megasporangia, (b)haploid megaspores, (c)female gametophyte **9.** Cycadophyta **10.** dioecious **11.** Ginkgophyta **12.** Gnetophyta **13.** vessel elements **14.** palms, grasses, orchids, irises, onions, lilies, etc. **15.** oaks, roses, mustards, cacti, blueberries, sunflowers, etc. **16.** (a)three, (b)one, (c)endosperm

17. (a)four or five, (b)two, (c)cotyledons   **18.** (a)sepals, petals, stamens, carpels; (b)stamens; (c)carpels; (d)perfect   **19.** (a)anther, (b)ovary   **20.** (a)megaspores, (b)embryo sac   **21.** (a)microspores, (b)male gametophyte (pollen grain)   **22.** zygote, endosperm tissue   **23.** fruit   **24.** (a)cross-fertilization, (b)genetic variation   **25.** leaves   **26.** (a)progymnosperms, (b)seed ferns   **27.** gymnosperms

Building Words: **1.** angiosperm   **2.** gymnosperm   **3.** monoecious   **4.** dioecious

Matching: **1.** a   **2.** e   **3.** f   **4.** i   **5.** k   **6.** m   **7.** h   **8.** l   **9.** b   **10.** c

Making Comparisons: **1.** herbaceous or woody   **2.** endosperm   **3.** cotyledons   **4.** mostly broader than in monocots   **5.** 1 cotyledon   **6.** floral parts in multiples of 4 or 5   **7.** parallel venation   **8.** netted venation

Making Choices: **1.** b,c   **2.** a   **3.** b   **4.** b   **5.** a,c,d,e   **6.** c,d,e   **7.** a,e   **8.** b,e   **9.** a-e   **10.** b,c

Chapter 29

Reviewing Concepts: **1.** 99   **2.** heterotrophs   **3.** sea (salt) water   **4.** fresh water   **5.** dehydrate   **6.** Cambrian radiation or Cambrian explosion   **7.** Hox   **8.** (a)anterior, (b)posterior, (c)dorsal, (d)ventral   **9.** (a)medial, (b)lateral, (c)cephalic, (d)caudal   **10.** germ layers   **11.** acoelomates   **12.** (a)pseudocoelomates, (b)coelomates   **13.** (a)mouth, (b)anus   **14.** (a)radial cleavage, (b)spiral cleavage   **15.** Eumetazoa   **16.** segmentation   **17.** choanoflagellates   **18.** (a)spongocoel, (b)osculum   **19.** spicules   **20.** (a)Hydrozoa, (b)Scyphozoa, (c)Anthozoa   **21.** cnidocytes   **22.** nematocysts   **23.** (a) epidermis, (b)gastrodermis, (c)mesoglea   **24.** medusa   **25.** polyp   **26.** planula   **27.** mesoglea

Building Words: **1.** pseudocoelom   **2.** ectoderm   **3.** mesoderm   **4.** protostome   **5.** deuterostome   **6.** schizocoely   **7.** enterocoely   **8.** spongocoel   **9.** gastrodermis

Matching: **1.** k   **2.** h   **3.** a   **4.** d   **5.** e   **6.** c   **7.** m   **8.** j   **9.** l

Making Comparisons: **1.** hydras, jellyfish, coral   **2.** biradial symmetry; diploblastic; gastrovascular cavity with mouth and anal pores   **3.** Platyhelminthes   **4.** biradial symmetry; triploblastic; simple organ systems; gastrovascular cavity with one opening   **5.** Mollusca   **6.** soft body with dorsal shell; muscular foot; mantle covers visceral mass; most with radula   **7.** some marine worms, earthworms, leeches   **8.** segmented body; most with setae   **9.** Rotifera   **10.** crown of cilia; constant cell number   **11.** cylindrical, threadlike body; pseudocoelom   **12.** centipedes, crabs, lobsters, spider, insects   **13.** segmented body; jointed appendages; exoskeleton; some with compound eyes; insects with tracheal tubes   **14.** Echinodermata   **15.** bilateral larva, pentaradial adult; triploblastic; organ systems; complete digestive tube   **16.** ring of cilia around mouth; three-part body; wormlike   **17.** Chordata   **18.** notochord; dorsal, tubular nerve cord; embryonic gill slits; segmented body

Making Choices: **1.** a,b,d   **2.** c   **3.** c,e   **4.** e   **5.** a   **6.** b   **7.** a,e   **8.** c,e   **9.** d   **10.** a,c,e

Chapter 30

Reviewing Concepts: **1.** 99   **2.** mesoderm   **3.** independent movement of digestive tract and body, coelomic fluid bathes coelomic cells and organs (transports food, oxygen, wastes), acts as hydrostatic skeleton, space for development and function of organs   **4.** (a)bilateral, (b)three, (c)ganglia   **5.** (a)Turbellaria, (b)Trematoda and Monogenea, (c)Cestoda   **6.** auricles   **7.** pharynx   **8.** scolex (head)   **9.** proglottid   **10.** everted from the anterior end of the body   **11.** (a)mantle, (b)Arthropoda   **12.** hemocoel   **13.** (a)trochophore, (b)veliger   **14.** eight   **15.** Polyplacophora   **16.** insects   **17.** torsion   **18.** (a)mantle, (b)calcium carbonate   **19.** adductor muscle   **20.** (a)tentacles, (b)ten, (c)eight   **21.** "head-foot"   **22.** (a)septa, (b)setae   **23.** parapodia   **24.** "many bristles"   **25.** (a)Annelida, (b)Oligochaeta, (c)*Lumbricus terrestris*   **26.** (a)crop, (b)gizzard   **27.** (a)hemoglobin, (b)metanephridia, (c)skin   **28.** (a)Brachiopods, (b)Poronida, (c)Bryozoans   **29.** flame cells   **30.** decomposers and predators of smaller organisms   **31.** (a)jointed foot, (b)exoskeleton, (c)chitin   **32.** tracheae   **33.** (a)Chilopoda, (b)one   **34.** (a)Diplopoda, (b)two   **35.** Chelicerata   **36.** (a)merostomes, (b)arachnids   **37.** (a)cephalothorax, (b)six   **38.** book lungs   **39.** spinnerets   **40.** Crustacea   **41.** (a)mandibles, (b)biramous, (c)two   **42.** barnacles   **43.** Decapoda   **44.** (a)maxillae, (b)maxillipeds, (c)chelipeds, (d)walking legs   **45.** (a)reproductive, (b)swimmeretes   **46.** (a)jointed, (b)possess tracheal tubes, (c)having six feet   **47.** (a)three, (b)one or two, (c)one, (d)Malpighian tubules

Building Words: **1.** exoskeleton **2.** bivalve **3.** biramous **4.** uniramous **5.** trilobite **6.** cephalothorax **7.** hexapod **8.** arthropod

Matching: **1.** c **2.** h **3.** a **4.** g **5.** f **6.** b **7.** i **8.** j **9.** e

Making Comparisons: **1.** organ system **2.** organ system **3.** moderate cephalization, (pair of ganglia) ventral nerve cords, sense organs **4.** brain, ventral nerve cords, well-developed sense organs **5.** closed system **6.** open system **7.** complete digestive tract **8.** complete digestive tract **9.** sexual: hermaphroditic **10.** herbivores, carnivores, scavengers, parasites

Making Choices: **1.** c **2.** b,c,e **3.** a **4.** d **5.** b,d,e **6.** a,d **7.** a,b,d **8.** b,c,e **9.** a,c,e **10.** b,d,e **11.** d **12.** b **13.** a,b **14.** b,c

Chapter 31

Reviewing Concepts: **1.** echinoderms and chordates **2.** indeterminate **3.** (a)bilateral, (b) pentaradial, (c)coelom **4.** Anura **5.** central disk **6.** tube feet **7.** (a)collect and handle food, (b)locomotion **8.** test **9.** locomotion **10.** transport oxygen and perhaps nutrients **11.** bilateral **12.** (a)notochord, (b)nerve cord, (c)pharyngeal gill slits **13.** tadpoles **14.** *Amphioxus* **15.** pharynx **16.** capturing prey **17.** (a)vertebral column, (b)cranium, (c)cephalization **18.** (a)6, (b)4, (c) Amphibia, Reptilia, Aves, and Mammalia **19.** ostracoderms **20.** fins **21.** placoid scales **22.** (a)lateral line organs, (b)electroreceptors, **23.** (a)oviparous, (b)ovoviviparous, (c)viviparous **24.** (a)ray-finned, (b)sarcopterygians **25.** swim bladders **26.** fishes and tetrapods **27.** (a)Urodela, (b)Anura, (c)Apoda **28.** (a) skin, (b)three, (c)mucous glands **29.** (a)two, (b)one **30.** (a)three, (b)ectothermic **31.** Testudines, Squamata, Rhynchocephalia, and Crocodilia **32.** *Archaeopteryx* **33.** endothermy **34.** (a)four, (b)endothermic, (c)uric acid, (d)feathers **35.** (a)crop, (b)proventriculus, (c)gizzard **36.** hair, mammary glands, differentiated teeth **37.** (a)therapsids, (b)Triassic period **38.** monotremes **39.** (a)marsupials, (b)marsupium

Building Words: **1.** agnathan **2.** Anura **3.** Apoda **4.** Chondrichthyes **5.** tetrapod **6.** Urodela **7.** echinoderm **8.** endoskeleton

Matching: **1.** h **2.** f **3.** i **4.** a **5.** l **6.** m **7.** k **8.** n

Making Comparisons: **1.** sharks, rays, skates, chimeras **2.** cartilage **3.** jawed, gills, marine and fresh water, placoid scales, well-developed sense organs **4.** Actinopterygii **5.** bone **6.** salamanders, frogs and toads, caecilians **7.** three chambered heart **8.** Aves **9.** four chambered heart **10.** light hollow bone with air spaces **11.** mammalia **12.** bone **13.** mostly tetrapods, possess hair and mammary glands, endothermic, highly developed nervous system

Making Choices: **1.** d **2.** d **3.** b,c **4.** c **5.** a,b **6.** c,e **7.** d **8.** a-e **9.** a,b **10.** b,e **11.** b,c,d **12.** c,e **13.** a,d **14.** d **15.** c,e **16.** d,e

Chapter 32

Reviewing Concepts: **1.** flowering plants **2.** roots, stems, leaves **3.** annuals **4.** life history strategies **5.** (a)tissue, (b)simple tissues, (c)complex tissues **6.** organs **7.** photosynthesis, storage, secretion **8.** support **9.** sclereids and fibers **10.** hemicelluloses and pectins **11.** (a)water and dissolved minerals, (b)parenchyma cells or xylem parenchyma, (c)fibers **12.** food **13.** (a)sieve tube elements, (b)sieve plates **14.** (a)epidermis and periderm, (b)cuticle, (c)stomata **15.** protection **16.** trichomes **17.** (a)meristems, (b)division, elongation, differentiation (specialization) **18.** (a)primary, (b)secondary **19.** (a)area of cell division, (b)area of cell elongation, (c)area of cell maturation **20.** (a)leaf primordia, (b)bud primordia **21.** (a)vascular cambium, (b)cork cambium **22.** bark

Building Words: **1.** trichome **2.** biennial **3.** epidermis **4.** stoma

Matching: **1.** o **2.** m **3.** a **4.** n **5.** c **6.** k **7.** b **8.** h **9.** f **10.** g

Making Comparisons: **1.** ground tissue **2.** stems and leaves **3.** sclerenchyma **4.** structural support **5.** xylem **6.** vascular tissue **7.** conducts sugar in solution **8.** extends throughout plant body **9.** dermal tissue **10.** protection of plant body, controls gas exchange and water loss on stems, leaves **11.** covers body of herbaceous plants **12.** periderm **13.** protection of plant body

Making Choices: **1.** d   **2.** a   **3.** e   **4.** a,d   **5.** d   **6.** c,d,e   **7.** c   **8.** a,b,e   **9.** c   **10.** a,b,c,e

Chapter 33

Reviewing Concepts: **1.** (a)blade, (b)petiole, (c)stipules   **2.** (a)simple, (b)compound   **3.** (a)alternate leaf arrangement, (b)opposite leaf arrangement, (c)whorled leaf arrangement   **4.** (a)mesophyll, (b)palisade mesophyll, (c)spongy mesophyll   **5.** (a)xylem, (b)phloem, (c)bundle sheath   **6.** parallel   **7.** (a)carbon dioxide, (b)oxygen   **8.** (a)turgid (swollen), (b)flaccid (limp)   **9.** (a)vacuoles, (b)guard, (c)open   **10.** close   **11.** light and darkness, CO_2, a circadian rhythm   **12.** (a)prevent plant from overheating, (b)distribute essential minerals throughout plant   **13.** hydrologic   **14.** guttation   (b)transpiration   **15.** ethylene   **16.** (a)carotenoids, (b)anthocyanins   **17.** (a)abscission zone, (b)fibers,   **18.** middle lamella   **19.** spines   **20.** tendrils   **21.** (a)passive, (b)active

Building Words: **1.** abscission   **2.** circadian   **3.** mesophyll   **4.** transpiration

Matching: **1.** b   **2.** j   **3.** g   **4.** c   **5.** h   **6.** f   **7.** d   **8.** m   **9.** e

Making Comparisons: **1.** allows for efficient capture of sunlight   **2.** helps reduce or control water loss, enabling plants to survive the dry terrestrial environment   **3.** stomata   **4.** allows light to penetrate to photosynthetic tissue   **5.** air space in mesophyll tissue   **6.** bundle sheaths and bundle sheath extensions   **7.** transports water and minerals from roots   **8.** phloem in veins

Making Choices: **1.** a   **2.** d   **3.** b,c,e   **4.** c   **5.** b,c,e   **6.** d   **7.** a,d,e   **8.** a,b,e   **9.** a-e   **10.** a,b,c   **11.** c   **12.** a,e   **13.** e

Chapter 34

Reviewing Concepts: **1.** roots, leaves, stems   **2.** support, conduction (internal transport), produce new stem tissue   **3.** (a)buds, (b)terminal bud, (c)bud scales   **4.** leaf-scar   **5.** vascular bundles   **6.** (a)xylem and phloem, (b)vascular cambium   **7.** (a)vascular cambium, (b)cork cambium, (c)periderm   **8.** (a)xylem (wood), (b)phloem (inner bark)   **9.** (a)rays, (b)parenchyma   **10.** cork cells   **11.** cork parenchyma   **12.** (a)heartwood, (b)sapwood   **13.** (a)springwood, (b)late summerwood   **14.** natural physical processes   **15.** (a)water potential, (b)less, (c)higher (less negative), (d)lower (more negative)   **16.** negative   **17.** tension-cohesion model   **18.** transpiration   **19.** (a)pressure flow, (b)source, (c)sink

Building Words: **1.** translocation   **2.** periderm   **3.** internode

Matching: **1.** a   **2.** g   **3.** h   **4.** b   **5.** c   **6.** l   **7.** m   **8.** j   **9.** i

Making Comparisons: **1.** secondary phloem   **2.** conducts dissolved sugar   **3.** produced by cork cambium   **4.** storage   **5.** cork cells   **6.** periderm   **7.** vascular cambium   **8.** produces secondary xylem and secondary phloem   **9.** a lateral meristem, usually arises from parenchyma   **10.** produces periderm (secondary growth)

Making Choices: **1.** c   **2.** d   **3.** a,b,c,d   **4.** a,d   **5.** a,b,d   **6.** b,d,e   **7.** a-e   **8.** a,b,d   **9.** a,c,e   **10.** b   **11.** e

Chapter 35

Reviewing Concepts: **1.** anchoring, absorption of water and minerals, conduction and storage   **2.** (a)taproot system, (b)fibrous root system   **3.** adventitious roots   **4.** root cap   **5.** root hairs   **6.** (a)endodermis, (b)Casparian strip   **7.** (a)pericycle, (b)vascular   **8.** (a)cell walls, (b)cellulose, (c)endodermis, (d)epidermis, (e)endodermis, (f)xylem   **9.** (a)storage, (b)pericycle   **10.** pith   **11.** (a)periderm, (b)cork cambium   **12.** (a)adventitious   **13.** (a)prop, (b)contractile   **14.** mycorrhizae   **15.** cell signaling   **16.** minerals, organic matter, air, water   **17.** (a)sand and silt, (b)clay, (c)humus, (d) air and water   **18.** castings   **19.** acid precipitation   **20.** (a)essential; (b)carbon, oxygen, hydrogen, nitrogen, potassium, phosphorus, sulfur, magnesium, calcium, silicon; (c)iron, boron, manganese, copper, zinc, molybdenum, chlorine, sodium, nickel.   **21.** hydroponics   **22.** nitrogen, phosphorus, potassium   **23.** soil erosion   **24.** salinization

Building Words: **1.** hydroponics   **2.** macronutrient   **3.** micronutrient   **4.** endodermis

Matching: **1.** i **2.** g **3.** d **4.** a **5.** c **6.** k **7.** j **8.** e **9.** m **10.** l

Making Comparisons: **1.** area of cell division that causes an increase in length of the root **2.** root hairs **3.** protects root **4.** storage **5.** endodermis **6.** pericycle **7.** xylem **8.** conducts dissolved sugars

Making Choices: **1.** a,c,e **2.** b,e **3.** c **4.** b,d **5.** a,b,d **6.** b **7.** a,b,d,e **8.** c **9.** c **10.** c **11.** c

Chapter 36

Reviewing Concepts: **1.** (a)sporophyte generation, (b)gametophyte generation **2.** flowering **3.** sepals, petals, stamens, and carpels **4.** (a)calyx, (b)corolla **5.** (a)microsporocytes, (b)microspores, (c)pollen grain, (d)sperm cells **6.** (a)anther, (b)stigma **7.** inbreeding **8.** self-incompatability **9.** (a)blue or yellow, (b)red **10.** flowers **11.** (a)style, (b)ovule **12.** endosperm **13.** suspensor **14.** (a)radicle, (b)cotyledons **15.** hypocotyls **16.** simple, aggregate, multiple, accessory **17.** (a)simple, (b)berries, (c)drupes, (d)follicle, (e)legumes, (f)capsules **18.** (a)aggregate, (b)multiple **19.** (a)accessory, (b)receptacle, (c)floral tube **20.** wind, animals, water, explosive dehiscence **21.** scarification **22.** coleoptile **23.** (a)rhizomes, (b)tubers, (c)bulbs, (d)corms, (e)stolons **24.** suckers **25.** dispersed **26.** sexual reproduction

Building Words: **1.** endosperm **2.** hypocotyl **3.** coevolution

Matching: **1.** d **2.** g **3.** l **4.** b **5.** a **6.** k **7.** j **8.** m **9.** c **10.** f

Making Comparisons: **1.** dry: does not open; nut **2.** stony wall does not split open at maturity **3.** single seeded; fully fused to fruit wall **4.** simple **5.** fleshy: berry **6.** accessory **7.** multiple **8.** formed from the ovaries of many flowers that fuse together and enlarge after fertilization **9.** aggregate **10.** many separate ovaries from a single flower **11.** simple **12.** fleshy, drupe **13.** hard, stony pit; single seeded

Making Choices: **1.** c **2.** e **3.** a **4.** b **5.** b,d,e **6.** d **7.** b,d,e **8.** b,c,d **9.** e **10.** d **11.** d **12.** e **13.** b,d,e **14.** a,c

Chapter 37

Reviewing Concepts: **1.** hormones **2.** photoreceptor **3.** (a)phototropism, (b)gravitropism, (c)thigmotropism **4.** auxins, gibberellins, cytokinins, ethylene, abscisic acid **5.** enzyme-linked receptors **6.** indoleacetic acid (IAA) **7.** (a)shoot apical meristem, (b)polar transport **8.** fungus **9.** flowering and germination **10.** cell division and differentiation **11.** senescence **12.** thigmomorphogenesis **13.** dormancy **14.** abscisic acid **15.** gibberellin **16.** brassinosteroids, jasmonates, salicylic acid, systemin, and oligosaccharins **17.** florigen **18.** photoperiodism **19.** (a)short-day (long-night), (b)long-day (short-night), (c)intermediate-day, (d)day-neutral **20.** phytochrome **21.** shade avoidance **22.** (a)red, (b)Pr, (c)Pfr **23.** (a)potassium, (b)turgor movements **24.** the time of day

Building Words: **1.** phototropism **2.** gravitropism **3.** phytochrome **4.** photoperiodism

Matching: **1.** b **2.** a **3.** j **4.** d **5.** c **6.** f **7.** l **8.** e

Making Comparisons: **1.** auxin **2.** cytokinin **3.** gibberellin **4.** ethylene **5.** gibberellin and cytokinin **6.** abscisic acid

Making Choices: **1.** a,c **2.** b **3.** a-e **4.** c **5.** a **6.** a,d **7.** a **8.** b **9.** e

Chapter 38

Reviewing Concepts: 1. number 2. (a)tissue, (b)organs, (c)organ systems 3. (a)sheets (continuous layers), (b)basement membrane 4. protection, absorption, secretion, sensation 5. (a)squamous, (b)cuboidal, (c)columnar 6. (a)simple, (b)stratified 7. (a)intercellular substance, (b)fibers, (c)matrix 8. intercellular substance 9. (a)collagen, (b)elastin, (c)collagen and glycoprotein 10. fibers and the protein and carbohydrate complexes of the matrix 11. macrophages 12. loose, dense, elastic, reticular, adipose, cartilage, bone, blood, lymph, those that produce blood cells 13. loose connective tissue 14. subcutaneous layer 15. collagen 16. (a)tendons, (b)ligaments 17. skeleton 18. (a)cartilage, (b)bone 19. (a)chondrocytes, (b)matrix, (c)lacunae 20. (a)matrix, (b)osteocytes, (c)vascularized 21. canaliculi 22. (a)osteons, (b)lamellae, (c)Haversian canal 23. plasma 24. transport oxygen 25. defend the body against disease-causing microorganisms 26. bone marrow 27. muscle fiber 28. (a)actin and myosin, (b)myofibrils 29. (a)skeletal, (b)smooth, (c)cardiac 30. (a)neurons, (b)glial 31. synapses 32. nerve 33. (a)cell body, (b)dendrite(s), (c)axon(s) 34. homeostatic mechanisms 35. steady state 36. thermoregulation 37. lower 38. six 39. (a)torpor, (b)hibernation, (c)estivation

Building Words: 1. multicellular 2. pseudostratified 3. fibroblast 4. intercellular 5. macrophage 6. chondrocyte 7. osteocyte 8. myofibril 9. homeostasis

Matching: 1. l 2. a 3. k 4. c 5. i 6. e 7. b 8. g 9. f 10. d

Making Comparisons: 1. simple squamous epithelium 2. air sacs of lungs, lining of blood vessels 3. epithelial tissue 4. some respiratory passages, ducts of many glands 5. secretion, movement of mucus, protection 6. connective tissue 7. food storage, protection of some organs, insulation 8. bone 9. support and protection of internal organs, calcium reserve, site of skeletal muscle attachments 10. connective tissue 11. within heart and blood vessels of the circulatory system 12. transport of oxygen, nutrients, waste product and other materials 13. walls of the heart 14. skeletal muscle 15. nervous tissue

Making Choices: 1 a 2. a,b,c,e 3. d 4. a,b,e 5. c 6. a,d 7. a,c,e 8. b,e 9. d 10. b 11. c 12. c,e 13. a,c,d 14. a,c 15. c 16. b 17. d 18. a,c,d 19. c

Chapter 39

Reviewing Concepts: 1. protect underlying tissues, exchange gasses, excrete wastes, regulate temperature, secrete substances such as poisons or mucous 2. skeleton 3. cuticle 4. skin 5. skin 6. (a)epidermis, (b)strata 7. stratum basale 8. (a)keratin, (b)strength, flexibility, waterproofing 9. (a)connective tissue, (b)subcutaneous 10. muscle 11. (a)hydrostatic skeleton, (b)longitudinally, (c)circularly, (d)outer, (e)inner 12. septa 13. exoskeleton 14. protection 15. molting 16. (a)endoskeletons, (b)chordates 17. echinoderms 18. (a)axial, (b)appendicular 19. (a)skull, (b)vertebral column, (c)rib cage 20. (a)pectoral girdle, (b)pelvic girdle, (c)limbs 21. (a)periostium, (b)epiphyses, (c)diaphysis, (d)metaphysis, (e)epiphyseal line, (f)compact bone 22. (a)endochondral, (b)intramembranous 23. (a)joints, (b)immovable joints, (c)slightly movable joints, (d)freely movable joints 24. (a)actin, (b)myosin 25. (a)smooth, (b)striated 26. asynchronous 27. tendons 28. (a)antagonistically, (b)antagonist 29. fascicles 30. sarcomere 31. (a)acetylcholine, (b)action potential 32. (a)calcium ions, (b)actin 33. center 34. ATP 35. creatine phosphate 36. glycogen 37. skeleton 38. (a)slow-oxidative fibers, (b)fast-oxidative fibers 39. (a)smooth, (b)cardiac

Building Words: 1. epidermis 2. periosteum 3. endochondral 4. osteoblast 5. osteoclast 6. myofilament

Matching: 1. h 2. f 3. j 4. l 5. g 6. b 7. d 8. m 9. i 10. a

Making Comparisons: 1. chitin 2. exoskeleton 3. external covering jointed for movement, does not grow so animal must periodically molt 4. echinoderms 5. calcium salts 6. cartilage or bone 7. endoskeleton 8. internal, composed of living tissue and grows with the animal

Making Choices: 1. c 2. a,b,c 3. d 4. c 5. a 6. e 7. c,d 8. a 9. d 10. c,e 11. a 12. a,b,d 13. b,c 14. a,c 15. b,e 16. b,e 17. d 18. a

Chapter 40

Reviewing Concepts: 1. stimuli 2. (a)endocrine, (b)nervous 3. (a)reception, (b)afferent, (c)transmit , 4. (a)interneurons, (b)integration, (c)efferent, (d)effectors 5. (a)dendrites, (b)cell body, (c)axon 6. (a)terminal branches, (b)synaptic terminals, (c)neurotransmitter 7. (a)Schwann cells, (b)myelin sheath, (c)nodes of Ranvier 8. (a)nerves, (b)tracts (pathways) 9. (a)ganglia, (b)nuclei 10. (a)astrocytes, (b)microglia 11. (a)resting potential, (b)-70 mV 12. inner 13. (a)inside, (b)outside 14. sodium-potassium pumps 15. (a)excitatory, (b)inhibitory 16. (a)action potential, (b)sodium and potassium ions 17. (a)voltage-activated ion channels, (b)threshold level, (c)-55 mV 18. spike 19. wave of depolarization 20. repolarization 21. node of Ranvier 22. all-or-none 23. synapse 24. (a)presynaptic, (b)postsynaptic 25. (a)electrical synapses, (b)chemical synapses, (c)neurotransmitter molecules, (d)synaptic cleft 26. (a)acetylcholine, (b)neuromuscular 27. (a)adrenergic neurons, (b)catecholamine (biogenic amine) 28. (a)ligands, (b)synaptic vesicles 29. (a)excitatory postsynaptic potential (EPSP), (b)inhibitory postsynaptic potential (IPSP) 30. graded potentials 31. (a)summation, (b)temporal summation, (c)spatial summation 32. reverberating

Building Words: 1. interneuron 2. neuroglia 3. neurotransmitter 4. multipolar 5. postsynaptic 6. presynaptic

Matching: 1. d 2. j 3. f 4. b 5. i 6. k 7. a 8. h 9. n

Making Comparisons: 1. -70mV 2. stable 3. sodium-potassium pump active, sodium and potassium channels open 4. threshold potential 5. voltage-activated sodium ion channels open 6. varies, but is about +35mV 7. wave of depolarization (rise: depolarization; fall: repolarization) 8. depolarization 9. neurotransmitter-receptor combination opens sodium ion channels 10. hyperpolarization

Making Choices: 1. b,c,e 2. b 3. c,d,e 4. c,d 5. b 6. a,d 7. c 8. b,c 9. d 10. c,e 11. a

Chapter 41

Reviewing Concepts: 1. (a)nerve net, (b)cnidarians 2. nerve ring 3. (a)ladder-type, (b)cerebral ganglia 4. ganglia 5. (a)central nervous (CNS), (b)peripheral nervous (PNS) 6. sympathetic and parasympathetic 7. neural tube 8. (a)cerebellum, (b)pons, (c)metencephalon, (d)medulla, (e)myelencephalon 9. brainstem 10. association 11. (a)superior colliculi, (b)inferior colliculi, (c)red nucleus 12. (a)thalamus and hypothalamus, (b)cerebrum 13. (a)hemispheres, (b)white matter, (c)cerebral cortex, (d)convolutions 14. (a)dura matter, arachnoid, pia mater, (b)choroid plexus 15. second lumbar 16. (a)white matter, (b)tracts 17. (a)sensory, (b)motor, (c)association 18. (a)frontal lobes, (b)parietal, (c)central sulcus 19. brain stem and thalamus 20. (a)electroencephalogram (EEG), (b)delta, (c)alpha, (d)beta 21. (a)rapid eye movement (REM), (b)non-REM 22. cerebrum 23. (a)memory, (b)implicit memory, (c) explicit memory 24. temporal lobe 25. (a)sense receptors, (b)sensory neurons, (c)motor neurons 26. (a)cranial, (b)spinal 27. (a)sympathetic, (b)parasympathetic 28. (a)preganglionic neuron, (b)postganglionic neuron 29. psychological dependence

Building Words: 1. hypothalamus 2. postganglionic 3. preganglionic 4. paravertebral

Matching: 1. g 2. d 3. j 4. c 5. h 6. a 7. e 8. k 9. b 10. i

Making Comparisons: 1. pons 2. midbrain 3. center for visual and auditory reflexes 4. diencephalon 5. relay center for motor and sensory information 6. hypothalamus 7. diencephalon 8. metencephalon 9. cerebrum 10. reticular activating system 11. limbic system

Making Choices: 1. a 2. a 3. b 4. a-e 5. a,d,e 6. b 7. a,b 8. d 9. a,c 10. b 11. b 12. a,b 13. b 14. c 15. a-e

Chapter 42

Reviewing Concepts: 1. echolocation 2. reception 3. (a)electrical, (b)receptor 4. (a)graded, (b)depolarizes, (c)action potential, (d)sensory 5. brain 6. (a)number and identity, (b)frequency and total number 7. sensory adaptation 8. (a)exteroceptors, (b)interoceptors 9. pit vipers and boas 10. hypothalamus 11. electromagnetic receptors 12. (a)spinal cord, (b)glutamate, (c)substance P 13. mechanoreceptors 14. Pacinian corpuscles 15. (a)Meissner corpuscles, Ruffini corpuscles, Merkel disks (b)pain 16. (a)muscle spindles, (b)Golgi tendon organs, (c)joint receptors 17. (a)statoliths, (b)hair cells 18. statocysts 19. kinocilium 20. vibrations 21. (a)canal, (b)sensory hair cells, (c)cupula 22. (a)otoliths, (b)saccule and utricle 23. (a)angular acceleration, (b)endolymph, (c)cristae 24. (a)cochlea, (b)mechanoreceptors 25. (a)tympanic membrane, (b)malleus, incus, stapes, (c)oval window 26. (a)basilar membrane, (b)organ of Corti, (c)cochlear nerve 27. sweet, sour, salty, bitter 28. glutamate 29. olfaction 30. pheromones 31. rhodopsins 32. eyespots (ocelli) 33. ommatidia 34. (a)sclera, (b)shape (rigidity) 35. cornea 36. (a)lens, (b)retina, (c)rods, (d)rhodopsin 37. (a)cones, (b)fovea

Building Words: 1. proprioceptor 2. otolith 3. endolymph 4. chemoreceptor 5. thermoreceptor 6. interoceptor

Matching: 1. l 2. b 3. j 4. i 5. n 6. k 7. o 8. d 9. a 10. c

Making Comparisons: 1. electroreceptor 2. exteroceptor 3. pressure waves (sound) 4. mechanoreceptor 5. interoceptor 6. receptor in the human hypothalamus 7. muscle contraction 8. interoceptor 9. thermoreceptor 10. pit organ of a viper 11. light energy 12. photoreceptor 13. chemoreceptor 14. mammalian taste buds 15. chemoreceptor 16. exteroceptor

Making Choices: 1. c 2. d 3. a 4. a 5. c 6. b 7. c 8. b,d 9. b,d 10. c 11. d 12. e 13. a,d

Chapter 43

Reviewing Concepts: 1. cholesterol 2. sponges, cnidarians, ctenophores, flatworms, nematodes 3. (a)blood, (b)heart, (c)vessels 4. (a)open circulatory, (b)hemocoel 5. (a)hemolymph, (b)interstitial 6. (a)arthropods, most mollusks (b)sinuses 7. (a)hemocyanin, (b)copper 8. (a)dorsal, (b)ventral, (c)five 9. plasma 10.(a)closed, (b)heart, (c)blood 11. (a)nutrients, oxygen, metabolic wastes, hormones, (b)maintenance of fluid balance, defense, distribution of metabolic heat 12.(a)red blood cells, white blood cells, platelets, (b)plasma 13. (a)interstitial, (b)intracellular 14. (a)fibrinogen, (b)gamma globulins, (c)plasma proteins 15. oxygen 16. (a)red bone marrow, (b)hemoglobin, (c)120 17. (a)neutrophils, (b)basophils and eosinophiles, (c)granular leukocytes 18. (a)lymphocytes, (b)monocytes, (c)macrophages 19. (a)thrombocytes, (b)platelets 20. (a)arteries, (b)veins 21. capillaries 22. (a)ventricles, (b)atria 23. (a)one, (b)one 24. (a)three, (b)sinus venosus, (c)conus arteriosus 25. (a)more, (b)higher 26. (a)pericardium, (b)pericardial cavity 27. (a)interventricular, (b)interatrial septum 28. (a)four, (b)atrioventricular valves, (c)tricuspid valve, (d)mitral valve, (e)semilunar valves 29. (a)sinoatrial (SA) node, (b)atrioventricular (AV) node, (c)atrioventricular (AV) bundle 30. (a)cardiac, (b)systole, (c)diastole 31. (a)lub, (b)AV valves 32. (a)dub, (b)semilunar valves 33. beta blockers 34. Starling's law of the heart 35. (a)cardiac output (CO), (b) stroke volume 36. five liters/minute 37. (a)hypertension, (b)increase, (c)salt 38. diameter of arterioles 39. low-resistance 40. baroreceptors 41. (a)pressure changes, (b)cardiac and vasomotor centers 42. (a)pulmonary circuit, (b)systemic circuit 43. (a)rich, (b)left 44. (a)aorta, (b)brain, (c)shoulder area, (d)legs 45. (a)superior vena cava, (b)inferior vena cava 46. lymph 47. (a)lymph, (b)interstitial fluids, (c)lymph nodes, (d)subclavian, (e)thoracic, (f)right lymphatic 48. edema

Building Words: 1. hemocoel 2. hemocyanin 3. erythrocyte 4. leukocyte 5. neutrophil 6. eosinophil 7. basophil 8. leukemia 9. vasoconstriction 10. vasodilation 11. pericardium 12. semilunar 13. baroreceptor

Matching: 1. h 2. i 3. g 4. j 5. d 6. k 7. b 8. l 9. a 10. m

Making Comparisons: 1. lipoproteins 2. plasma 3. cell component 4. transport of oxygen and carbon dioxide 5. blood clotting 6. cell component 7. defense; differentiate to form phagocytic macrophages in tissue 8. albumins and globulins 9. plasma 10. defense, principle phagocytic cell in blood

Making Choices: **1.** d   **2.** a   **3.** d   **4.** c,e   **5.** a   **6.** b,c   **7.** d   **8.** a,b,d   **9.** a,c,e   **10.** b,d,e   **11.** b
12. a,d,e   **13.** c,e   **14.** a,b,c,e   **15.** c   **16.** c   **17.** d

Chapter 44

Reviewing Concepts: **1.** immunology   **2.** nonspecific immune responses, specific immune responses
3. antigen   **4.** antibodies   **5.** danger signals   **6.** phagocytosis   **7.** antimicrobial peptides
8. (a)lymphocytes, (b)lymph nodes   **9.** (a)outer covering, (b)acid secretions and enzymes, (c)mucous
membranes   **10.** neutrophils and macrophages   **11.** viral replication   **12.** vasodilation, increased
capillary permeability, increased phagocytosis   **13.** (a)stem, (b)bone marrow   **14.** (a)cytotoxic, (b)helper
15. immunocompetent   **16.** antigen-presenting cells   **17.** MHC antigens   **18.** perforins and granzymes
19. (a)immunoglobulins, (b)Ig, (c)antigenic determinants (epitopes), (d)binding sites   **20.** IgG, IgM, IgA,
IgD, IgE   **21.** (a)IgGs, (b)IgAs, (c)IgD, (d)IgE   **22.** phagocytic cells   **23.** clonal selection
24. monoclonal antibodies   **25.** IgM   **26.** IgG   **27.** DNA vaccines   **28.** temporary   **29.** DNA
microarrays   **30.** protein malnutrition   **31.** protease inhibitors   **32.** (a)sensitization stage, (b)effector
stage   **33.** erythroblastosis fetalis   **34.** (a)allergen, (b)IgE   **35.** histamine   **36.** autoreactive

Building Words: **1.** antibody   **2.** antihistamine   **3.** lymphocyte   **4.** monoclonal   **5.** autoimmune
6. lysozyme

Matching: **1.** b   **2.** o   **3.** d   **4.** n   **5.** e   **6.** g   **7.** m   **8.** l   **9.** i   **10.** k

Making Comparisons: **1.** plasma cells   **2.** memory B cells   **3.** B-cells and macrophages   **4.** helper T
cells   **5.** cytotoxic T cells   **6.** memory T cells   **7.** natural killer cells   **8.** macrophages   **9.** neutrophils

Making Choices: **1.** a,d   **2.** a,c,d   **3.** b   **4.** b   **5.** b,c,e   **6.** a-e   **7.** a,c,e   **8.** b,c,d   **9.** a,b,c,e
10. a-d   **11.** b,c,e

Chapter 45

Reviewing Concepts: **1.** respiration   **2.** (a)more, (b)faster   **3.** ventilation   **4.** (a)thin walls, (b)moist,
(c)blood vessels (or hemolymph)   **5.** body surface, tracheal tubes, gills, lungs   **6.** surface-to-volume
7. surrounding water   **8.** (a)tracheal tubes, (b)spiracles   **9.** tracheoles   **10.** filaments
11. countercurrent exchange system   **12.** (a)body surface, (b)body cavity   **13.** book lungs   **14.** ciliated
cells   **15.** (a)three, (b)two   **16.** (a)pharynx, (b)larynx, (c)trachea, (d)bronchi, (e)bronchioles, (f)alveoli
17. (a)increases, (b)decreases   **18.** (a)tidal volume, (b)500 ml   **19.** vital capacity   **20.** pressure (tension)
21. Dalton's law of partial pressures   **22.** (a)100 mm Hg, (b)40 mm Hg   **23.** (a)heme (iron-porphyrin),
(b)globin   **24.** (a)iron, (b)heme, (c)oxyhemoglobin (HbO_2)   **25.** (a)lower, (b)Bohr effect
26. (a)bicarbonate (HCO_3^-), (b)carbonic anhydrase   **27.** medulla   **28.** chemoreceptors   **29.** blood
pressure   **30.** hypoxia   **31.** decompression sickness   **32.** the metabolic rate decreases 20%, breathing
stops, bradycardia occurs   **33.** bronchial constriction

Building Words: **1.** hyperventilation   **2.** hypoxia   **3.** ventilation   **4.** oxyhemoglobin

Matching: **1.** n   **2.** a   **3.** f   **4.** l   **5.** c   **6.** j   **7.** b   **8.** d   **9.** i   **10.** g

Making Comparisons: **1.** gills   **2.** some, tracheal tubes; some book lungs   **3.** lungs   **4.** lungs
5. body surface   **6.** tracheal tubes   **7.** dermal gills   **8.** gills   **9.** lungs

Making Choices: **1.** e   **2.** c,d   **3.** a-e   **4.** b,c,e   **5.** a,e   **6.** a,b,e   **7.** a,b,d   **8.** a   **9.** b,d,e   **10.** e

Chapter 46

Reviewing Concepts: **1.** heterotrophs (consumers)   **2.** egestion   **3.** herbivores   **4.** (a)carnivores,
(b)shorter   **5.** omnivores   **6.** (a)single, (b)mouth   **7.** (a)mouth, (b)anus   **8.** (a)pharynx (throat),
(b)esophagus, (c)small intestine   **9.** (a)incisors, (b)canines, (c)molars and premolars, (d)enamel, (e)dentin,
(f)pulp cavity   **10.** salivary amylase   **11.** peristalsis   **12.** bolus   **13.** pepsin   **14.** (a)simple columnar
epithelial, (b)parietal, (c)chief, (d)pepsin   **15.** (a)duodenum, jejunum, ileum, (b)duodenum   **16.** (a)villi,
(b)microvilli   **17.** (a)glycogen, (b)fatty acids and urea   **18.** gallbladder   **19.** trypsin and chymotrypsin
20. (a)pancreatic lipase, (b)pancreatic amylase   **21.** monosaccharides   **22.** amino acids
23. monoacylglycerols, diacylglycerols, fatty acids and glycerol   **24.** (a)endocrine cells, (b)digestive system
25. villi   **26.** chylomicrons   **27.** (a)cecum, (b)ascending colon, (c)transverse colon, (d)descending colon,

(e)sigmoid colon, (f)rectum, (g)anus   **28.** (a)excretion, (b)elimination   **29.** (a)Calories, (b)kilocalories
30. fiber   **31.** lipoproteins   **32.** (a)high-density lipoproteins (HDLs), (b)low-density lipoproteins (LDLs)
33. (a)9, (b)essential amino acids   **34.** (a)fat-soluble, (b)water-soluble, (c)B and C   **35.** fat-soluble
36. sodium, chloride, potassium, magnesium, calcium, sulfur, phosphorus   **37.** trace elements   **38.** free
radicals   **39.** (a)fruits and vegetables, (b)fats   **40.** (a)basal metabolic rate (BMR), (b)total metabolic rate
41. (a)increases, (b)decreases   **42.** essential amino acids   **43.** adipose tissue

Building Words: **1.** herbivore   **2.** carnivore   **3.** omnivore   **4.** submucosa   **5.** peritonitis
6. epiglottis   **7.** microvilli   **8.** chylomicron

Matching: **1.** h   **2.** e   **3.** j   **4.** i   **5.** o   **6.** l   **7.** a   **8.** c   **9.** k   **10.** n

Making Comparisons: **1.** splits maltose into 2 glucose molecules   **2.** small intestine   **3.** pepsin
4. stomach   **5.** trypsin and chymotrypsin   **6.** pancreas   **7.** RNA to free nucleotides   **8.** pancreas
9. pancreatic lipase   **10.** dipeptidase   **11.** duodenum   **12.** lactose to glucose and galactose

Making Choices: **1.** a,b,d   **2.** c,d,e   **3.** d   **4.** a,d   **5.** a   **6.** a,b   **7.** a,b,c   **8.** a,b,c   **9.** d,e   **10.** a-d
11. b   **12.** c   **13.** a   **14.** d   **15.** c,e   **16.** e   **17.** c,d   **18.** a   **19.** a,b,e   **20.** b,e

Chapter 47

Reviewing Concepts: **1.** water   **2.** osmoregulation and excretion   **3.** osmoregulation   **4.** ammonia,
urea, uric acid   **5.** uric acid   **6.** (a)ammonia, (b)carbon dioxide   **7.** (a)nephridiopores, (b)protonephridia,
(c)metanephridia   **8.** gut wall   **9.** hemolymph in the hemocoel   **10.** diffusion and active transport
11. skin, lungs or gills, digestive system   **12.** (a)ammonia, (b)urea   **13.** drink sea water   **14.** (a)uric acid,
(b)urea   **15.** (a)ureters, (b)urethra   **16.** (a)Bowman's capsule, (b)renal tubule, (c)Bowman's capsule,
(d)loop of Henle, (e)collecting duct   **17.** (a)cortical, (b)juxtamedullary   **18.** filtration, reabsorption,
tubular secretion   **19.** Bowman's capsule   **20.** renal tubules   **21.** (a)descending, (b)ascending,
(c)countercurrent mechanism   **22.** vasa recta   **23.** urine   **24.** (a)antidiuretic hormone (ADH),
(b)posterior lobe of the pituitary gland   **25.** hypothalamus   **26.** (a)adrenal cortex, (b)sodium

Building Words: **1.** protonephridium   **2.** podocyte   **3.** juxtamedullary

Matching: **1.** e   **2.** a   **3.** f   **4.** n   **5.** o   **6.** j   **7.** m   **8.** l   **9.** i   **10.** b

Making Comparisons: **1.** kidney   **2.** Malpighian tubules   **3.** metanephridia   **4.** protonephridia

Making Choices: **1.** a,c,e   **2.** b   **3.** a,b   **4.** a,d,e   **5.** c   **6.** c,d   **7.** b,d   **8.** d   **9.** c   **10.** e   **11.** c,e
12. c   **13.** a,c,d   **14.** b,d   **15.** a,d,e   **16.** e   **17.** b,c,d   **18.** a,c

Chapter 48

Reviewing Concepts: **1.** hormones   **2.** exocrine glands   **3.** ducts   **4.** (a)neurohormones, (b)nervous
5. (a)negative feedback systems, (b)homeostasis   **6.** steroids, peptides or proteins, fatty acid derivatives,
amino acid derivatives   **7.** cholesterol   **8.** (a)axons, (b)interstitial fluid and the blood   **9.** autocrine
signaling   **10.** (a)receptor up-regulation, (b)receptor down-regulation   **11.** hormone-receptor complex
12. cell-surface receptor   **13.** (a)cyclic AMP, (b)calcium ion   **14.** neurons   **15.** (a)BH or ecdysiotropin,
(b)prothoracic   **16.** ecdysone   **17.** (a)overstimulated, (b)deprived of needed stimulation   **18.** seven
19. antidiuretic hormone (ADH)   **20.** oxytocin   **21.** tropic   **22.** prolactin   **23.** (a)somatotropin,
(b)anterior pituitary   **24.** (a)gigantism, (b)acromegaly   **25.** (a)thyroxine, (b)iodine   **26.** (a)anterior
pituitary, (b)thyroid-stimulating hormone (TSH)   **27.** cretinism   **28.** (a)bones, (b)kidney tubules
29. (a)thyroid, (b)calcitonin   **30.** (a)islets of Langerhans, (b)beta, (c)insulin, (d)alpha, (e)glucagon
31. liver, muscle, and fat cells   **32.** diabetes mellitus   **33.** epinephrine, norepinephrine   **34.** (a)sex
hormone precursors, mineralocorticoids, glucocorticoids, (b)testosterone, (c)estradiol   **35.** aldosterone
36. cortisol (hydrocortisone)   **37.** (a)corticotropin-releasing factor, (b) adrenocorticotropic hormone
38. (a)pineal, (b)melatonin   **39.** thymosin   **40.** atrial natriuretic factor

Building Words: **1.** neurohormone   **2.** neurosecretory   **3.** hypersecretion   **4.** hyposecretion
5. hyperthyroidism   **6.** hypoglycemia   **7.** hyperglycemia   **8.** acromegaly

Matching: **1.** i   **2.** d   **3.** l   **4.** a   **5.** g   **6.** c   **7.** e   **8.** n   **9.** m   **10.** k

Making Comparisons: **1.** steroid **2.** maintain sodium and potassium balance; increase sodium reabsorption and potassium excretion **3.** amino acid derivative **4.** stimulates metabolic rate; essential to normal growth and development **5.** ADH **6.** glucocorticoids (cortisol) **7.** steroid **8.** prostaglandins **9.** fatty acid derivative **10.** stimulates secretion of adrenal cortical hormones **11.** amino acid derivative **12.** help body cope with stress; increase heart rate, blood pressure, metabolic rate; raise blood sugar (glucose) level **13.** oxytocin **14.** peptide

Making Choices: **1.** c **2.** a,c,e **3.** b **4.** a,d,e **5.** b **6.** a,d **7.** a,d **8.** b,c **9.** b,c **10.** a,c,d **11.** d

Chapter 49

Reviewing Concepts: **1.** sperm and eggs **2.** budding **3.** fragmentation **4.** unfertilized egg **5.** external fertilization **6.** hermaphroditism **7.** sexual **8.** (a)spermatogonium, (b)primary spermatocyte **9.** acrosome **10.** (a)scrotum, (b)seminiferous tubules **11.** inguinal canals **12.** (a)epididymis, (b)vas deferens **13.** urethra **14.** (a)200 million **15.** seminal vesicles and prostate gland **16.** (a)shaft, (b)glans, (c)prepuce (foreskin), (d)circumcision **17.** (a)cavernous bodies, spongy body (b)blood **18.** Testosterone **19.** (a)FSH and LH, (b)testosterone **20.** (a)testosterone, (b)interstitial **21.** (a)oogenesis, (b)oogonia, (c)primary oocytes **22.** (a)follicle, (b)FSH, (c)secondary oocyte **23.** corpus luteum **24.** wall of the oviduct **25.**(a)endometrium, (b)menstruation **26.** cervix **27.** receptacle **28.** vulva **29.** clitoris **30.** lactation **31.** prolactin **32.** ovulation **33.** (a)estrogen (estradiol), (b)FSH and LH, (c)LH **34.** estrogens and progesterone **35.** (a)hypothalamus, (b)ova, (c)infertile **36.** reabsorbed **37.** vasocongestion, increased muscle tension (myotonia) **38.** excitement phase, plateau phase, orgasm, resolution **39.** human chorionic gonadotropin or hCG **40.** estrogen and oxytocin **41.** progestin and synthetic estrogen **42.** 90 million **43.** sexually transmitted diseases (STDs) **44.** progestin pills **45.** vasectomy **46.** tubal sterilization **47.** prostaglandins **48.** (a)human papillomavirus or HPV, (b)20 million, (c)75%

Building Words: **1.** parthenogenesis **2.** spermatogenesis **3.** acrosome **4.** circumcision **5.** oogenesis **6.** endometrium **7.** postovulatory **8.** preovulatory **9.** vasectomy **10.** vasocongestion **11.** intrauterine **12.** contraception

Matching: **1.** m **2.** j **3.** k **4.** b **5.** g **6.** c **7.** e **8.** d **9.** i **10.** l

Making Comparisons: **1.** anterior pituitary **2.** gonad **3.** stimulates development of seminiferous tubules, stimulates spermatogenesis **4.** stimulates follicle development, secretion of estrogen **5.** stimulates interstitial cells to secrete testosterone **6.** stimulates ovulation, corpus luteum development **7.** testes **8.** testosterone **9.** ovaries **10.** estrogen (estradiol)

Making Choices: **1.** c **2.** a,e **3.** a,c **4.** a,d, **5.** a,c **6.** b,d **7.** a-e **8.** a,c,e **9.** c **10.** c,d,e **11.** a **12.** d **13.** a,c,e **14.** e **15.** e **16.** d **17.** a,b **18.** b,d **19.** c,e **20.** c **21.** c **22.** a,b,c,e

Chapter 50

Reviewing Concepts: **1.** increases **2.** cellular differentiation **3.** morphogenesis **4.** (a)diploid, (b)sex **5.** bindin **6.** (a)capacitation, (b)acrosome reaction **7.** polyspermy **8.** cortical reaction **9.** (a)increases, (b)active, (c)protein **10.** microtubules **11.** totipotent **12.** blastula **13.** (a)invertebrates and simple chordates, (b)holoblastic, (c)radial, (d)spiral **14.** (a)meroblastic, (b)blastodisc **15.** unequal distribution **16.** surface proteins **17.** gastrulation **18.** (a)germ layers, (b)endoderm, ectoderm, mesoderm **19.** blastopore **20.** primitive streak **21.** (a)notochord, (b)neural groove, (c)neural tube, (d)central nervous system **22.** (a)pharyngeal pouches, (b)branchial grooves **23.** chorion, amnion, allantois, yolk sac **24.** amnion **25.** allantois **26.** (a)zona pellucida, (b)morula, (c)blastocyst **27.** trophoblast **28.** (a)implantation, (b)seventh **29.** embryonic chorion and maternal **30.** second and third **31.** end of first trimester **32.** (a)seventh, (b)convolutions **33.** ultrasound **34.** amniocentesis **35.** carbon dioxide **36.** 8 **37.** decline

Building Words: **1.** telolecithal **2.** isolecithal **3.** blastomere **4.** blastocyst **5.** blastopore **6.** holoblastic **7.** meroblastic **8.** archenteron **9.** trophoblast **10.** trimester **11.** neonate

Matching: **1.** a **2.** e **3.** j **4.** n **5.** c **6.** f **7.** k **8.** l **9.** i **10.** h

Making Comparisons: **1.** gastrulation **2.** 24 hours **3.** fertilization **4.** 0 hour **5.** 7 days **6.** blastocyst begins to implant **7.** fertilization **8.** organogenesis **9.** late cleavage **10.** 3 days

Making Choices: **1.** c **2.** b,c,e **3.** d **4.** c **5.** c,e **6.** a **7.** d **8.** b **9.** c,e **10.** d **11.** c **12.** b,c **13.** b,d

Chapter 51

Reviewing Concepts: **1.** stimuli **2.** learning **3.** behavioral ecology **4.** direct fitness **5.** (a)innate behavior, (b)learned behavior **6.** nervous, endocrine **7.** sign stimulus **8.** experience **9.** habituation **10.** recognizes **11.** (a)conditioned stimulus, (b)unconditioned **12.** (a)positive reinforcement, (b)negative reinforcement **13.** insight learning **14.** exercise, learning to coordinate movements, learning social skills **15.** circadian rhythms **16.** suprachiasmatic nucleus (SCN) **17.** directional orientation **18.** optimal foraging **19.** society **20.** pheromones **21.** vomeronasal **22.** dominance hierarchy **23.** testosterone **24.** home range **25.** intrasexual selection **26.** lek **27.** courtship rituals **28.** (a)polygyny, (b)polyandry **29.** parental investment **30.** kin selection **31.** sentinel behavior **32.** reciprocal altruism **33.** dance **34.** (a)culture, (b)social learning **35.** natural selection

Building Words: **1.** sociobiology **2.** polyandry **3.** polygyny **4.** monogamy

Matching: **1.** b **2.** e **3.** i **4.** m **5.** l **6.** g **7.** h **8.** a **9.** c **10.** f

Making Comparisons: **1.** learned **2.** classical conditioning **3.** operant conditioning **4.** habituation **5.** imprinting **6.** insight learning

Making Choices: **1.** d **2.** e **3.** b **4.** b **5.** a,b **6.** a **7.** a,b,d **8.** c,d

Chapter 52

Reviewing Concepts: **1.** (a)biotic factors, (b)abiotic factors **2.** population density, population dispersion, birth and death rates, growth rates, survivorship, age structure **3.** population dynamics **4.** population density **5.** (a)random dispersion, (b)uniform dispersion, (c)clumped (aggregated) dispersion **6.** natality and mortality **7.** (a)change in number of individuals in the population, (b)change in time, (c)natality, (d)mortality **8.** (a)immigration, (b)emigration **9.** intrinsic rate of increase **10.** exponential population growth **11.** carrying capacity **12.** S-shaped **13.** predation, disease, competition **14.** high **15.** disease **16.** (a)interference competition, (b)exploitation competition **17.** density-independent **18.** (a)semelparous, (b)iteroparous **19.** (a)*r* strategists, (b)*K* strategists **20.** survivorship **21.** (a)source habitats, (b)sink habitats **22.** (a)birth rate, (b)death rate **23.** (a)doubling time, (b)less **24.** (a)replacement-level fertility, (b)lower, (c)higher **25.** (a)total fertility rate, (b)higher **26.** number and proportion **27.** (a)former, (b)latter

Building Words: **1.** intraspecific competition **2.** interspecific competition **3.** biosphere

Matching: **1.** o **2.** i **3.** l **4.** b **5.** e **6.** p **7.** h **8.** a **9.** f **10.** n **11.** g **12.** m

Making Comparisons: **1.** density-independent **2.** environmental factor that affects the size of a population but is not influenced by changes in population density **3.** density-dependent **4.** dominant individuals obtain adequate supply of a limited resource at expense of others **5.** density-dependent **6.** as density increases, encounters between population members increase as does opportunity to spread disease **7.** density-independent **8.** random weather event that reduces vulnerable population

Making Choices: **1.** g **2.** h **3.** d **4.** f **5.** b **6.** b **7.** b **8.** a,e **9.** d **10.** c **11.** a **12.** b,c **13.** a,d **14.** c

Chapter 53

Reviewing Concepts: **1.** community **2.** ecosystem **3.** competition, predation, symbiosis **4.** niche **5.** (a)fundamental, (b)realized **6.** limiting resource **7.** intraspecific competition **8.** competitive exclusion principle **9.** resource partitioning **10.** character displacement **11.** coevolution **12.** (a)aposematic coloration, (b)warning coloration **13.** Batesian mimicry **14.** mutualism, commensalism, and parasitism **15.** pathogen **16.** (a)keystone, (b)dominant **17.** less **18.** edge effect **19.** (a)species richness, (b)species diversity **20.** primary **21.** secondary **22.** climax community **23.** organismic **24.** individualistic

Building Words: **1.** carnivore **2.** ecosystem **3.** pathogen **4.** intraspecific competition **5.** herbivore **6.** interspecific competition

Matching: **1.** m **2.** l **3.** j **4.** g **5.** n **6.** i **7.** e **8.** b **9.** o

Making Comparisons: **1.** commensalisms of species 1 and species 2 **2.** beneficial **3.** no effect **4.** predation of species 2 by species 1 **5.** beneficial **6.** adverse **7.** parasitism of species 2 by species 1 **8.** beneficial **9.** adverse **10.** parasitism of species 2 by species 1 **11.** beneficial **12.** adverse **13.** beneficial **14.** beneficial **15.** adverse **16.** adverse **17.** commensalism of species 1 and species 2 **18.** beneficial **19.** no effect

Making Choices: **1.** b **2.** d **3.** c **4.** b **5.** a,c,e **6.** e **7.** a **8.** a,d,e **9.** e **10.** a **11.** c

Chapter 54

Reviewing Concepts: **1.** ecosystem **2.** biosphere **3.** (a)radiant, (b)chemical, (c)heat **4.** food web **5.** (a)producers, (b)primary consumers (herbivores), (c)secondary consumers (carnivores) **6.** (a)numbers, (b)biomass **7.** energy **8.** gross primary productivity **9.** net primary productivity **10.** biological magnification **11.** biogeochemical cycles **12.** carbon, nitrogen, phosphorus, water **13.** (a)photosynthesis, (b)respiration **14.** (a)nitrogen fixation, (b)nitrification, (c)assimilation, (d)ammonification, (e)denitrification **15.** (a)rocks, (b)soil **16.** runoff **17.** groundwater **18.** (a)limited nitrogen and phosphorus, (b)few trophic levels and low species richness **19.** water and temperature **20.** long-wave infrared (heat) energy **21.** (a)high, (b)low, (c)Coriolis Effect **22.** Pacific **23.** (a)right, (b)clockwise, (c)left, (d)counterclockwise **24.** climate **25.** precipitation **26.** rain shadows **27.** (a)free nutrient minerals essential for plant growth locked in dry organic matter, (b)remove plant cover and expose the soil, thereby stimulating the germination and establishment of seeds requiring bare soil, as well as encouraging the growth of shade-intolerant plants, (c)increase soil erosion, leaving the soil more vulnerable to wind and water. **28.** deforestation

Building Words: **1.** biomass **2.** atmosphere **3.** hydrologic cycle **4.** microclimate

Matching: **1.** m **2.** i **3.** e **4.** g **5.** k **6.** l **7.** d **8.** f **9.** c **10.** j

Making Comparisons: **1.** carbon **2.** carbon dioxide **3.** nitrogen **4.** nitrogen **5.** hydrologic **6.** enters atmosphere by transpiration, evaporation; exits as precipitation

Making Choices: **1.** c **2.** a,c **3.** a,c **4.** a-e **5.** d **6.** b,d,e **7.** a,d,e **8.** c,e **9.** b **10.** c

Chapter 55

Reviewing Concepts: **1.** climate **2.** permafrost **3.** short growing season and mineral-poor soil **4.** industrial wood and wood fiber **5.** lumber and pulpwood **6.** 30-50 inches (75-126 cm) **7.** tall grass prairies **8.** Mediterranean **9.** estivation **10.** African **11.** South America and Australia **12.** 80-180 inches (200-450 cm) **13.** vegetation **14.** salinity, light, nutrient minerals, oxygen **15.** (a)plankton, (b)nekton, (c)benthos **16.** (a)phytoplankton, (b)zooplankton **17.** strength of the current **18.** (a)littoral, (b)limnetic, (c)profundal, (d)littoral **19.** (a)thermal stratification, (b)thermocline, (c)fall turnover, (d)spring turnover **20.** (a)marshes, (b)swamps **21.** (a)prairie potholes, (b)peat moss bogs **22.** salt marshes **23.** intertidal zone **24.** neritic province **25.** euphotic **26.** oceanic province **27.** boundries **28.** center of origin **29.** cosmopolitan **30.** Palearctic, Nearctic, Neotropical, Ethiopian, Oriental, and Australian

Building Words: **1.** phytoplankton **2.** zooplankton **3.** intertidal **4.** thermocline

Matching: **1.** n **2.** i **3.** j **4.** a **5.** c **6.** l **7.** d **8.** e **9.** h **10.** g

Making Comparisons: **1.** long, harsh winters, short summers, little rain **2.** permafrost **3.** low-growing mosses, lichens, grasses, and sedges adapted to extreme cold and short growing season; voles, weasels, arctic foxes, grey wolves, snowshoe hares, snowy owls, oxen, lemmings, caribou, etc. (few plant and animal species) **4.** taiga **5.** seasonality, high rainfall **6.** rich topsoil, clay in lower layer **7.** broad-leaf trees that lose leaves seasonally, variety of large mammals **8.** temperate grassland **9.** deep, mineral-rich soil **10.** wet, mild winters; dry summers **11.** thickets of small-leafed shrubs and trees, fire adaptation; mule deer **12.** desert **13.** low in organic material, often high mineral content **14.** savanna **15.** low in essential mineral nutrients **16.** high rainfall evenly distributed throughout year

17. mineral-poor   18. high species diversity, distinct stories of vegetation, epiphytes in organic material, often high mineral content

Making Choices: 1. b   2. d   3. b,d   4. a,d   5. e   6. a,d   7. d   8. b,d   9. e   10. b,e   11. b

Chapter 56

Reviewing Concepts: 1. declining biological diversity, deforestation, global warming, and ozone depletion in the stratosphere   2. environmental sustainability   3. genetic diversity, species diversity, and ecosystem diversity   4. (a)endangered, (b)threatened   5. habitat destruction   6. habitat fragmentation   7. biotic pollution   8. nature (the wild)   9. human controlled settings   10. animal and plant habitats   11. landscape ecology   12. (a)restoration ecology, (b)*in situ* conservation   13. (a)artificial insemination, (b)host mothering   14. protect endangered and threatened species in the U.S. and abroad   15. endangered animals and plants in international wildlife trade   16. subsistence agriculture, commercial logging, and cattle ranching   17. industrial wood and wood fiber   18. CO_2, CH_4, O_3, N_2O, chlorofluorocarbons (CFCs)   19. greenhouse effect   20. changes in seal level, changes in precipitation patterns, effects on organisms, and effects on agriculture   21. 10   22. ultraviolet radiation   23. chlorofluorocarbons (CFCs)   24. September and November   25. cataracts, skin cancer, and weakened immune system   26. (a)50, (b)CFCs   27. overpopulation   28. consumption overpopulation

Building Words: 1. deforestation   2. stratosphere

Matching: 1. g   2. o   3. b   4. e   5. h   6. a   7. j   8. n   9. c   10. l

Making Comparisons: 1. deforestation   2. permanent removal of forests for agriculture, commercial logging, ranching, use for fuel   3. reduce current practices   4. global warming   5. atmosphere pollutants that trap solar heat   6. prevent build-up of gases, mitigate effects, adapt to the effects   7. chemical destruction of stratospheric ozone   8. reduce use of chlorofluorocarbons and similar chlorine containing compounds